Species Richness

Patterns in the Diversity of Life

Jonathan Adams

Species Richness

Patterns in the Diversity of Life

Springer

Published in association with
Praxis Publishing
Chichester, UK

PRAXIS

Dr Jonathan Adams
Assistant Professor in Biological Sciences
Department of Biological Sciences
Rutgers University
Newark
New Jersey
USA

SPRINGER–PRAXIS BOOKS IN ENVIRONMENTAL SCIENCES
SUBJECT *ADVISORY EDITOR*: John Mason M.B.E., B.Sc., M.Sc., Ph.D.

ISBN 978-3-642-09363-0 e-ISBN 978-3-540-74278-4

Springer is part of Springer-Science + Business Media (springer.com)

Cover design: Jim Wilkie
Project management: OPS Ltd, Gt Yarmouth, Norfolk

Printed in Germany on acid-free paper

Contents

To my girls
Mei Ling, Tera, Kendera, and Morgy
Beauty within diversity

Preface

When I began this book, I already knew that I was attempting the impossible. The Earth's biological diversity is much too vast a subject for any one person to really grasp. Yet I could see that there is the urgent need for a textbook which is both accessible enough, and rigorous enough, to provide a start for anyone interested in the serious study of biodiversity.

So far, the published literature on this subject has either been too dense and inaccessible, or much too cursory and simple, for students starting out in research or in conservation management. This book is intended to provide a bridge, between the simple and the complex.

The task of pulling so much information together and putting it into a straightforward form has involved more effort than I ever thought I was capable of. Even so, it has been an immensely rewarding experience, opening up and clarifying many areas of knowledge for me. This is knowledge which I now hope can be passed on to others. I am aware that there are gaps, and inevitably specialists will find inaccuracies somewhere in this book. Yet, at a time when we are on the brink of losing so much, I cannot allow fear of nit-picking to prevent me from explaining the story and the mysteries of biological diversity, to as many people as will listen.

This book has greatly benefited from comments by my friends and colleagues. Among them, Ellen Thomas, James Steele, Michael Rampino, Matt Huber, and David Currie gave particularly thorough and helpful reviews of chapters. I would also like to thank all those who have provided photographs for this book: I have acknowledged each of them in turn in the figure captions. Special thanks to Mark June-Wells for the butterfly photos, and Maile Nielle for the fynbos photos, on the front cover of the book. And in particular, my wife and best friend Mei Ling Lee has been a wonderful source of support and inspiration for me throughout.

Newark, New Jersey

Figures

* *See also* Color section.

Tables

Abbreviations and acronyms

AET	Actual EvapoTranspiration
BCI	Barro Colorado Island
FRIM	Forest Research Institute of Malaysia
FACE	Free Air CO_2 Experiment
GCM	General Circulation Model
MVP	Minimum Viable Population
IUCN	International Union of Conserving Nations
MAR	Minimum Area Required
NPP	Net Primary Productivity
NOAA	National Oceanic and Aerospace Administration
CDIAC	Center for Dissemination of Information on Atmospheric Change
SLOSS	Single Large Or Several Small?
UNEP	United Nations Environment Program
UNESCO	United Nations Educational, Scientific, and Cultural Organization
WCMC	World Conservation Monitoring Center

1

Local-scale patterns in species richness

1.1 LOCAL-SCALE TRENDS IN SPECIES RICHNESS

To my mind, there is nothing prettier than an English chalk meadow on an early summer's day. Scattered through the closely nibbled turf are yellow, purple, blue, and red flowers of many different hues, sparkling there like little jewels. In the most diverse spots within chalk grassland, there can be more than 50 species of plants packed into a single square meter.

I still remember the warm, sunny afternoon many years ago during my undergraduate days, when I sat down in the meadow on the top of a round chalk hill to eat the sandwiches from my backpack, and looked at what was around me. I wondered, why are there so many kinds of flowers here, and how do they all fit in together? In that moment I first saw nature as an intricate and delicate machine: so much more than just the wild jumble of struggling creatures that I had previously imagined it to be. There and then I decided to abandon my goal of specializing in molecular biology like almost everyone else, and instead to try to understand nature—the nature that ecology sees, in all its beauty and its glory.

Though I did not know it at that time, many others have asked similar questions about the diversity of life. In fact these questions are among the deepest and most consuming in ecology, and their answers have remained a riddle for more than a hundred years. Now at last, with the sum total of the efforts of many thousands of ecologists around the world, we may be getting close to solving these puzzles.

1.2 WHAT IS SPECIES RICHNESS?

Before we jump into the intricate world of species richness, it is important to make clear what is meant by the term itself. The very word "richness" is apt, for it conveys the idea of a bounty of nature—in this case, the bounty of different types, the diversity. That is what this book is essentially about: the exuberance of life's diversity, and what creates and destroys it.

In a science, though, precise definitions are helpful, because they can cut down on confusion. Strictly speaking, in ecological terms "richness" refers to the number of biological types—of species—that can be found in a particular area. In ecology, the word richness is also often used interchangeably with the word "diversity". Yet strictly speaking, diversity means something rather different. Diversity is a term that mixes total species richness with some measure of how evenly species are represented. There are various ways of weighing up the two aspects of species richness and evenness (or equability as it is called) (MacGurran, 1988). In this book I will concentrate on richness in the strict sense, although it is impossible to completely avoid talking about equability when trying to explain patterns in richness. To avoid letting the writing become monotonous, I often use the word "diversity" in place of "richness". However, it is important to realize that what I am talking about here does not necessarily take into account the even-ness aspect.

Species richness can only be talked about meaningfully if we know exactly how it is being sampled. Two different samples that record numbers of species can only be compared if they are taken in roughly the same way. The actual scale that the samples are taken on is not so important, so long as the scale is the same. For example, we might record the numbers of plants in meter-square areas of different meadows and compare them. Or we might compare numbers of species found in hectare-sized plots in different locations. Some studies have even used data from published floras to compare sample areas tens or hundreds of kilometers across (e.g., Currie and Paquin, 1987; Adams and Woodward, 1989). They are all perfectly valid scales for studying species richness, though the explanations needed to understand the patterns we see may be quite different from one scale to another.

Another important thing for studying species richness is generally to sample only wild populations. Trees and flowers from other places around the world can be planted out and survive, yet not actually form a breeding population. A large botanic garden, such as Kew in London, may have hundreds of species of trees planted on its grounds and is perhaps comparable with an area of tropical rainforest in its total plant species diversity. Yet most of these species are not native and they are each present as only one or a few individuals, often surrounded by mown grass. In many cases, they would not be able to reproduce in the local climate even if they were given the chance to establish a population. Their presence then has no real ecological meaning; they are not part of any ecological system that we might study and understand.

In-between these totally artificial populations—which we should ignore in any study of richness—and the native species that we should certainly study, are exotic introduced species that have run wild after they were planted or released. These may behave in many ways as if they were native populations, even though we know that they are not. In some cases introduced species have become part of the natural heritage of nature, as far as the humans in those places are concerned. An example would be many of those chalk grassland species that enchanted me as a student. They probably only exist in England because of farming practices that accidentally brought them in from the mainland of Europe thousands of years ago. Now they are so well established and integrated, that we can really study them as if they were a natural system. In a general way it can also be interesting for its own sake to study the richness of invading exotic species and the factors that seem to affect it, and in some later chapters in the book we will examine this too.

1.3 WHAT IS MEANT BY "LOCAL" VARIATIONS IN SPECIES RICHNESS?

Local-scale variations in species richness, like those in English meadows, are on a far more human scale than the broad and long-term patterns described in other chapters of this book. As I have chosen to define them here, these local patterns may occur across anything between a few centimeters and several tens of kilometers. Really though, the different scales fade into one another, and sometimes it becomes a subjective issue whether a pattern is actually local or geographical in scale.

These local-scale differences in diversity may occur between one type of habitat and another (e.g., if we compare the number of species of plants or insects in a patch of forest compared with a patch of grassland right next to it). Or they may occur within the same general habitat in different localities (e.g., between distinct patches of grassland on different soil types, perhaps within a few hundred meters of one another).

What causes such local species richness patterns? One thing that can be said from the outset is that most of these local differences are on too small a scale to have been produced by variations in climate, or by the history of drifting continents that I will often talk about in other chapters. The only exception to this on the relatively local scale are altitudinal gradients in richness, which are probably very much affected by climate. Generally, we have to look for a distinct set of causes apart from climate that are more suited to the smaller scale. Something to bear in mind though, is that the causes of local-scale species richness patterns might also help us understand the broad-scale patterns. Perhaps if we look hard enough, it will turn out that there is only one underlying mechanism behind all diversity patterns, at all of these different scales? It would be very satisfying to find a sort of "grand unified theory" that explains all variations in species richness. Yet knowing that the natural world is complex, we should not really expect to find everything fitting together so neatly: it would be satisfying enough to find just a few common strands across different scales.

Understanding the reasons for local-scale differences in species richness also has practical uses for those who care about nature. It helps conservationists reach decisions about how best to manage nature reserves, in order to encourage and hold onto high levels of species richness (Chapter 8).

1.4 LOCAL-SCALE PATTERNS ARE MOST NOTICEABLE IN ORGANISMS THAT DON'T MOVE AROUND MUCH

Local-scale patterns in diversity tend to be more noticeable in organisms that stay in one place through most of their life cycle, because they are literally fixed to the spot. These are known as "sessile" organisms. They include land plants rooted into soil, and marine seaweeds glued onto rock, plus many aquatic animals such as corals, barnacles, sponges, and bryozoans that are also fixed in place. Local patterns in richness *are* also found in organisms that can move around very freely, such as birds, mammals, and flying insects. But the patterns tend to be weaker, and they are found less often. For this reason, more of the attention in this chapter is given to things like plants and corals, than to the many other types of creatures that move around throughout their life. It also makes sense that understanding what is happening with photosynthesizers (and that includes corals, because they have algae in their cells) might help us to understand any patterns we might see in the local diversity of animals—since animals depend on the plants that power the ecosystem and also provide the structure of the habitat.

First, let's take plants, the most abundant sessile organisms on the land and in the shallowest seas. For plants, some of the patterns that have often been noticed on the local scale include

- A change in species richness as vegetation recovers from a big disturbance event (succession).
- A difference in species richness across areas with differing intensities of disturbance.
- A difference in species richness between different levels of nutrients in the soil (on land) or in the water (in aquatic plant communities)

It is much the same with corals, which are in a sense half-way to being plants.

So, to get anywhere in explaining local-scale patterns, we need to pay particular attention to how and why species richness varies with disturbance, and with nutrient levels.

1.5 A RELATED AND IMPORTANT QUESTION: HOW DO SPECIES COEXIST *ANYWHERE*?

The question of what causes local place-to-place variation in species richness is closely interwoven with another question that really applies to species richness at any scale: What is it that enables multiple species to coexist? If we can understand how they manage this when they are side by side, then variation in this mechanism (whatever it is) could explain differences in species richness on both a local and a geographical scale.

It ought to be difficult for species to coexist if they are similar in almost everything they do. Imagine that two species are competing for exactly the same resources (be it food, water, light, nutrients, or whatever). One of the two species will probably turn out to be just slightly better at grabbing some key resource, and eventually it will starve the other species out to extinction. This is often called Gauss' Principle, after a Russian biologist who came up with it based on observing how microbes behave in cultures in the lab—he found that he could not get two similar species to keep going in the same culture medium. This general sort of situation—two similar species competing—was first put into mathematical terms by Alfred Lotka, a German, in 1925 and independently by an Italian, Vito Volterra, in 1926 (science has always been such a refreshingly international business, with ideas, and people, bouncing around between different countries). The equations that express the idea are called the *Lotka–Volterra competitive equations* (by the way, Lotka and Volterra also produced some famous equations on population cycles between predators and prey, and that is where their names are more often heard). The only way that two competing species can survive will be if each of them is better at taking only a certain part of some important resource for itself. For example, if one animal species is better at taking insects only from the upper part of trees (but cannot hunt effectively lower down the tree), while the other can take them better from the middle part (and because it cannot climb so well cannot feed in the upper branches), they might both coexist forever in the same patch of forest. Or if a plant species is specialized in growing in patches of waterlogged soil, while another plant growing nearby prefers drier patches, again they can both potentially coexist. The important thing is that each species has its *forte*, its own particular strength, that no other species can push it out from. This forte is known as a "niche", the slot in the ecosystem that it—and only it—occupies.

But very often we see lots of very similar-looking species all living side by side and apparently living in just the same way, without any obvious differences in what they feed off. The contrast between the logical reasoning of the Lotka–Volterra mathematics and what is actually seen in nature was expressed by the great biologist Evelyn Hutchinson as "the paradox of the plankton". It is paradoxical, he pointed out, that there can be so many different types of microscopic plants floating in the waters of the same lake at the same time—using the same nutrients and sunlight as one another. Surely, with them all being so similar in what they require, one species should push out all the others just as the equations predict. Equally we might ask, how can there be so many similar fish species in a coral reef, when they are so often eating similar food? Or how can there be 200 species of trees in a hectare of tropical rainforest, when they are rooted into the same soil and competing for the same sunlight?

It is possible that we are just not thinking and looking hard enough for ways in which niches can be different, even if the species do look very similar to one another at first glance. With more imagination one might see the potential for a huge number of different niches in a diverse community such as a coral reef or a rainforest, with each species being so effective at living in its own particular way that it cannot be pushed out by competitors. Over the years, ecologists have documented many aspects of the behavior, anatomy, or physiology of organisms that seem to fit particular species into their own unique niche. Some years after that moment of epiphany with the chalk grassland, when I worked in the Plant Sciences Department at Cambridge University (on satellite images of rainforest, not chalk grassland by the way), there was a whole bunch of industrious grad students working with the ecologist Peter Grubb, spending their time finding and documenting slight differences in the niche requirements of chalk grassland plants—often the exact conditions that prevailed just after the seeds sprouted (those of us in the room down the corridor named their activities "Grubbing around"). These differences seemed to help explain how so many different types of wild plant could coexist packed together in chalk grassland.

In communities of sedentary organisms such as plants or corals, the important thing for coexisting may not be so much what the adult does, as what the young do. Once a young tree or coral has successfully established itself and grown big, its place in the reef or the forest is more or less assured. Instead, it is in the early stages that life is most hazardous, when there is most to gain and most to lose—for the vast majority of any species die young. Take the situation in a tropical rainforest. It is possible that the seedlings of each species of tree in the forest are adapted to do best in a very particular set of soil conditions, light levels, or water content in the soil—or some precise combination of all three. At this small scale, there might be patches of forest floor which have different soil or light conditions. Each little patch might suit the seedlings of some particular species of rainforest tree—and in this place a tree of that species will establish itself, growing slowly up towards the forest canopy. In a sense then, a seedling has its own very particular "niche", that might be very different from the adult's niche. The idea that it is differences in niches at the seedling stage that allows the coexistence of plants was first put forward by Peter Grubb in 1977 (Grubb, 1977). He called the niche of the seedling its "regeneration niche"—regeneration being the process by which the forest replaces the old trees that die with young ones in their place. There could be similar regeneration niches in other plant communities (such as the chalk grassland that his students worked on), and in coral reefs where larvae settle down after drifting with the plankton.

A different explanation for how species can coexist also involves an important role for disturbance, emphasizing that no single species could ever do best in an environment that gets disturbed every once in a while. When a disturbance comes along, it favors one set of species that do well in quickly colonizing the open ground and multiplying up. In a northern temperate forest in Europe, North America, or China, these are typically the cherries (*Prunus*) and birches (*Betula*), plus all the varied wild flowers and shrubs that tend to appear first after a big tree has blown down, or after a fire has swept through. In a grassland that has been damaged or killed after a disturbance event such as a landslide, severe drought, or overgrazing, a particular set of grassland plant species adapted to the bare ground will get in fast. So, for example, in the North American prairies a wide range of species will get in and colonize a disturbed patch of ground (e.g., the grasses *Agrostis scabra* and *Agropyron repens*—along with many others).

In ecology these fast-colonizing species are known as "r" selected species, named after the "r" for potential rate of increase in equations used to describe population growth. Their populations can grow fast because from early on in their lives they put a lot into producing offspring, with each of these offspring taking a relatively short time to reach maturity. Only with more time can slower growing species get into the patch and then relentlessly overtop and shade out the "r" species. These are often called "K" species, with the "K" coming from "carrying capacity" (ecologists are not good at spelling), the maximum capacity of population that can be held in a place. "K" species spend a lot of their time competing in populations that are already filled up, no longer growing, near carrying capacity: hence the term. In a northern temperate forest these will often be species of oaks (*Quercus*) and beech (*Fagus*)—slower dispersing and slower growing trees.

In a varied, disturbed forest environment, there is no way that the oaks will ever drive the cherries to extinction, because cherries and oaks are each good at exploiting different circumstances—they have different niches for exploiting distinct situations, even if they may prefer exactly the same types of soil. The cherries may be pushed out of the forest in any one spot as it closes over and the oaks dominate, but by this time they have already moved on elsewhere to another gap in the forest. Eventually (maybe centuries later) another disturbance will hit the patch of big mature oaks and kill many of them, and the cycle will be repeated.

In the North American prairies, the slower colonizing competitor species that slowly get in are two grasses—little and big bluestem (*Schizachyrium scoparium* and *Andropogon gerardii*). In experiments on prairie land in Minnesota, David Tilman (Tilman, 1994) found that given between 11 and 17 years without disturbance, they will take over and push everything else out. Tilman has suggested that essentially the diversity of the prairie (with a total list of more than a hundred species occurring in this plant community) may be dependent on enough disturbances to allow many different faster colonizing species to get in and exploit gaps, before they can be pushed out. The key to coexistence might be not the fact of disturbance itself (which is what is emphasized by others such as Connell, 1972), but the range of speeds with which different species can get in to the bare ground created: the speed of colonization by each species is in effect part of its "niche".

If coexisting in nature is all about having distinct niches—during at least some part of the life cycle—perhaps the reason that there are more species in some places is just that there are more niches available? For example, if there is a greater range of soil types, or a wider range of food types, this could explain why we see more plant or animal species there. Or if the exact types of disturbance that occur differ, maybe some types of disturbance give more niche opportunities?

Or, on the other hand, perhaps the *width* of niches is for some reason narrower in a more diverse community, so more species in total can be squeezed into it? To clarify how niche width can differ, imagine a forest where some bird that eats a whole range of different berries and insects. It has a very wide niche in terms of its diet, and if there are many species like that around there are not likely to be many niches available so there will not be many species there in total because of the way they divide up the resources available to them. Imagine another forest where there are no birds that have such broad diets. Instead we have a species that eats only hairy caterpillars on trees, another that eats only mosquitos, another that eats only dragonflies (I am making this up), and so on. The narrower niches will mean more space for cramming in other extra niches and thus more different species can exist there.

The tropics have always been famous for their incredible levels of species richness (Chapter 2). There might be something about a tropical, moist environment that both allows and rewards finer adaptation to niches, each species becoming very good at doing one narrow thing. Certainly, that is the impression one often gets: the adaptations are often quite exquisite. For instance, some plants have colonies of ants living inside and defending them. Some types of insects have wings that closely resemble leaves, for camouflage (Figure 1.1). Neither of these is ever seen in temperate forests, as far as I know. Perhaps the reason for this precise adaptation is the relative constancy of the climate in the tropics, without a sharp winter or dry season? This constancy would make the environment very predictable, and predictability is likely to be something that favors specialization. For example, a temperate zone plant must generally cope with temperatures ranging anywhere between freezing and around 40°C, and that is just during its growing season (temperatures will dip much lower in winter when it is dormant). A tropical rainforest plant only has to cope with the upper end of this scale, never the colder temperatures at the lower end. In the tropics, the lack of strong seasons means that trees can fruit at any time of year, and this provides

Figure 1.1. Exquisite adaptations seem to be more common in tropical organisms, and they may help them to fit into narrower niches than in the temperate zone. This leaf insect (*Phyllodes eyndhovii*) closely resembles a fallen dead rainforest leaf, and it even has false "leaf miner" damage on each of its wings to make its disguise look more plausible (photo: Author, northern Borneo).

the constancy that will support specialized fruit-eating birds (such as toucans and hornbills) or bats (such as the flying foxes of Asia and Australia). In the temperate zones, by contrast, fruit is gobbled up by all sorts of different creatures. None of these are specialized to the role because fruit only appears during the warmer times of year—and these animals must live by other means when there is no fruit around. Variation in niche width would affect how many different niches can be packed into the ecosystem: how many different species can coexist. This then could be a part of the explanation for latitudinal gradients in species richness (Chapter 2).

However, some ecologists wonder how it is really possible for hundreds of similar species living side by side all to find their own niche. When we have, for example, a single hectare of rainforest with 150 or 200 species of trees all living together, how can each and every one of those trees find a unique way of living, where no other species can outcompete it? Or the untold thousands of species of insects in that same hectare of tropical forest—how can they each have their own niche in the rainforest? Or take the example of the hundreds of species of heathers in the South African fynbos vegetation (Chapter 4), or the hundreds of species of cichlid fishes in Lake Tanganyika (Chapter 4)? Granted, sometimes the niche differences are obvious (e.g., the difference between a herbivorous and a carnivorous species of cichlid fish in a lake). And more niche differences can often be found on closer examination—slight differences in soil preferences between different types of rainforest tree, for example. But niche differences are not always found, even when they are looked for carefully. When ecologists study the distribution and the behavior of species in very diverse natural communities such as rainforests or coral reefs, they are often at a loss to explain what could mark each species off as having a unique niche—because species seem to behave in an interchangeable way, with overlapping preferences in terms of where they will grow or what they can eat. For example, in the Barro Colorado Island forest in Panama which has probably been studied in more detail than any other patch of rainforest in the world, John *et al.* (2007) found only fairly limited evidence of trees specializing on particular soil conditions. Many tree species in the tropical forest seemed to grow randomly scattered across any type of soil.

Probably if we looked closely enough at nature, we could always find some slight difference in niche requirements between any two species, no matter how similar they look—just because some of their genes happen to be different from one another. The relevant question is whether these differences are big enough to enable these species to coexist indefinitely. Imagine, for example, we have two species of birds living in the same forest—if we were to find that they take the same type of food 95% of the time. Even though we have found a difference 5% of the time, it is surely not going to be enough to allow them to coexist.

So when we look at communities that have many hundreds, even thousands, of rather similar species living together in small areas, it becomes hard to explain how they could all manage to have different niches. Surely there are only just so many ways in which to be different.

Many ecologists suspect that the reason we cannot always find evidence for distinct niches for the species in diverse communities, is that they *do not* all have different niches. In fact it is believed that a lot of those species living side-by-side effectively have almost exactly the same niche as one another, with only minor differences that are not possibly enough to explain how they avoid getting pushed out by competition. This view is especially widespread among ecologists who study tropical forests, which have a huge diversity of trees all apparently growing in the same soil and needing the same things (water and sunlight)—and also among many who study diverse tropical fish, coral, or Mediterranean shrub communities. An articulate and thoughtful proponent of this view is Stephen Hubbell of UCLA (Hubbell, 2001). But if these species are in effect living in the same niche, how is it that they can still all coexist without one strong species outcompeting the others, as Lotka and Volterra predicted? There are certain mechanisms that could prevent this pushing-out process, though we are not yet sure how effectively they work in nature.

For example, one mechanism allowing different species to coexist, despite a lot of niche overlap, could operate when selective enemies of each species prevent it from becoming too abundant. Imagine that there is a strongly competitive species of tree (or coral, or whatever) in the community, that has the potential to take over by outcompeting most of the other species. It starts to multiply up its population, becoming more and more abundant by pushing out members of the other species that surround it. But as it is doing this, specialized pests, parasites, and diseases that are adapted to attack *only this one type of tree* may begin to increase in numbers, able to spread more easily through the densely packed population of their host. The increased damage and death rate from these tiny attackers either directly reduces the population of the host, prevents its seedlings from surviving, or weakens the adult trees so much that they can no longer push out their weaker competitors in the community. The result: an even-ing up of the scores, and coexistence of many species instead of just one. There may be various species living in exactly the same niche in the community, but none of them can become too abundant because of the attack that soon follows as they start to multiply up. With these little enemies helping out, there is always room for many different species—even if they are living in exactly the same niche. This idea was put forward by Daniel Janzen and Joseph Connell independently in 1970 (Janzen, 1970; Connell, 1970), hence it is called the Janzen–Connell hypothesis. More on this in Chapter 2, which deals with latitudinal gradients in species richness.

Another idea that may explain how species could exist in overlapping or identical niches involves disturbances: disruptive events that strike the community (Huston, 1993). These either kill individuals of nearly every species in that locality, or they seriously weaken them by damaging them. What exactly are these disturbances? Sometimes they are fairly dramatic (e.g., a fire burning vegetation and killing most of the plants, or a landslide that creates a jumble of bare soil). When a river floods, it may deposit a thick layer of silt that smothers and kills much of the vegetation along its banks. In the upper parts of a coral reef, a big storm can scour away most of the coral. But often, disturbances are gradual, subtle processes such as the grazing of a meadow by sheep, which damages and weakens most of the plants. Mowing a lawn or meadow can have a similar effect to natural grazing or fire. The disturbances that are important in a community may even be minuscule events that we would barely notice unless we were studying it closely. For example, the scuff mark left as a cow's foot slips on the wet turf of a meadow can kill or cripple the plants across a few square centimeters. It may not seem much to us, but to the plants that were growing in this little bit of meadow, it was a catastrophe—but perhaps also an opportunity for their offspring starting out from seeds on the bare soil. Likewise, parts of a big coral colony might be accidentally damaged by a fish that grazes algae off the reef surface, giving a space that could be colonized by another young coral establishing itself.

In the view of most ecologists, the important thing about disturbance events is that they are often followed by a free for all in which the normal hierarchy of competitors does not have any importance. Plants re-establish from seed or sprout back from fragments of roots, but at first they will be spaced widely enough apart that they are not competing strongly for either nutrients, water, or light. Also, after a disturbance event the young colonizing plants are often helped in getting rooted and established by an abundant supply of nutrients in the soil. This can be due to the decay of the older plants that were killed, from the fertile silt left behind by a flood, or the mineral-rich ash left behind by a fire. An abundance of sunlight—with the shading of previous vegetation cover now removed—can also help the plants to grow quickly. If such disturbance is frequent enough, the order of stronger vs. weaker can never properly assert itself before everything is jumbled up again.

Let's take the case of a forest that has been damaged by a powerful storm blowing down a swath of trees. After the devastation, populations of plants start to recover from scattered seeds on the ground and from a few damaged survivors. In the open ground left by the death of most of the adult trees, there is little competition and so it does not really matter if the first tree seedlings coming back have overlapping niches—they are not going to get pushed out anyway. Only as more trees fill in and the forest finally closes over does competition between trees really begin to bite, and the weaker species with overlapping niches begin to get pushed out slowly in a process that will take centuries. Eventually what will cause their populations to die out is the failure of the adults of the weaker species to replace themselves in equal numbers after they die, because their seedlings and juvenile trees get outcompeted before they can reach up into the canopy. But before this pushing-out process can get very far, another big disturbance will likely hit the forest, resetting the clock of competition. Here again, as in the meadow, in this situation all species are more or less equal, regardless of how they would fare if they were matched up in lengthy close competition. The result is that a potentially dominant tree species, which has the ability to push out everything in the forest that occupies the same niche, will never actually get to succeed in doing that (note that this is different from the idea I mentioned above, of species coexisting because some are better dispersers than others: they can all be equally abysmal dispersers, but the jumbling-up caused by disturbance can give all of them a good chance).

Even if species are not quite precisely matched in their ability to compete, and not different enough to avoid being outcompeted eventually, the pushing out of the weaker competitors might take a long time. With long-lived trees in a forest, it might take thousands of years. Before that can happen, everything will have been shaken up again by another disturbance that resets the competition clock, and allows young of the weaker species to establish themselves. What we could be seeing, then, is the coexistence of so many species in essentially the same niche because they are almost identical to one another in terms of competitive ability: so similar that even a bit of disturbance just once in a while will allow the weaker ones to persist.

And thus, due to disturbance, many more species may be able to coexist in the community, crammed into the same few niches. This is known as the "intermediate disturbance mechanism"—involving disturbances that are frequent enough to help species to coexist, and yet not too frequent. If they were too frequent, hardly any species would be able to cope with the relentless damage, so that would actually mean fewer species around. Differing intensities of disturbance are often suggested as the cause of differences in species richness on both the local and geographical scale (see below). For instance, my former colleague Michael Huston (Huston, 1993) has been a strong advocate of this point of view.

Still another idea, which builds from the disruptive role of disturbance, is based around an often-overlooked prediction of the Lotka–Volterra competitive equation. In the equation, if two species have exactly the same growth rate and carrying capacity in that one niche, then neither of them can push the other one out. Since neither can win, they will simply coexist while squarely occupying the same niche—with nothing more than luck determining which is more abundant than the other, in that particular spot. It could be the same for a group of 50 species, or 100, if they all do just about as well in competition.

This idea is known as the "neutral theory", and it has been emphasized by Hubbell (2001) as an explanation for what happens in the world's forests, and in other diverse communities (and it explains Hubbell's assertion that species are mostly existing in the same niches as one another, as I outlined earlier). In Hubbell's theory, it is basically a "lottery" whether a seedling of one tree species rather than another establishes in a particular spot in the forest. Their relative abundance in any particular place will drift up and down randomly, with no species pushing any other out by competition. If the environment allows all the species to grow just about as well as one another, and survive just about as easily, where is the basis for any one species to be a stronger competitor than the others? With almost no competitive hierarchy, many different species can coexist in essentially the same niche.

This seems odd, considering what I wrote earlier, about "being different" enabling species to coexist. In theory then, there is a knife edge–thin zone of extreme similarity, where the differences between species would be too small to give an advantage to any one of them. Beyond this zone of very close similarity, with greater mismatches between species, the only way to coexist is by being *different* enough.

Hubbell's view of tropical forests as a big lottery, with every tree species having about the same chance of getting established as any other, seems to be borne out by at least some studies. If every tree had its own specialized niche (e.g., a "regeneration niche" as Grubb would suggest), then there should be a fair amount of predictability about which species manage to get in to colonize a gap in the forest left when a big tree falls. Extensive comparisons of more than a thousand gaps in the rainforest on Barro Colorado Island in Panama shows that it is essentially impossible to predict which of the many tree species in the forest will establish in a gap, even if we know the gap's size, light level, soil conditions, etc. It looks a lot like it is just chance, a matter of whose seeds happen to fall in there first, which determines which tree species grows where in the forest (Hubbell *et al.*. 1999). This, by the way, is bad news for Grubb's (1977) regeneration niche theory which says that seedlings are specialized to a precise set of conditions, and should be very choosy about which gap they will grow in.

Another slightly different way of visualizing how species might coexist has been put forward by Axel Kleidon of BGC–Jena and a group of his colleagues (including myself) (Kleidon *et al.*, 2009). In this scenario, species do tend to push one another out by competition, but they can never do it completely—even if they are in the same niche. That is because it is all a lottery, rather like in Hubbell's theory above, as to which species gets in. But in this case it is a weighted lottery, and the better at growing a species is, the more likely it is to establish itself in a particular spot and then grow big. Even so, there is still a chance the species that are less well adapted to growing in that climate will get in there sometimes—they just won't get in so often and they won't grow as big. The better adapted species in the forest or coral reef (in terms of basic ability to grow well in that environment) win their places more often, and are the biggest and most abundant. But if there are several well-adapted species around, they cannot win out over others if the differences in their competitive ability are slight, and all of them will be just about equally abundant. If the differences in competitive strength between two species are greater, then yes they will in a sense push out the weaker ones—but not necessarily to complete extinction. Often these stronger species will just push the weaker ones down to low levels of abundance by reducing the *probability* that they will establish. But at very low levels of abundance, their populations are likely to go extinct by chance anyway—because in small populations the numbers often dip down to nothing just out of bad luck. The idea is basically rather like Hubbell's in some ways, but not entirely neutral, for there can still be relative winners and relative losers. Maybe we can call it the "nearly-neutral" theory.

So these then are some scenarios that can be used to explain how so many species can be crammed into a diverse community: precisely and finely honed niches might allow many species to coexist. Or perhaps species are actually so evenly matched in competition that none can entirely push out the others. Or, maybe variations in the amount of disturbance or selective pest pressure might allow niche overlap—and because of this many more species can fit in. Disturbance might also widen the range of different niches available by creating opportunities for rapid colonizers, allowing more species in.

The various explanations for how species can coexist basically come down to two different types of mechanism. In one set of explanations, a key requirement for diversity is the availability of many distinct niches. If this is so, then the more opportunities there are for forming different niches, the more diverse the community can become. The other set of explanations involve mechanisms which allow niches to overlap without species getting pushed out by competition. In such cases, the more strongly these factors operate, the more completely the species can overlap with one another—and the more species can fit into the community.

While such coexistence mechanisms may be necessary for high species richness, they may not be *sufficient* to bring it about. For all we know, it could be that just as many species could fit into (say) a temperate forest as a tropical rainforest. The thing that limits diversity in the high-latitude forest might be lack of supply of species through evolution. If species have not appeared in that environment then it will not fill up with species, no matter how much room there is for more species to fit in there. Why would there be a shortage of supply of species in some environments? Perhaps because these environments have not existed for a long enough geological time (Chapters 3, 4), or because they recently suffered a mass extinction (Chapter 3), or because something about that environment is just not conducive to species formation.

1.6 DIFFERENT SCALES OF SPECIES RICHNESS, FROM LOCAL TO GEOGRAPHICAL

Ecologists find local-scale patterns in richness occurring at slightly different scales, and in slightly different ways. If we compare one meadow or patch of woodland with another, the species richness pattern we see is at the habitat level. One habitat in one place has a different species richness (e.g., in terms of species per square meter, or per hectare) from another patch of habitat nearby. The diversity *within* a local bit of habitat is called "alpha diversity".

Apart from just knowing the number of species counted in each local habitat, we may also be interested to know the amount of difference in species *composition* between different habitats or patches of the same habitat. This is really additional species richness that occurs because the species lists do not completely overlap. If different samples turn out to have a very different set of species within them, we can say that in the landscape in general there are probably a lot more species present than we would guess just from a single local sample. The additional species richness which turns up when one compares the lists of different local samples is called "beta diversity". Beta diversity, then, adds to the total number of species we would see if we zoomed out from the landscape and counted up all the species in all the local patches of habitat. This local or regional total of species that we count from stepping back and counting up all the lists from alpha diversity (which differ from one another through beta diversity) is known as "gamma diversity".

This slightly broader scale gamma diversity is more difficult to estimate than alpha diversity, because it occurs scattered across landscapes over distances of tens to hundreds of kilometers. It is not likely that we would be able to sample every square meter in the landscape to know exactly what is there. The best way to estimate this landscape-scale diversity is to start taking local samples and see how quickly new species are added in to the list as we add more samples. Or instead of taking more samples of the same size, we might just expand the size of our first sample and see how fast extra species are found when we take larger and larger samples. As the data accumulate, we can plot a line, showing how quickly the total goes up as we add more samples or look on ever-larger scales. From seeing where this line seems to be headed we can then guess roughly how many species could be found in total if we sampled every last bit of that region. Only when the line of total species found begins to level off—known as the inflection point (Figure 1.2) can we be reasonably sure that we

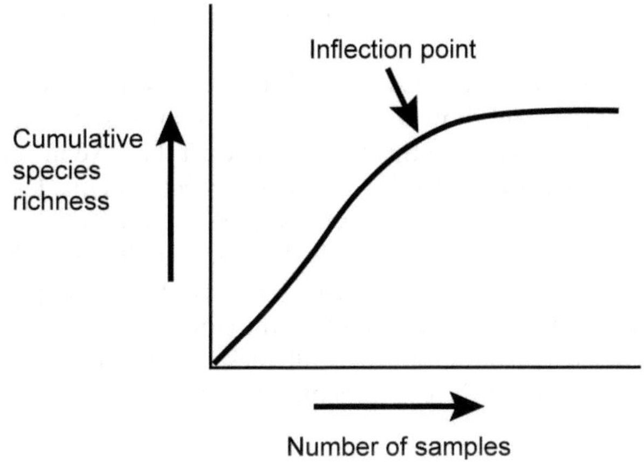

Figure 1.2. Accumulation curve for estimating the species richness "out there" in the area being sampled. The numbers of species found increases as more samples are added. When the line starts to bend (the inflection point) this suggests one is nearing the true total number of species in the habitat or region being sampled.

have found most of the species. A graph like this that plots the build-up of species in samples is known as an *accumulation curve*. In effect, this method detects species that have not yet been found in the study—not by psychic sense but by simple mathematical principles!

Depending on the locality and the group of organisms being sampled, the rising total of species richness in the accumulation curve can show one of two trends. It may go up fairly slowly, because even a very local sample captures most of the diversity present in the region. My guess is that a single valley of temperate forest in New Jersey would actually capture the majority of the tree species richness in the whole state, maybe even the whole northeastern United States, because most species are fairly evenly distributed throughout. Only occasionally as we increase the size of area sampled in the northeastern United States would we be lucky enough to find another tree species to add to the total. This slow-rising trajectory in terms of adding more species is called a Type 1 trend. On the other hand, the total of species richness can go on rising steeply as more samples are added because the local-scale list of species actually differs a lot from one place to another. Tiger beetles in Scotland show this sort of trend (Koleff and Gaston, 2001), varying a lot from one valley to another, and so as we expand the sample size we take in more and more new valleys with all their tiger beetles. This is known as a Type 2 trend, and we would also expect to see the same steep trend in the plants or insects of a tropical rainforest where each species is thinly scattered throughout the forest and there are a lot of species in total.

Looking at local-scale species richness and zooming out, there is no definite point at which one moves up from the "local" to the "geographical". The distinction between the two scales is convenient for trying to make sense of the range of species richness patterns in nature, and for dividing a book up into manageable chapters. But it is important to bear in mind that they actually blend as part of a continuum. For instance, we might ask, are the different climate zones on a mountain (each with their own levels of species richness) something that exists on the "geographical" or the "local" scale? Are the differences between two patches of grassland still "local" if they are 100 km apart? Nature has not been made in order for us to classify it into neat categories, so we should not be surprised if sometimes it frustrates us when we try to do this.

If we look across the full range of scales, from a few tens of centimeters to the whole world, how does the rising trend of species richness look? Smith *et al.* (2005) tried this with the tiny phytoplankton that live in surface waters of ponds, lakes, and seas. Starting with tiny artificial lab ecosystems derived from pond water (known as microcosms), they zoomed out and out, counting species richness at every scale until at the upper end of the scale they took in all of the phytoplankton species known from all the world's ocean basins. The comparison spanned 15 orders of magnitude in scale. What they found is that the slope of rising species richness in the phytoplankton stayed constant, through all these many orders of magnitude. What this actually means about the controls on diversity is not clear, but it is an interesting observation anyway!

1.7 A VARIED ENVIRONMENT TENDS TO ALLOW MORE SPECIES IN

Even if species tend to overlap in their niches, no one species can live everywhere. Different species each have their own strengths in terms of making a living, or at least different requirements for establishing themselves. For example, an area that has several different soil types or a wide range of moisture levels is likely to offer opportunities to several different sets of plants, each of them specialized to certain soil conditions. So a diversity of environments will tend to increase the species richness of the area. This is going to be the case even if many of the species have identical or overlapping niches *within* each set of soil conditions. There is a great variety of plants out there specialized for either wet or dry soils, alkaline or acidic, salty or non-salty, regardless of whether they each grow alone in those soils.

For example, along a river in the rainforest of the Amazon Basin there may be several different zones above the mean river level, each zone flooded with differing frequencies and for different lengths of time in an average year. At the lowest levels, on the fringes of the river itself, is *varzea* forest which is flooded several meters deep throughout the wettest season of the year. At higher levels above the river there are other zones, less and less likely to be flooded, up to the *terra firme* forest that never floods. Each zone has its own characteristic set of species—for example, the rubber tree *Hevea braziliensis* is a part of the varzea forest, adapted to withstand flooding for several months, and with seeds that are actually dispersed by fish after they drop into the water. Having this varied set of environments will add in several different sets of species to the landscape, adding to the overall species richness.

Along tropical shores, mangrove trees establish themselves in different zones according to distance above average tide level—each zone gets flooded by saltwater with a particular frequency. Some zones are flooded with every tide, others only get saltwater around their roots when there is an exceptionally high tide (Figure 1.3). Each zone has its own set of mangrove species, adapted to the amount of saltwater and flooding that it must endure. Having these several zones adds in more species, and thus more species richness.

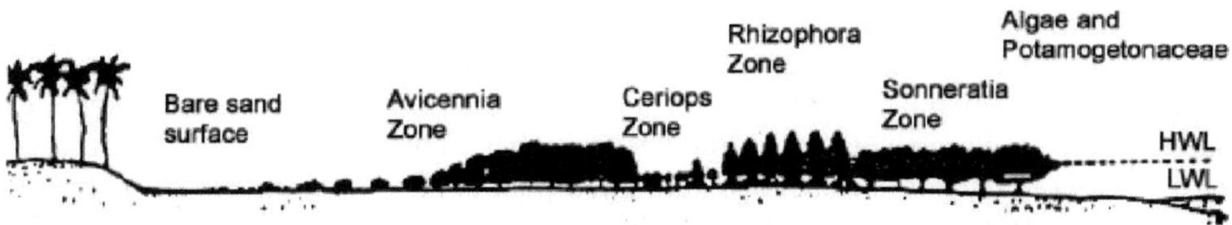

Figure 1.3. The various species zones of mangroves found on an East African coast. The range of environments, brought about by different distances above mean high tide level, give opportunities for different species of mangroves to coexist in the same area (redrawn from Huston, 1993).

There are many rather more subtle examples of the same principle. In a temperate forest, some trees tend to occur on a sandy soil, others on a heavy clay—so having a mosaic of soil types gives more variety. For animals, a patchwork of open grassland and areas of forest is likely to give a chance for species that prefer open environments a chance to live in the same area as those that require closed forest, again raising the overall species richness.

Generally, the more varied the topography and geology of any given area, the more species of plants and animals will be found there. On a mountainside a range of different outcropping rock types gives different soil conditions, each associated with different species of plants (see below, on mountains). Heterogeneity of slope, aspect, and drainage conditions in a landscape is also important. On some hills I know near the Mediterranean coast in southern France, moist temperate forest trees such as beech (*Fagus sylvatica*) grow on the cooler, damper north-facing side. Just over the brow of the hill, on the south-facing slope is another completely different set of species: Mediterranean shrubs (such as rockroses, *Cistus*) that cope with the much drier and hotter sunlit environment there. Because of the differences in environment caused by aspect, both sets of species can grow on the same hill.

In savanna environments in the tropics, local variations in soil drainage conditions affect the moisture supply to plants—which is all-important in this semi-dry environment. This leads to a wider range of plant communities, and greater overall species richness.

Even very small-scale topographical variation can make an important difference in terms of allowing more species to coexist, each in its own specialized niche. For example, in meadows in England there is often a pattern of low ridges that is a legacy of a medieval cultivation system known as "ridge and furrow". The high parts of these ridges are better drained than the lower parts, and this allows species of plants specialized on each set of soil conditions to coexist. For example, on the beautiful Port Meadow by the Thames in Oxford, where I spent many lazy and sunburned afternoons as an undergraduate, three species of buttercups (*Ranunculus*) coexist—one (*R. repens*) on the top of the ridges, one (*R. bulbosus*) at the bottom, and one (*R. acris*) in-between—half-way up the ridges. If Port Meadow were completely flat, presumably only one species of buttercup would be present there.

Another example of a community that contains greater species diversity due to environmental patchiness, is the "limestone heath" that occurs locally in England and

Wales. Normally limestone yields quite an alkaline soil, which supports plants that are specialized to grow at high pH. But in the dismal and rainy British climate, organic debris that builds up above the soil surface (from dead leaves and suchlike) can get leached of lime by the rain, and become acidic. These patches of acidic organic matter support the roots of acid soil–adapted plants. Also in some places there is a thin layer of sand over the top of the chalk, which also forms a surface layer of acidic soil that acid-loving plants can root down into, while side by side with them are deeper rooted limestone-loving plants that tap into the high-pH soil beneath. So the total community is much more species-rich because it contains the plants of two totally different soil types, growing right next to one another.

For animals, it is not so much soil types as the number of different physical environments to clamber over or hide in that is important in determining local differences in species richness. For example, Pianka (1973) found this in lizard species diversity patterns in the deserts of the southwestern U.S.A.—more structurally complex vegetation with different layers and closed and open patches tended to support a lot more lizard species. Presumably, the higher species richness is due to complex vegetation providing lots of niches in which individual species can thrive, without being eliminated by competition. Likewise, Pianka also found that from the point of view of bird species richness in the American southwest, the local structural diversity of the scrub vegetation that birds live and feed in is more important than anything else. MacArthur and MacArthur (1961) also found a trend of increasing bird species richness with structural complexity of vegetation across the northeastern U.S.A. and Canada (Figure 1.4).

Habitats with a diverse range of growth forms of plants turned out to be much richer in bird species than those that had a more limited range of forms. This seemed to be independent of the number of plant species present—as long as the plants present had a wide range of shapes and forms among them, bird species diversity could be high. Structural complexity of vegetation also seems to control variation in bird species richness among the California Channel Islands (Diamond, 1969). However, this trend is not always present: in Patagonia at the southern tip of South America, Ralph (1985) found that species richness of birds actually decreases with greater structural complexity of vegetation.

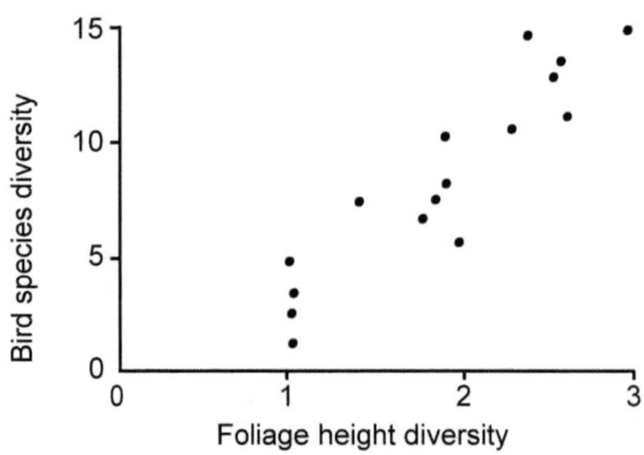

Figure 1.4. Bird species richness in habitats across North America is greater in vegetation with a more complex structure, here called foliage height diversity (after MacArthur and MacArthur, 1961).

On the upper reaches of a mountain, small alpine cushion plants may help to provide a diversity of small sheltered environments—microclimates—in which other plant species can establish themselves. Thus species richness can multiply itself by providing still more opportunities from diverse growth forms.

Locally, a big disruptive disturbance such as a landslide, a fire, or a volcanic explosion can wipe out almost every mature individual of the plants that were present before. I will talk later about the response of species richness overall to different *frequencies* of disturbance, but we can first think about how it varies in relation to an individual disturbance event.

After the dust has settled or the ground has cooled, plant species diversity increases as seeds or spores disperse in. They may be carried on the wind and aided by animals such as birds—or begin to grow back from seeds and roots near the surface. Usually, after a fire or landslide it takes no more than a week or two to see the first new life sprouting. Yet after a big volcano the first signs of recolonization can take several months: following the big volcanic explosion on Mount St. Helens in 1980, the first weeds were poking up from the moonscape of ash only after about six months. On the volcanic island of Krakatoa in Indonesia—sterilized by a vast explosion—and on the newly formed island of Surtsey near Iceland, the first plants to appear were mosses and lichens. Within a year or two flowering plants began to appear, and their diversity took off rapidly as more and more species arrived from nearby lands. On these volcanic islands, plant species richness seems to have risen especially fast as plants that had already established started to provide a more varied environment for others to establish successfully when their seeds arrived, by giving different levels of shade, for example.

Generally, as vegetation recovers back toward its original state—a process known as succession—species come and go, pushed out by the bigger and more competitive species that take over from them (often after a couple of decades in temperate-zone forests). The greatest species richness tends to be in the middle stages of recovery, when early fast-growing colonizers (the "r" species, see below) grow alongside the slower growing competitor species that will later grow and shade them out. Some of the competitors will have arrived in the early years too, but they remain small for a while as they slowly put on growth which they will eventually use to overtop their neighbors and dominate the community.

Birds too may follow a successional trend in species richness as vegetation grows back and becomes taller. Even though individual birds can fly in and out of the recovering habitat at will, their diversity recorded in observations reflects the number of species that choose to spend their time in the place. In the case of birds in a landscape of abandoned fields returning to forest, species richness tends simply to increase with succession, until it reaches the full levels of species richness of closed forest, with all its specialists (Figure 1.5).

Different amounts of heterogeneity in the local environment can explain some of the small-scale differences in species richness that are found across landscapes. But it cannot explain all of them. Some fairly uniform patches of habitat are just richer than others, even if they all occur on the same soil or rock type. This often has something to do with disturbance.

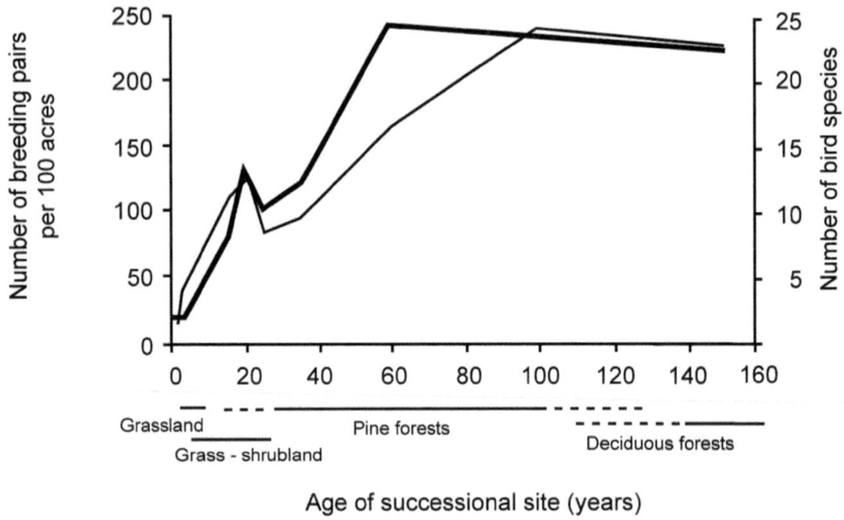

Figure 1.5. Bird species richness (right-hand axis) increases during the successional process in abandoned old fields (after Johnston and Odum).

As I mentioned above, disturbances to vegetation or a coral reef may explain why so many species manage to coexist, even though they have just about the same niche. The disturbances knock back the competition between the species, by destroying a large part of all of their populations every once in a while, and creating a temporary free for all. Now what would be the effect of different frequencies of disturbance?

Let's take a plant community, a meadow or a forest, for example. If there were no more disturbances at all, eventually the plants that had established on bare ground would grow bigger and start to play out a slow contest. Aboveground, the shoots and leaves of the fastest growing species would reach for the sun, overtopping other plants in a battle for light. Belowground, roots would intertwine and outdo one another in sucking up nutrients and water. Some would produce poisons that slowly killed their neighbors. Eventually, plants of certain species would begin to win out and smother the weaker types around them. The plant community would be left with only a few dominant species which would win the contest every time.

Disturbance that is sufficiently frequent tends to prevent this stage from ever being reached, allowing the weak to continue to coexist with the strong. We would expect that the more disturbance events there are, the less competition between species there will be, and the more species that are weak competitors can survive.

But there can only be so much of a good thing, even from the point of view of the species that benefit from the disturbance. If the plants or corals are too often trampled, uprooted, burned, or whatever, they will not be able to recover enough to build up a healthy population again, before the next disturbance hits. Beyond a certain point, relentless hard-hitting disasters such as storms or heavy grazing will damage plants or corals to the point where most species cannot recover. Only a few species, those that just happen to grow fast enough and reach maturity soon enough, will be able to survive. The result will be a local community with lower levels of species richness—because there is too much disturbance, and just a few species that can survive that level of disturbance. So it is rather like the state of the porridge in the *Tale of Goldilocks and the Three Bears* ("one bowl of porridge was too hot, the second was too cold—but the third was just right"). In terms of disturbance, there is such a thing as a happy medium, "just right" for high species richness. At this level, along a graph of disturbance frequencies, there will be a peak or a "hump" of species richness, rather like an old humpbacked bridge. Hence the predicted pattern is called a "humpback diversity curve". A certain amount of disturbance, then, may be good for species richness. By knocking back the likely winners as well as everything else, such disturbance events may level the field of play, allowing weaker species a chance to survive.

So, to recap, there are two different factors at work here in producing a humpback diversity trend along a disturbance gradient. At low levels of disturbance, the competitive game can easily run to its conclusion, and a few strong competitor species win out to give a community that is poor in species (Figure 1.6, left-hand side).

At intermediate levels of disturbance, the easing of competition allows more species to survive without strong competitors pushing the rest out. The result is a high point in species richness (Figure 1.6, middle).

Yet as the rate of disturbance increases further, another factor enters. This is the ability of plants to recover from severe disturbance. Most species of organisms do not have time to grow back after they are squashed, torn, burnt, or otherwise damaged. Or they do not have time to mature and scatter their seeds (or if they are sessile invertebrates, their larvae) before the next disturbance event kills nearly all the adults. There are fewer species around that can manage to cope with this amount of disturbance and keep their population going. As a consequence, species richness declines as disturbance increases beyond a certain point. Eventually things will reach a point where, at a certain level of disturbance, no species can survive and the species richness will reach zero (Figure 1.6, right-hand side). The result then is a "humpback curve" along a gradient in disturbance.

The idea that differences in the amount of disturbance can explain differences in species richness was first clearly put by Joseph Connell in 1972. It is often called Connell's Intermediate Disturbance Theory, because it explains the importance of intermediate levels of disturbance in promoting species richness. It focuses on "sessile" organisms, those which stay in one spot for most of their life, and includes not only plants but also many aquatic invertebrates such as corals, barnacles, and certain shellfish.

1.8 USING MODELS TO VALIDATE THE LOGIC OF THE ROLE OF DISTURBANCE IN ALLOWING COEXISTENCE

This idea of a balance between disturbance and competition is intuitively quite reasonable. But in science we prefer to use more than just our intuition—we like to take the imagined mechanism apart and look at it step by step to see if things could

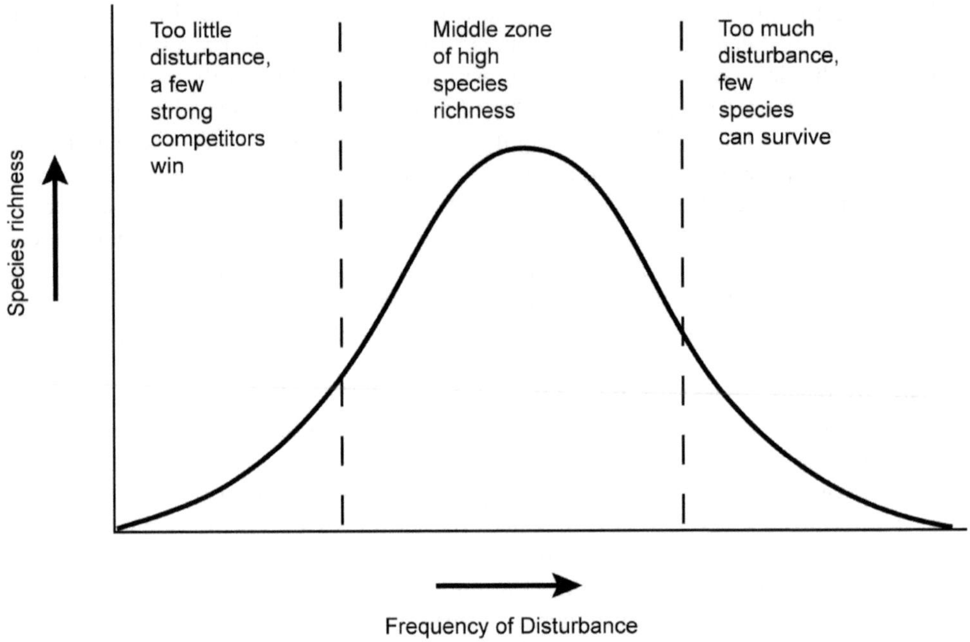

Figure 1.6. The "humpback curve" in species richness comparing habitats with differing frequencies of disturbance (Author).

really work that way. How then might we show more rigorously that disturbance could work as Connell originally suggested, allowing the weak to coexist with the strong? Usually, a good way to attack a problem is to reduce it to simple mathematical principles—a model, which can run on a computer. Mathematical logic adds an undeniable rigor to the reasoning process, and as long as the steps are based on reasonable assumptions it can be very useful in checking how well everything in a theory fits together (Huston, 1993).

In our model, we can set up a "virtual" plant community with some commonsense attributes of growth rate, birth rates, and death rates, and assign some species as stronger competitors and others as weaker. Then we let the plants grow and fight it out, following them through several generations of birth, growth, and death.

If we leave things undisturbed, eventually the stronger species will win and dominate, and the weaker species will be pushed out (Huston, 1993). If, however, we run the models with a disturbance event every so often, randomly killing a proportion of plants and then leaving open space for plants to come back from seed or scattered survivors—we find that the weaker ones manage to make a comeback each time. If the disturbances are frequent enough, the stronger species will never quite manage to push out the weaker ones. At least, the time it takes them to win the contests can take an almost infinite amount of time, the time scales on which the evolution of new species and mass extinctions take over more as the controls on species richness. The fact that one can make the same mechanism work in a model suggests that our intuition may be correct, and disturbance really *could* help prevent the stronger species from pushing everything else out.

1.9 DO HUMPBACK CURVES REALLY OCCUR ALONG DISTURBANCE GRADIENTS?

It sounds reasonable that humpback curves should exist in response to the amount of disturbance. Both commonsense and mathematical models support the idea. But it would be even more convincing to see the pattern actually occurring out in nature, comparing different localities that differ both in species richness and in levels of disturbance. Some of the first sets of observations that supported it were from rocky shores. Lubchenco (1978) looked at algal communities in New England in small rock pools around the tideline, subject to different amounts of disturbance in the form of grazing from marine snails (Figure 1.7). As the theory predicted, pools that had very little grazing had very few species of algae (seaweeds and suchlike) but a dense overgrowth of just a few species. Those with intermediate levels of grazing had the highest algal species richness, while rocks with the most grazing had low levels of richness.

Lubchenco did not find this humpback curve everywhere, though. On more wave-battered rocks of the shorefront, the species richness of algae simply declined with more grazing from the snails, instead of going over a hump. This she explained in terms of the total load of disturbance—from waves and snail grazing combined—being too much for the algae. With adequate amounts of disturbance from the waves,

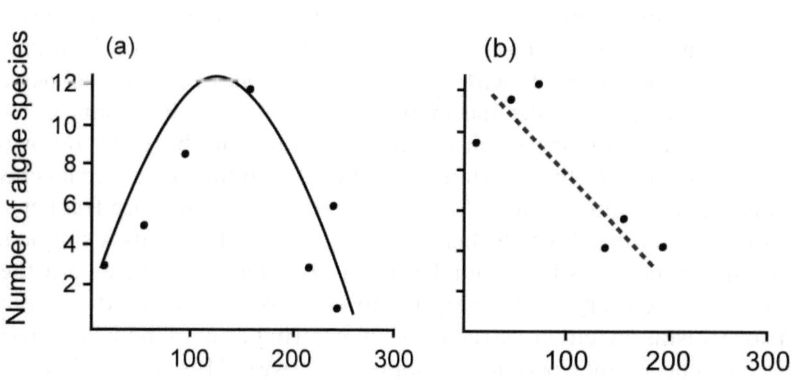

Figure 1.7. (a) Intermediate levels of grazing disturbance by snails on shoreline algae promote the greatest species richness (after Huston, after Lubchenco, 1978), (b) except in places on the shore which already have a heavy background amount of disturbance by waves. In these places the additional snail grazing is too much for the algae, and species richness declines with more grazing.

competition was not enough anyway to suppress the species richness of the algae, so adding snail grazing simply meant that fewer algal species could survive.

So, how about if we compare forests with different disturbance rates? When it comes to forests there is not as much data to compare, unfortunately. Certainly at one extreme, very high rates of disturbance keep forests species-poor and dominated by fast-growing colonizer species (Huston, 1993). For example, there are the "storm forests" of Kelantan in Malaysia, frequently smashed by storms coming in off the sea, and low in species richness. There are other rather mysterious forests in parts of the moist tropics where just a single tree species dominates over stands that can extend for many kilometers—surrounded by other types of forest that have all the usual tropical diversity. For example, there are the monotonous forests of *Gilbertiodendron*, a tree in the pea family Leguminosae, that extend over large parts of the Congo Basin in central Africa. There is nothing obviously different about the soils of these one-species forests compared with surrounding areas of forest that are very diverse, and the small plants that live in the shade under the trees are just as diverse in the *Gilbertiodendron* forests as elsewhere. Michael Huston suggests that these unusual low-diversity tropical forests are simply a product of lack of disturbance, in effect proving the point that without disturbance one strong competitor will push everything else out and dominate. This is a fascinating idea, and plausible, but it still remains to be properly tested.

There is no expectation among ecologists that they will *always* find a humpback curve of richness along a gradient of disturbance. For example, it might be that some environments are never actually disturbed often enough to produce the predicted decline in species richness at high disturbance intensities. In places where plants or corals grow and recover from disturbance very fast, there might never happen to be enough storms, or enough landslides, or enough fires to produce the decline that is supposed to happen with too much disturbance. Likewise, some environments might always end up getting so much disturbance that the pushing out of species by competition is not really an issue. In such cases, we might see just part of the humpbacked curve—either the left-hand side with its rise up to a peak of species richness, or just the right-hand side with its decline in richness with more disturbance. The more disturbed boulder environments that Lubchenco studied on shorelines (above) would be one example of this. In these instances, half of the humpback curve has been "cut off" because of the limited range of environments and disturbance frequencies that actually occurs in nature. Indeed, there are many cases where only a simple increase or decrease in diversity has been found along a disturbance gradient. But these tend to be in areas where everything is very highly disturbed, or very little disturbed, or where the organisms are either extremely slow-growing or extremely fast-growing, or at the extreme end of the spectrum in some other way. Huston (1993) has made a strong case that in a sense, these are the exceptions that prove the rule.

So humpback curves are not always found along disturbance gradients, nor should we expect them to be. Just finding "complete" humpback curves part of the time is fair enough evidence that the mechanism Connell described is actually at work in nature, and there are plenty of instances where they have indeed been found.

1.10 GRIME AND TILMAN: DISTURBANCE CREATES OTHER SORTS OF OPPORTUNITIES FOR COEXISTENCE, AND VARIATION IN LEVELS OF DISTURBANCE AFFECTS THESE OPPORTUNITIES

There are other explanations besides Connell's, for why we might get a humpbacked curve of richness along a gradient in disturbance. Some of these have already been mentioned above in the section on how species manage to coexist. The explanations of Grime (2001), Tilman (1994), and Grubb (1977) all invoke the role of a moderate amount of disturbance in opening up a wider range of niches for different species to exist in a plant community. Yet each of them is quite different from Connell's theory.

According to Grubb's (1977) explanation, disturbance gives a more varied environment of different light and soil conditions, and that gives more niches for plants to occupy—especially when they are young and first establishing. Imagine how the forest can be disturbed in so many different ways: large gaps and small ones, severe fires and mild ones, gaps left when a big tree dies and loses its leaves but stays standing, and ones left when a healthy tree blows down and uproots ... Each of these gaps has its own particular combination of soil and light conditions, and this combination may give opportunities for a particular set of species that can do well there (Grubb, 1977). However, if the environment is too heavily disturbed, all one has is essentially a uniform bare expanse of soil—with very little in the way of a range of niches for regenerating species. This would tend to mean fewer species in the community if there is too much disturbance.

In contrast, in the mechanisms which Grime (2001) and Tilman (1994) invoke, disturbed open environments in general are available to a range of species that are good at colonizing them. These are species that get in quick and multiply up, before any serious amount of competition starts to bite. Grime and other ecologists have called these "r" species—the "r" label referring to the "natural rate of increase" in the logistic equation that describes population growth toward a steady carrying capacity (I introduced these terms earlier in this chapter, in the context of "how does anything coexist with anything else?"). They are species whose natural rate of increase is very high, aided by their good dispersal abilities—due to a short generation time and investment in large numbers of small offspring. In a temperate forest, the most extreme r-selected species include weeds such as hawksbeard (*Hieracium*) and goldenrod (*Solidago*) whose seeds drift in on the wind. If disturbances in a forest, a shrubland, grassland, or coral reef are reasonably frequent, there will be patches where the "r" species are temporarily abundant, but intermixed with other areas where the community has not been disturbed recently and is dominated by long-lived competitor species. These strong competitors are often known in ecology as "K" species, after a term in the logistic equation that labels the stable carrying capacity for a population—the reason being that these species spend a lot of their time up near carrying capacity. In a temperate forest, the most extreme "K" species may be, for example, oaks (*Quercus*) and especially beech (*Fagus*), large bulky trees that live a long time. Grime, adding more finesse to his theory, actually divides up the "K" species into "C" species that do not colonize as early on but can grow relatively fast and push other species out once they get established, and "S" species that hang in under stressful conditions after the most readily available supply of nutrients has run out. For the sake of simplicity (this is not after all a book about general community ecology), I'll lump the later successional "S" and "C" species together into the category of "K" species.

So we have a mosaic of areas that have been colonized by vegetation after disturbance, some of them full of "r" species still living in gaps that were made recently, others occupied by "K" species that have pushed out the gap-colonizing ones. This sort of landscape will have the greatest number of species packed into it. Take another landscape, where disturbances hardly ever happen. Here, the only things found are likely to be the "K" species—and so that area is deprived of all the "r" species that could have been there. This is one end of the disturbance scale, and it has relatively few species. Now take a landscape where disturbances happen not just every once in a while but all the time, after just a few years a storm sweeps away most of the trees leaving the ground bare again to be colonized. In this situation, competitor species like oaks cannot gain a foothold and keep their population going, even though the "r" species do manage to get in and reproduce in time. Here then the landscape has lost out on its "K" species. So at this other extreme on the disturbance scale there are also relatively few species.

So, according to what Grime or Tilman (1994) invoke, along a gradient of disturbance a moderate amount of disturbance gives the greatest range of opportunities by providing a mosaic of different stages of recolonization of gaps. Too little disturbance means you lose the fast-growing "r" species, and too much disturbance means you lose the slower growing "K" species.

In fact, both this theory and that of Connell could operate in parallel, in the same places. It could be that even within the group of "K" species, there is actually a hierarchy with some being stronger competitors than others—and given too little disturbance, just one of these species will eventually push the other ones out. So a little disturbance both adds the "r" species and allows more "K" species to coexist by loosening up the competitive hierarchy.

1.11 WHEN STRATEGIES MIX—THE HUMPBACK CURVE WITH SUCCESSION, AFTER A DISTURBANCE

As a result of the different strategies invoked by Grime or by Tilman (1994), following a single disturbance event there can be a humpback diversity curve over *time* (not space) in that same spot. Looking at it simply, suppose we have only two sets of species in the forest, the "r" species and the "K" species. At first, following the devastation, mostly there will just be the fast-colonizing, fast-growing "r" species. Over time, however, the slower moving "K" species start to appear, and there is a stage when both "r" and "K" species are growing mixed together. This is the time when species richness is at its highest, because both types are present. However, gradually the "K" species push out the less competitive "r" species until there are none left. Because there are now only "K" species, diversity has declined from the middle stage.

In fact, a more accurate way of looking at it is to recognize that each species is not "all-r" or "all-K". Each can be placed on a spectrum in terms of how good it is at either colonizing fast or pushing out others after it arrives. Because of this, with succession there is often quite a complex sequence of species replacements, the whole community becoming more "K- like" over time. So for example, in a destroyed patch of forest in North America, the first plants on the scene are typically weed-like members of the daisy family Compositae and cabbage family Cruciferae. They are then followed by fast-growing trees such as cherry (*Prunus*) or birch (*Betula*), and later on by slower growing but longer lived species like oak (*Quercus*) and maple (*Acer*). If the forest is left undisturbed for a very long time, beech (*Fagus*) may start to dominate. The various different stages—each stage is in effect a niche—allow more species to exist in a community that is subject to disturbance. Zoom out to look at the landscape, and it is a mosaic of different species in patches at different stages in the recovery process. Even within a little localized patch of forest a few meters or tens of meters across, succession can give a mix of species as the leftovers from the previous stages are still growing side by side with those of later stages. The closer one is to the "middle" part of the succession, the more different species from different stages are likely to be present.

1.12 THE "OTHER" HUMPBACK CURVE: ALONG GRADIENTS IN NUTRIENT LEVELS

Some of the most consistent trends in local-scale species richness occur in relation to soil nutrient levels. For example, among sheep-grazed meadows in England (Figure 1.8), or patches of prairie in North America, areas with either very nutrient-poor or very nutrient-rich soils have few species. But areas with *intermediate* nutrient levels have the *highest* levels of species richness.

Plant species richness in various other ecosystem types such as shrublands has also been found to show this pattern. Puerto *et al.* (1990) found it in grasslands and shrublands in Spain. Often, rather than measuring the nutrient levels themselves, ecologists measure the growth rate of the vegetation (the net primary productivity) because it is known that plants on high-nutrient soils tend to grow faster. So in a meadow, for example, from harvesting and weighing the hay that grows in a certain area each summer, one has a rough proxy for nutrient levels. Another way that soil fertility can be guessed at indirectly is from the overall area of living leaves about each unit area of ground (the leaf area

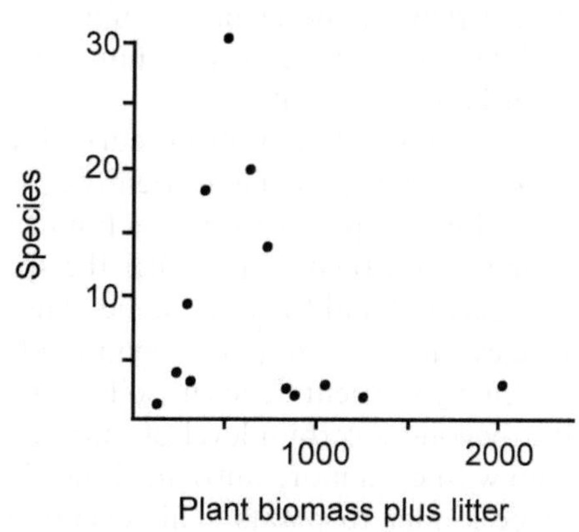

Figure 1.8. Species richness against soil fertility in herbaceous plant communities in Britain (after Al Mufti *et al.*, 1977).

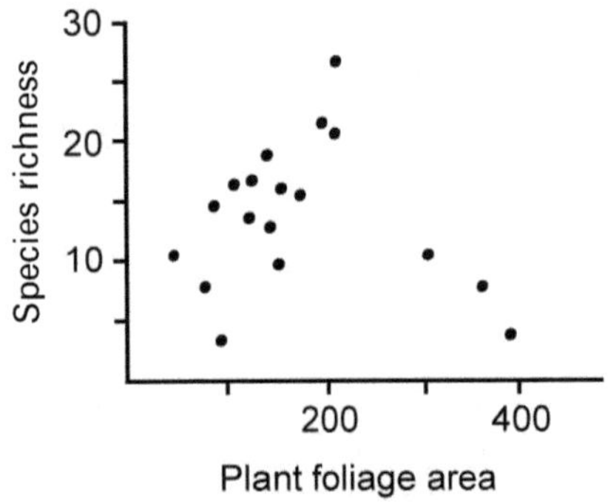

Figure 1.9. Humpback curve in species richness against an indirect measure of soil fertility, in South African fynbos scrubland (after Bond, 1983).

index): more leaves tend to require more soil nutrients to build and maintain them (Figure 1.9). The water-holding capacity of the soil can also be used as a predictor of soil fertility—in this case water itself is rather like a nutrient.

The same general trend of species diversity vs. nutrients has also been found in the phytoplankton species richness of lakes, which differ from one another in the nutrient levels in their waters. Very nutrient-poor lakes tend to have few species of phytoplankton, and so do very nutrient-rich ones. But waters with intermediate levels of nutrients seem to be "just right", with the highest species richness. It is the same out in the open ocean: the very nutrient-poor centers of ocean gyres (great whirls of water circulating around ocean basins) have higher species richness of phytoplankton (Huston, 1993). In contrast the nutrient-rich shallow-shelf seas close to the continents are much poorer in phytoplankton species, even though they have a much higher biomass of the species which do occur.

Although I have concentrated here on plant communities, a similar gradient could exist along gradients in the nutrient or food supply among animal communities (e.g., coral reefs). Indeed, Huston suggests that the differences in coral diversity with depth in reefs is partly a product of this. In reefs around the world, the highest coral species richness is found at some depth—below about 10 meters—although it declines at greater depths. This peak of diversity corresponds to intermediate amounts of sunlight—and intermediate coral growth rates as a result.

So very often with the plant communities in a particular region, if you arrange species richness along an axis of nutrient supply, there will be a peak or a hump in the middle. So it too is called a "humpback diversity curve", just like the one that follows disturbance gradients.

This sort of humpback curve initially puzzled ecologists. After all, we would intuitively expect that a nutrient-rich environment would be easy for plants to survive in, so lots of species would be found living in those conditions. And the greater the supply of nutrients, the higher the species richness ought to be. The expected trend *does* seem to hold true to some extent, going up from very low levels of nutrients; species richness in plant communities definitely increases compared with most extremely nutrient deficient soils (e.g., the "white sand" soils of parts of the tropics). But beyond a certain level of nutrients the number of species starts to level off, and then with even more nutrients it declines—even though the environment would seem to be kinder to plants. This decline in species richness despite increasing nutrient supply was named the "paradox of enrichment" by Michael Rosenzweig (1971). As well as being seen by just observing different patches of habitat, it can also be seen experimentally when nutrients are added. For example, if a lake or pond becomes polluted with fertilizer runoff—or if a mischievous ecologist adds nutrients just to see what happens—there is usually a sharp decrease in species richness of the tiny plants floating in the plankton. It tends to be the same if fertilizer is thrown onto a species-rich grassland. I saw this time and again when I worked in nature conservation in my early twenties, helplessly watching the last species-rich meadows in my home county vanishing under a mat of lush ryegrass, as farmers added artificial fertilizer for the first time in centuries. The same thing has been done on a more systematic and scientific level in a study at the Rothampstead Experimental Station in eastern England. A plot of traditional old species–rich meadow had chemical fertilizer added to it each year, and was kept mown, for 90 years from 1856 to 1949 (Huston, 1993). Species richness among the meadow plants progressively plunged, until by the end of the experiment only a very few types remained. As an aside, it is interesting that the scientists who set up this experiment were working well before the days of modern ecology, and had no clear picture of what they expected to see or why it would be beneficial to record every species and its abundance for nearly a century. It is hard to imagine any modern-day grant review panel approving something like that nowadays. Quite apart from the length of time taken to get results, they would slam it all as having "no clear hypothesis".

Adding to the confusion about the controls on species richness, in the same place different groups often respond differently to an increased supply of nutrients or water to the plants in that ecosystem. Moving eastwards from the dry central U.S.A. into moister climates, the number of species of herbaceous plants declines (Huston, 1993), while diversity of woody plants increases (Currie and Paquin, 1987).

In the deep oceans, less supply of food falling from above in the nutrient-poor open oceans apparently means *more* species of the tiny forams which feed off dead organic matter (just as there are more species of phytoplankton up above, supplying it), yet there are *fewer* species of fishes in these impoverished regions.

There are some types of plant communities in which the humpbacked curve has not as often been found along primary productivity gradients. For example, it has not so often been found among forest communities—at least in the tropics. Often it seems that the more nutrient-poor the soil that a tropical forest is growing upon, the richer it is in species, with no sign of species richness declining towards a minimum on the least fertile soils.

However, the humpbacked diversity pattern *has* sometimes been recorded in tropical rainforest in Southeast Asia, the Americas, and Africa (Figure 1.10), where sites with moderately low soil nutrient levels have much higher tree species richness than sites with either very low or high levels of nutrients. At the extreme low end of both the nutrient and diversity scale are places where the soil is very leached, old, white sand, consisting of almost pure quartz. At the opposite end of the scale are silty soils deposited by rivers—very rich in nutrients, but also poor in species.

Sometimes, soil nutrient analyses are in short supply, so indirect clues have to be used. Across the small Central American country of Costa Rica a rough way of estimating the fertility of the soil is to look at tree height—because trees tend to grow bigger in fertile soils. It turns out that forests of intermediate height, and thus intermediate soil fertility, have the greatest species richness (this is the index of

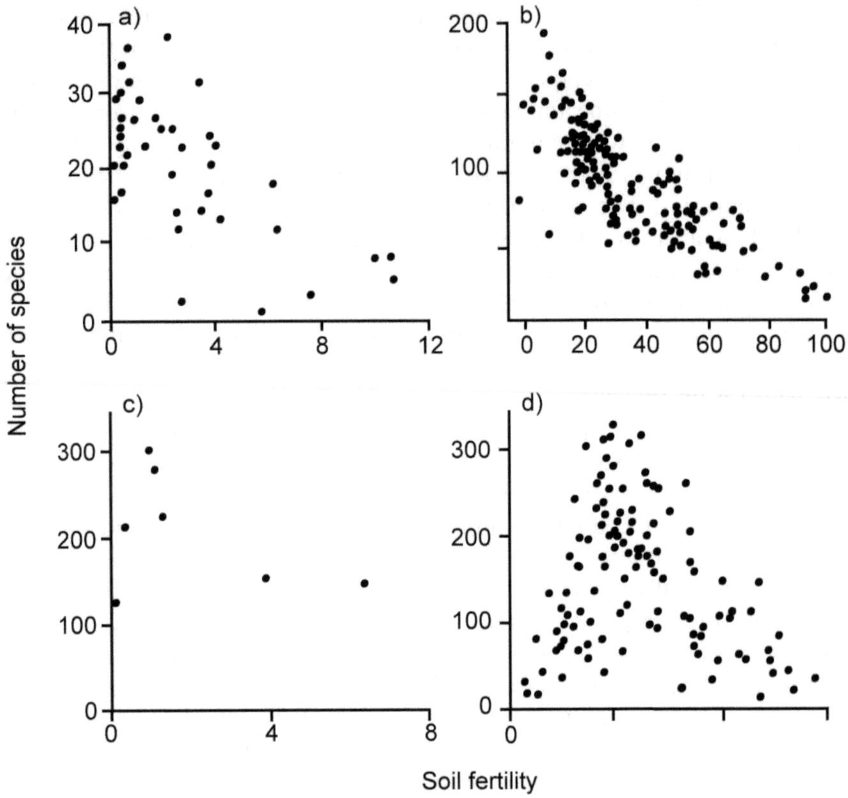

Figure 1.10. Species richness in tropical rainforests against soil fertility: (a) Costa Rica, (b) Ghana, (c) Malaysia/Indonesia, (d) Amazon Basin (after Huston, 1993).

fertility used to arrive at Figure 1.10a). In some cases these differences in soil fertility are really on a geographical scale, and they are actually due to differences in the rainforest climate over a few tens of kilometers. Generally, in places that are extremely rainy (even by tropical standards), soils are poorer because most the nutrients have been leached out of them—and the forests in these areas tend to be most species-rich.

So what could cause the "paradox of enrichment"? Eventually after much scratching of heads, ecologists came up with some possible explanations for what could cause this humpbacked pattern. One set of theories focuses on the interaction of growth rates (affected by nutrient levels) on a factor I have already mentioned, and the role of disturbance in shaking up plant communities. According to this idea, nutrient levels affect the extent to which a certain level of disturbance prevents strongly competitive types of plants from growing fast and pushing out weaker competitors. Suddenly, we find that the role of nutrients is tied in with the "intermediate disturbance" theories I mentioned above. So this is how it is generally thought to work:

— At *low-nutrient levels*, few plants are able to survive the difficult conditions. They need special adaptations to capture and store nutrients. Because low-nutrient environments are relatively rare, few plants have adapted to the difficulties presented by these environments ... so they are poor in species.
— At *intermediate-nutrient levels*, most types of plants have enough nutrients to survive and grow. But because nutrients aren't super-abundant, plants grow slowly relative to the rate of disturbance events which wipe out all or most of the plants irrespective of which species they belong to. These disturbance events reset the "clock" of competition, the process whereby the best competitors would otherwise wipe out the weaker ones so that relatively few species were left.
— At *high-nutrient levels*, however, most plants can grow well and grow fast, but they are all growing fast relative to the rate of disturbances. So between those disturbance events the plants can grow dense and tall and the better competitors win out over the weaker competitors. Eventually a relatively small number of the best competitor species push out all the weaker ones; so the site has low species richness.

Whereas in the scenarios mentioned earlier with the intermediate disturbance hypothesis, we had the vegetation growing at the *same* rate in different frequencies of disturbance, in this instance we have vegetation growing at *different* rates under the same frequency of disturbance. From the point of view of plants getting pushed out by competition, the effect is more or less the same—little effect from disturbance at one end of the scale, increasing to a strong influence of disturbance at the other.

As with the intermediate disturbance theory, we should not necessarily see a full "humpback" curve along the range of different soil nutrient conditions found in any particular area. It might be that all the grasslands or shrublands in that area have fairly nutrient-poor soils, so we will only ever see the "left-hand side" (see Figure 1.11) of increasing species richness with nutrient levels. On the other hand, in another area, soils might all be quite fertile, so we only ever get to see a decline in species richness with increasing nutrient levels. Just finding the expected "humpback" curve once in a while seems good enough evidence that the theory itself might be right.

In his book Michael Huston (1993) made a wide-ranging and thorough case for thinking that variations in vegetation growth rate, due to either nutrients or water supply, are a major cause of differences in species richness on the local scale. I must admit I like Michael's ideas, and not just because he used to work in the office next door to mine. His book, while it is heavy-going, makes a very good, coherent case for the role of certain factors in affecting species richness. He also suggests that if we look across different climate zones, biomes, and geographical regions, what controls differences in species richness is very often nutrient supply as well—relative to a fairly constant background rate of disturbance (see below). Here then we may be seeing an unexpected unity of scales: basically the same processes controlling variation in diversity between the local and the geographical scale.

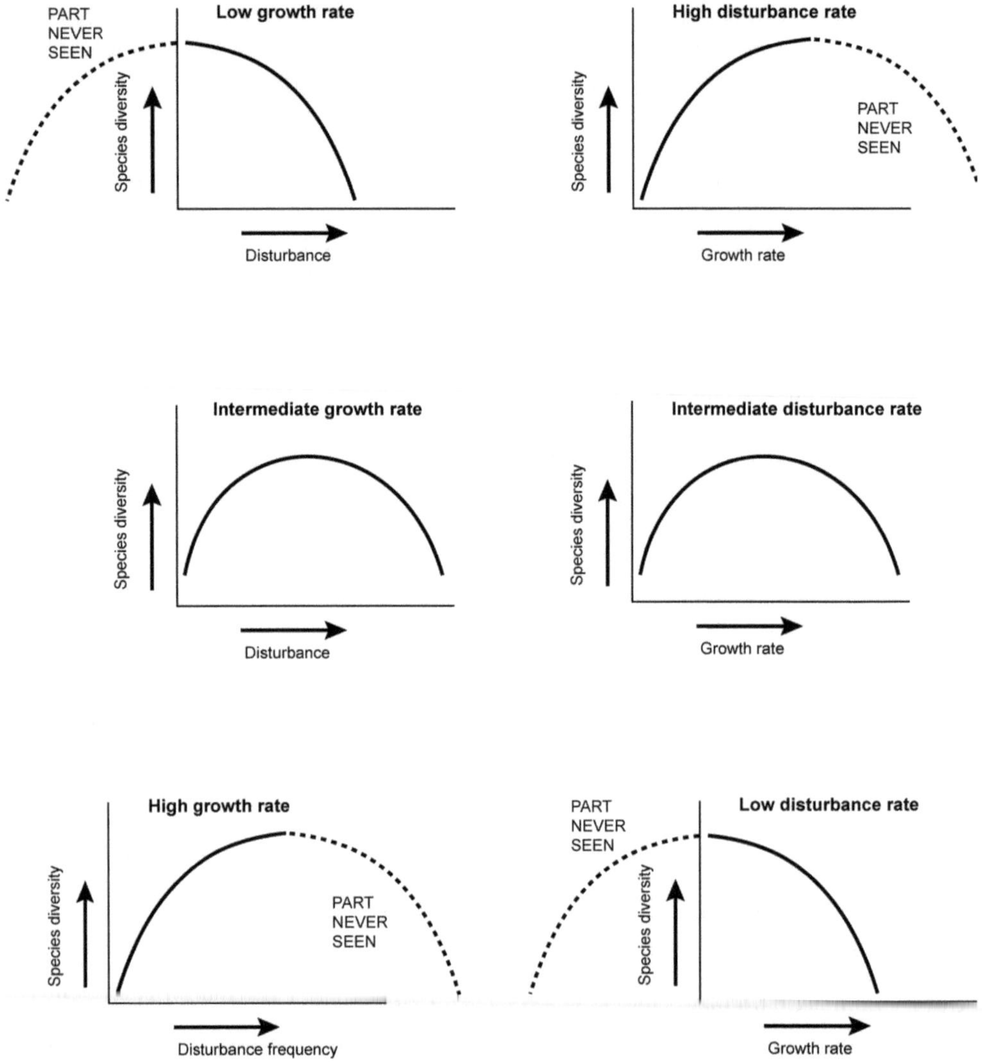

Figure 1.11. A summary of the possible circumstances leading to humpbacked species richness curves, and one-directional rising or falling species richness in other cases (modified from Huston, 1993). If background levels of disturbance or nutrients are either too low or high, we do not see a humpback curve as richness responds to the other factor. The "hidden" part of the humpback curve is rather like the hidden part of a rainbow that hits the side of a mountain range—we know that it potentially exists, but it is just not present due to the circumstances.

Only tropical (not temperate) forests show a humpback curve in relation to soil nutrients. In tropical forests, there seems to be a humpback curve in tree species richness, with a rise and then a decline along a gradient of soil fertility (Huston, 1993). So, for instance, we have those various examples of site-to-site differences in species richness in the tropics, that I mentioned above. Yet in the temperate zone there seems to be only increasing tree species richness as the soil becomes more fertile, without a decline at the higher levels. This is the conclusion that Peter Grubb (Grubb, 1987) reached, based upon his widespread observations of forests in the mid-latitudes of Europe, Asia, and North America. In the temperate zone, it seems to be the most nutrient-rich soils of flat river bottomlands (not really swamp forests, but forests that have moist soils enriched frequently by silt from floods) that have the most tree species packed in. Why is it that the temperate forests show only one trend: the left-hand part of the humpback curve? Could it be that disturbance rates are somehow greater in the temperate zone, preventing tree species from being pushed out, even when they are growing fastest?

We can probably better explain it in terms of temperate-zone trees growing more slowly relative to disturbance rates. In the cooler climate of the mid-latitudes, trees generally grow more slowly, so disturbance events always seem "very frequent" relative to the slow growth rate and long life cycle of these trees. So competitive exclusion of many species due to lack of disturbance is not a factor, even when the forest is growing fairly fast on a fertile soil. A forest then tends to be quite different from a temperate grassland, which can easily grow too fast for the disturbance rate if the soil is fertile: forests perceive their environment in centuries, not months or years as a grassland does.

In forests in the tropical zone, it seems to be a different story. What we tend to see, comparing sites with one another, is a humpbacked curve in local-scale tree species richness with increasing soil fertility. The tropics have a year-round growing season, and it is always very warm. By contrast, in the temperate zone, trees stop for the winter—so even if disturbances are happening at about the same rate in the tropical and temperate zone they would have more effect on the slower growing temperate trees. With warm temperatures and year-round growth in the tropics, there can easily be outcompetition by the stronger competitors, causing the decrease in diversity that is seen on more fertile soils in the tropics.

1.13 SPECIES RICHNESS IS A BALANCING ACT BETWEEN THE EFFECTS OF DISTURBANCE AND NUTRIENTS

So the humpbacked species richness curve has been found in relation to nutrient supply in many different ecosystem types, including grasslands and shrublands on land, plus coral reefs and deep-sea animals. It can occur in forests, but it seems to be less frequent there—only occurring in the tropics. In the case of forests it is perhaps rarer because the trees in the forest are growing so slowly relative to the frequency of disturbance that even rare disruptions are always enough to prevent dominant species from outcompeting others. Different-sized organisms will "perceive" disturbances very differently according to the time frame on which they live their lives. In the warm, moist tropics, the growth rate of the forests may be sufficiently fast that it takes a nutrient-poor soil to slow down growth enough to avoid the pushing out of species by competition.

There is a complex interaction then between disturbance and growth rate. Increasing the growth rate, while keeping the same disturbance frequency, suits the basic survival requirements of more species. But it also means more pushing out by competition, and beyond a certain point this means that fewer species will actually survive. Increasing disturbance while keeping growth rate constant means less pushing out, but if the damage is too severe it can translate into fewer species.

The two types of humpback curves are really different sides of the same thing: a balance between disturbance and growth rate. Michael Huston (1993) has combined them into one diagram, which expresses the importance of the *ratio* between the two. Along the middle he predicts a diagonal bar where this ratio is "just right", and where species richness reaches its maximum (Figure 1.12). This he suggests will explain many of the local-scale patterns that one sees in species richness. He goes on to suggest that the broader scale trends of latitudinal gradients and geographical "hot-spots" in species richness (Chapters 3 and 4) could also be the result of differences in this balance between growth rate and disturbance.

Huston also emphasizes that in any particular set of environments, we might not see the full humpbacked curve because it may be cut off either on the right- or the left-hand side—simply because that set of disturbance levels or growth rates doesn't occur in that region (Figure 1.11). I suppose it is rather like when a rainbow hits the side of a mountain—if the mountain was not there, the rainbow we see would continue round to form its full arc. Even though we do not see it, the full humpback diversity curve can still potentially be there, just not seen because there is not a wide enough range of environments.

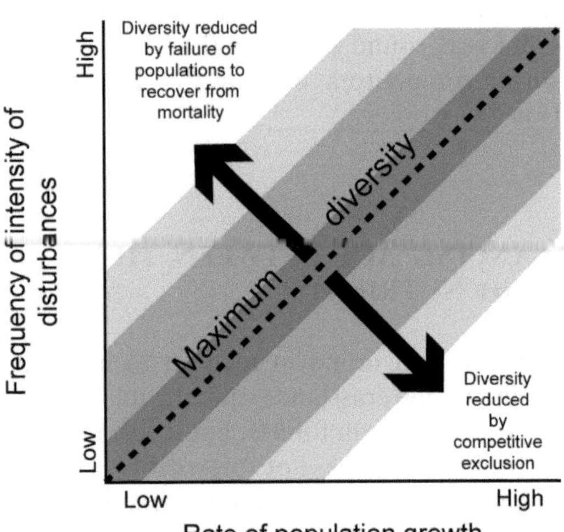

Figure 1.12. The interacting effects of growth rate and disturbance frequency on the species richness of communities. Maximum species richness is all about having the right balance between growth rate and disturbance (after Huston, 1993).

Some interesting experiments suggest that Huston has got it about right, when he talks of an interaction between growth rate and disturbance rate affecting species richness. Some of these experiments have succeeded in creating the diagonal bar of maximum richness which he predicted, where the ratio between disturbance and growth is just right.

One ingenious set-up used small chambers (known as "microcosms") of stirred growth medium (Rashit and Bazin, 1987) containing multiple species of bacteria or protozoa. The idea was to simulate different combinations of disturbance vs. growth rate. The microbes were floating in a liquid growth medium that was flushed out (and discarded) at a fixed rate, as new empty medium was added in to replace it. This represented a disturbance to the growing populations of microbes—loss and dilution of medium inside the chamber is like the mass death that follows a fire or a storm in land vegetation, for instance. The other side of things, growth rate of the microbes, was modified by varying either the temperature, or concentration of nutrients in the new empty medium that was added. The warmer the temperature or the more nutrients available, the faster the microbes grew, and the greater the chance that they would be able to start seriously competing with one another. More disturbance though was capable of counteracting this effect, by knocking back the growing populations frequently enough to prevent severe competition. The result was just as the theory predicts—two humpback curves of diversity, one along a gradient of disturbance (if growth rate is constant) and the other along a gradient of growth rate (if disturbance frequency is kept constant). Combining all the results gave a middle bar of high diversity of the type shown in Figure 1.12.

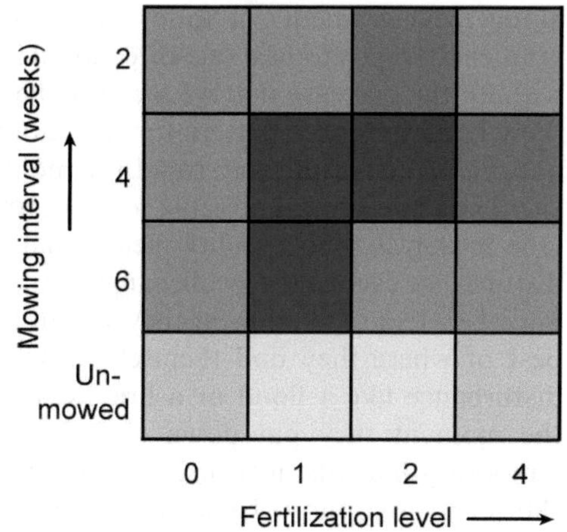

Figure 1.13. Huston's experiment with different intensities of mowing and differing degrees of soil fertility on plots of grassland. Darker colors equal more plant diversity. Maximum species richness occurs where the balance between the two factors is right, regardless of their absolute level.

Another very interesting and clever experiment was carried out by Huston himself (Huston, 1980), using plots of vegetation in an abandoned agricultural field at the Ann Arbor Campus of the University of Michigan. The grid of small plots was arranged to provide differing combinations of growth rate of plants (due to different amounts of fertilizer added each season) and disturbance (in terms of different frequency of mowing the vegetation). What he found was as expected: a high diversity band in the middle following the "optimal" ratio between growth rate and frequency of disturbance (Figure 1.13).

1.14 WHY HASN'T A HUMPBACKED DIVERSITY CURVE BEEN FOUND FOR ANIMALS?

The humpback diversity curve in relation to nutrients or disturbance frequencies has mostly been found for land-based and aquatic plant communities, and a few types of aquatic animal communities that are dominated by "sessile" animals—those rooted to the spot like plants are. However, for most animal communities (e.g., for the birds, mammals, fishes, and insects of the world), there is no sign of any humpback pattern along local gradients of food supply or disturbance frequency. When there is a gradient, it tends to be a one-directional increase in species richness with food supply, without the decrease that we see at higher productivity levels for plants. Why should there be a humpback pattern for sessile organisms like plants and corals, but either no pattern or just a one-way trend for mobile ones like birds or mammals?

There are several possible reasons, and they might all be correct to some extent. One reason is that mobility means that animals are free to move in response to a disturbance event, or a local change in the availability of resources. Plants are stuck in one place from the moment they sprout from a seed, and they must simply make the best of where they find themselves. They also cannot move out of the way of a disturbance like a flood or a fire. Sessile animals are similarly stuck in place from the moment they put down as a larva. Mobile animals, however, can flee the temporary hazards in their local environment (such as a fire, a tornado, or a landslide) and then move back in afterwards. They can also move out when resources become scarce, and go and find themselves another patch of habitat. This mobility tends to jumble up the relationship between the local populations of each species, and blur out the sorts of factors that would produce a humpbacked curve.

A very few examples of humpbacked curves in animals *have* been found on land though (e.g., along a net primary productivity, NPP, gradient in the Middle East deserts of Israel and Jordan). NPP, remember, is the growth rate of plants which ultimately supplies all terrestrial ecosystems with food (Chapter 2). Most species of rodents and lizards are found in moderately arid areas, which have intermediate productivity. Too moist or too arid, and fewer species are to be found in that area. On the other hand, this is more a geographical than a local-scale gradient, because it is ultimately controlled by climate, not soil. A strong climate gradient that occurs over the horizontal distance of a few tens of kilometers is unusual, but can occur on the edge of desert. Even a distance of a few tens of kilometers is likely to be outside the migration range of individuals of many small animals. In a sense this is actually a geographical-scale gradient that may tap into different mechanisms that normally operate at the broader scale (see Chapters 1 and 2, and below), because we do not have movements of individuals messing up the pattern. It is indeed a humpback curve though, even if it does not operate on a strictly local ecological scale.

Another set of humpback curves in animal diversity may be seen within a very different environment, the oceans. Some of these actually do include mobile animals, although they are small invertebrates on the ocean floor that probably do not move far during their lives. More about that later in this chapter, though.

Michael Huston (1993) suggested another general reason why animals less often show the same humpback trend as photosynthesizers such as plants and corals along a gradient of increasing primary productivity. Essentially, animals feed off scraps—the small proportion of the energy supply and materials originally fixed by photosynthesizers but is passed along the food chain to them. Given how sparse their supply of food is, there is not much chance of the animals becoming so abundant all the time that species crowd one another out by competition. Before they can become too abundant, bad-luck events such as droughts or other extreme weather events will tend to whack their numbers back down.

Thus we should expect that animals will only ever respond positively in terms of species richness, to the easier conditions of increased food supply from faster growing and lusher plant life. Are there any observations to either back up or refute Huston's idea? Supporting the idea are observations that show species richness of mammals in tropical forests tends to increase with soil fertility, even while the forest tree diversity itself is declining along the same gradient. If food is in short supply, few species can survive. So at high levels of plant growth, the mammals are benefiting from the abundance of edible leaves and fruits and perhaps insects too. But they do not reach such high densities—relative to population setbacks in disturbance events—that species start to push one another out because of their overlapping niches.

Animals, then, may never reach the peak of abundance that would begin forcing species diversity down as food supply increased, giving the characteristic humpbacked curve found in plants. And the higher up the food chain, the scrappier the remnants they receive—so predators should be especially unlikely to show a humpbacked curve in relation to resource supply, only an increase.

I must admit I am a bit skeptical of this idea, because I think that there is plenty of evidence for intense competition in nature among members of animal species (e.g., the territoriality of many mammals, birds, and other animals), and if members of the same species are competing then members of different species will be probably competing quite intensely too. But granted, the local-scale patterns in diversity seem to fit in with ideas Connell and Huston have on intermediate disturbance.

1.15 A QUITE DIFFERENT EXPLANATION FOR HUMPBACKED DIVERSITY CURVES IN PLANTS, IN RELATION TO SOIL FERTILITY

Another type of explanation for these humpback curves is that the level of nutrients affects how finely plants can divide up their world into "niches"—and the more niches there are, the more species can coexist in a stable arrangement. We can call these "niche-partitioning" hypotheses. They contrast with the explanation that I have just outlined above from Connell's theory, where the balance between growth rate and disturbance allows varying degrees of niche overlap—with most of the extra species in diverse communities being there because their niches are able to overlap.

One of these niche-partitioning ideas, to explain the humpbacked curve of plant species diversity relative to NPP, comes from David Tilman (1982). This is quite aside from his later (Tilman, 1994) explanation of how species can coexist in the same community, based on how quickly they can disperse after a disturbance. So in his 1982 book, Tilman suggested that each plant species may be specialized to exploit a particular *ratio of nutrient levels*. In a very nutrient-rich soil the ratios can't differ as much from one local patch within the soil from another—any variations are swamped out by the high overall concentrations. So because of this, the nutrient-rich environment is less heterogenous, giving fewer niches, so fewer different species of plants can coexist. This swamping-out effect is rather like an overexposed picture, where the

Figure 1.14. The effect of too high a background level of something (in this case reflected light) on the amount of patchiness. (Top) A scene at normal exposure shows clear contrasting spatial patterns. (Bottom) Overexposed photo "flooded" with light shows less heterogeneity. Tilman suggested that a rather similar effect occurs with heterogeneity in nutrient ratios—at high overall background levels of nutrients the patchiness of nutrient ratios is weaker.

whiteness overwhelms everything so that there is no contrast in tone across the picture (Figure 1.14).

So basically, Tilman suggested that there could simply be more opportunities in the relative concentrations of different nutrients in an environment that had moderate levels of nutrients overall.

At low nutrient levels, Tilman agrees with Connell, Huston and others that few types of plants have managed to adapt to survive on those soils. Tilman suggested that the interaction of these two different factors produces the humpback curve with increasing nutrient concentrations: there is less opportunity for niche differentiation and specialization at higher nutrient levels.

Tilman's nutrient heterogeneity explanation differs from the more generally accepted view of Connell and Huston (above) that plants in the same community generally have strongly overlapping niches, so that stronger competitors will push most of them out if disturbance doesn't even up the score. In Tilman's theory, the plants actually don't have strongly overlapping niches and can each "hold their ground" in their own niche if the environment is varied enough (in terms of nutrient ratios) to give them a little local environment that is just right for them.

Tilman's theory began from experiments on diatoms—microscopic algae which live in lakes, rivers, and seas. Individual species of diatom were grown in vessels with precisely controlled ratios of concentration of nutrients, and it was found that each species did best at absorbing certain nutrients from the water down to very low concentrations, but not so good at absorbing other nutrients. This meant that usually when they were grown in combinations of more than one diatom species, at a particular combination of concentrations of nutrients, one population would draw down the concentrations of some nutrient that was in especially short supply and drive the other population extinct.

So Tilman's idea partly rests on a principle of how species successfully compete. The species that will keep on fighting after there is very little left to fight over, is the one that wins. Tilman suggested that in a varied environment, there would be very local differences in terms of which nutrients were in shortest supply—and that would affect which species did best there in that patch of water or (by analogy) in a patch of soil. No one species would be able to win out everywhere, so this allowed a mosaic of species to coexist in that place. How then does one get to a humpbacked curve between different local environments, some of which are richer overall in nutrients and some poorer in nutrients?

I will set out the theory again just briefly, because I know that it is hard to grasp. At very low nutrient levels, it is hard for any plant to survive. At very high nutrient levels, there is not much opportunity for local differences in nutrient ratios to arise because everything is swamped out by high overall amounts of nutrients. Even plants locally sucking up nutrients in a very nutrient-rich soil cannot make much difference to the ratios in concentration. But at moderate nutrient levels, there is enough heterogeneity for many plant species to occupy their different niches and coexist.

Tilman's (1982) theory has not been well-received by the world of ecology in general (unlike his later suggestion put forward in 1994 that "r" and "K"-type differences in niches are very important in allowing coexistence). Most ecologists are skeptical that plant niches and plant communities could ever be as "fine-tuned" to precise nutrient ratios as Tilman has advocated. I myself feel that Tilman's mechanism is a little too elegant, and too dependent on precise adaptations, to really work out in nature which usually seems to be rather a messy place. Tilman has at least some evidence from laboratory experiments on diatoms (microscopic algae from lakes) that apparently show this mechanism at work. But there does not appear to be any such evidence for larger plants managing to coexist in this way. So perhaps Tilman's mechanism is important in producing the humpbacked curve in some circumstances (e.g., plankton), but not in others. There is no reason that all humpback curves everywhere have to be produced by the same mechanisms.

1.16 "POISONED" AND "EXTREME" ENVIRONMENTS ARE USUALLY POOR IN SPECIES

Serpentinite is a greenish rock that forms when older rocks like granite are squeezed and partially melted by the forces that build mountains. It occurs scattered in small outcrops around the world, with California having more than its fair share of these. The soils that form from serpentinite have an unusual chemistry that reflects the composition of the rock. They are unusually low in calcium but very rich in magnesium, and have high concentrations of poisonous metals like cadmium and nickel. An outcrop of serpentinite in California is usually strikingly different from the areas of other rocks around it, because of the sparse, stunted-looking vegetation. Within this vegetation, there are far fewer species than adjacent areas. The low species richness of the poisonous nutrient-poor environment over serpentinite rock is typical of a general pattern that goes with localized extreme environments, having far fewer species than other adjacent communities. Another classic example is the lead and copper spoil heaps in north Wales, in the U.K. Abandoned many decades ago, the debris that was dug out from the mine has accumulated an open cover of vegetation. This tends to be far poorer in species than adjacent areas of pasture, apparently because of the high concentrations of lead or copper in the soil. Whereas a nearby unpolluted meadow may contain 30 or 40 species of plants, a spoil heap often has only two or three. Environments that have an unusual chemistry tend to have lower species richness. This seems to be because relatively few species have evolved the mechanisms to cope with such rare environments. If we study the plants growing in such places in detail, it often turns out that they are distinct genetic strains, even distinct sub-species, that have adapted to this extreme environment. In the serpentine areas, endemic subspecies of more widespread plants are often present. Maybe if we left the Welsh spoil heaps for a hundred thousand years, they would gradually accumulate more species that had evolved to cope with them. But also bear in mind that as these rocks and sediments are eroded away or covered with ice every few tens of thousands of years, they vanish in ecological terms—so in this unstable world there is not much chance of them accumulating a full set of specially adapted species. As with the plants, soils that have high concentration of poisonous metals also seem to be far poorer in terms of their diversity of microbial life. In some experiments looking at the diversity of types of bacteria in soils using DNA analysis, "species" richness plunged to about one thousandth after heavy-metal salts were added to the soil.

In the aquatic realm, extreme environments are also species-poor. For example, hot springs above 70°C, such as those in Yellowstone National Park, cannot support any multicellular animals or protozoa: they are entirely inhabited by bacteria, the simplest forms of life. Even the specialized microbial flora in such springs is less diverse than in more "normal" aquatic environments.

Why are extreme environments less species-rich? Generally, the rarity of an extreme and "difficult" environment—in space and in time—may tend to mean that few species end up adapting to it. Furthermore, some extreme environments may preclude whole classes of organisms because they are incompatible with their growth form or the intricate cellular structures within them. The same argument might also be applied to the most nutrient-poor environments (above) that help form humpback diversity curves: these too are relatively rare, yet require specialized adaptations for survival. Their lack of nutrients may make them unsuitable for large organisms such as trees, and the paucity of material passed down the food chain may mean that few species of animals can live there.

1.17 MOUNTAIN-SCALE PATTERNS IN SPECIES RICHNESS

On the borderline between the local and geographical scale are the effects of elevation on diversity. On a mountain, strikingly different climates can occur within a few hundred meters in altitude—a distance that might be covered on foot in an afternoon. Not only does temperature decline with altitude—mostly because the blanket of greenhouses gases above is thinner—but there also tend to be big changes in rainfall. Air rising over the higher slopes of mountains tends to cool and then form clouds and rain, so mountains tend to be cloudier and wetter environments. However, the highest mountains which extend above the main cloud layer get drier towards the top (Adams, 2007). Adding to the complexity of mountain environments, they often tend to have a wide range of different rock types, including both folded-up layers of sedimentary rocks and also igneous rocks intruded from molten magmas inside the Earth. This geological complexity tends to give a range of soil types, which interact with the effects of climate variation on a mountain to give a broad diversity of environments for plants.

In general, species richness declines with increasing elevation on a mountain. For example, the trend has been noted (Chapter 2) in the diversity of trees on the slopes of tropical mountains. Among plots of 0.1 ha in size, there is a linear decrease in tree species richness with altitude in tropical Mexico and in the Andes mountains at the western edge of Amazonia (Silman, 2005). Jared Diamond also recorded how bird species richness declines with altitude in New Guinea and the Americas. The altitudinal decline in diversity has often been considered as a miniature version of the great latitudinal trends, where species richness also declines towards colder climates. Many of the same explanations have been put forward for both the altitudinal and latitudinal trends (e.g., the amount of energy available to partition up niches, or the rarity of such cold environments in the geological past making it less likely that species would have adapted to them).

Mountains have also often been considered as islands—small areas of cold climate cut off by warm lowlands. Since they are often so isolated and restricted in area, the upper parts of mountains might be subject to the same sorts of ecological processes that determine the diversity of remote oceanic islands (Chapter 4). According to the theory of MacArthur and Wilson (1967), isolated environments have difficulty accumulating species partly because few arrive, and partly because those that are there tend to die out frequently. It has been suggested that part of the reason diversity decreases towards the top of mountains is that they tend to slope up towards a point, which means that there is progressively less area in each climate zone for species at higher altitudes—and thus smaller populations and more frequent extinctions.

However, it may be worth considering the drawbacks to MacArthur and Wilson's theory, which are mentioned in the chapter on hotspots and coldspots in diversity (Chapter 4). Species diversity on mountains seems likely to be a lot more complex than a simple balance between chance arrival and random extinction. For one thing, we ought to consider recent geological history. During the numerous Quaternary ice ages over the last couple of million years, the whole world was colder, with altitudinal zones often as much as 1,000 meters lower. In some cases, plants that are now confined to mountains could spread all the way across the surrounding lowlands to the neighboring mountain areas (e.g., the conifer *Podocarpus* in many areas of the lowland tropics during the last glacial, and the arctic rose *Dryas* across the mid-latitude lowlands of Eurasia). Often, the ice-age lowering of altitudinal belts on mountains must also have increased the area of mountain slope on which each species occurred, so that the size of the mountain "island" that each species presently occupies is not really representative of most of the history of the last two million years.

Many high-mountain species have very widely scattered geographical ranges (e.g., several species of alpine plants that live in the mountains of Tasmania and southeastern Australia also occur thousands of kilometers away in New Zealand). It seems that they are often particularly good dispersers, probably helped by having been more widespread at lower altitudes during the ice ages. At least sometimes (just as on remote islands in the ocean) species richness has been added to by endemic mountain species that are confined to just one area of highlands. A couple of examples are the peculiar large fleshy *Senecio* and *Lobelia* species of the Ruwenzori and Virunga mountains in east Africa, relatives of small weedy plants that grow in the lowlands of other parts of the world.

Not all groups of animals and plants decline in species richness towards higher altitudes. Several studies have found a mid-elevation "peak" in diversity several thousand meters up a mountain. For instance, Watkins *et al.* (2006) found a peak in diversity of epiphytic ferns at 1,000 m in the mountains of Costa Rica, and a somewhat higher elevation peak in richness of ground-living ferns. A study of 2,800 bird species in tropical South America found that there is a peak in diversity some way up the Andean mountains (Colwell and Lees, 2000). Small-mammal species richness in the Philippines peaks at 1,500 m to 2,200 m altitude. Frogs and toads peak in diversity at about 1,200 m in the western Ghats of southern India (Naniwadekar and Vasodevan, 2007).

In some cases, these patterns might be explained in terms of characteristic features of the mountain climate. The very rainy, damp climate of the cloud layer (Figure 1.15) up in tropical mountains seems to favor high species richness of frogs, and epiphytic ferns. In areas that have a dry season in the lowlands, rainfall tends to

Figure 1.15. Cloud layer at around 2,300 m in a tropical mountain area in northern Borneo. Areas regularly bathed by clouds are much damper and likely to support more species of ferns and amphibians (photo: Author). *See also* Color section.

be more constant through the year at higher elevations. Usually, with further increasing altitude the diversity declines once more, perhaps due to less rainfall or maybe because conditions become too extreme in terms of temperature to favor many species, despite the damper climate. Rosenweig (1995) looked at the mid-elevation peak in species richness from the same point of view as he looked at latitudinal trends in species richness (Chapter 2), suggesting that if you randomly scatter species ranges at various altitudes on a mountain, they will tend to overlap in the middle, producing a "peak" in species richness at mid-elevations. However, this does not seem to account for the continuous decline in species richness with altitude, which is seen in many groups.

There is no doubt that mountainous areas with low valleys among them tend to be richer than surrounding areas of flat land. This is partly because there are more different environments, each with its own characteristic set of species. For one thing there are different climate zones on a mountain, but only one climate in a flat lowland area. For example, in the Santa Catalina Mountains of Arizona (Huston, 1993), many different plant species occur in the same mountains but at distinct altitudes, each species at its own climatic optimum. A similar diversity of species composition with elevation is found in the Siskiyou Mountains of Oregon, and in fact on almost any set of high mountains, simply because there is a wide range of climates there. Even on a very local scale, a varied landscape can have microclimatic differences (Adams, 1997) adding to species richness, as I mentioned briefly above.

As well as diverse topography, mountains tend to have diverse geology, with a range of different soil types that can allow a greater variety of species. Whittaker (1956) found evidence for the effect that a combination of altitude and rock type can have, in his classic study of the tree communities of the Great Smokey Mountains in the U.S.A. He found a wide range of communities, and species, each of them tending to be characteristic of a certain combination of climate and soils.

1.18 PATTERNS OF SPECIES RICHNESS WITH DEPTH IN THE OCEANS

How about if we go in the other direction, below sea level rather than above it? The general picture that emerges is a peak in diversity of marine animal life at intermediate depths in the oceans (at around 2,000–4,000 m), on the long gentle slope between the shallow-shelf seas at the edges of continents, and the abyssal plain that underlies the majority of the world's ocean area. This is a humpback diversity curve, resembling the ones along disturbance and nutrient gradients on land, and those at much shallower depths in the coral reefs. But is it coincidence, or does it have the same underlying cause? Could this be another effect of a changing balance between growth rate and disturbance? Huston (1993) suggests that it does, pointing out that the supply of dead material falling from the plankton layer (the nekton, as it is called) decreases with increasing depth. From the point of view of the animals that feed off this stuff, it is a resource just like the nutrients in soil. If there is plenty of material to be found, they will grow fast and crowd one another out quicker than disturbances can cut back their numbers. This, he suggests, is what happens on the shallower shelf seas with their abundant productivity. Even though disturbance rates from storms and currents are fairly high, they are not enough to keep pace with the rapid growth of well-fed populations. On the continental slope, Huston suggests, growth rate is slower because food is in shorter supply. Even though disturbance is not as frequent as in the shallow seas, sliding of sediments and shifting ocean currents are able to keep populations down enough that they cannot crowd one another out. This then gives the maximum species diversity, the product of the "right" balance between disturbance and food supply. At greater depths, food is too sparse to support many animals and their growth rate is slow. Under these conditions, it may simply be the extreme scarcity of food which means that fewer types can exist. Or it may be that the few disturbances which do occur cut back the populations too drastically for many species to survive.

Individual groups of marine organisms have peaks of diversity at different depths. Small detritus eaters such as bivalve mollusks and crustaceans reach their greatest diversity at 3,000 m to 4,000 m, and the single-celled forams reach their peak at 4,000 m to 5,000 m depth. Huston suggests that forams reach their peak at deeper depths because they feed low in the food chain, mostly on dead plant material, which is fairly abundant—allowing fast growth and outcompetition at shallower depths. Only with great depth does the food supply for forams diminish, giving less competition and allowing greater species richness. As we move up the food chain away from the primary producers, there is less food to go around and so less chance of competitive crowding because the food is patchy and difficult to find—so it is very much a matter of luck (and not so much a matter of competitive ability) which species happens to come across each patch of food. Predators such as polychaete worms, then, seem to show their peak at 1,000 m to 3,000 m depth. In the case of top predators such as fish—which themselves feed off smaller predators—food is so sparse that they reach a peak in diversity at just a few hundred meters depth, and then show a decline below this.

Near the ocean surface, the phytoplankton grows fastest just below the surface where sunlight is abundant. Yet this area has relatively low diversity—compared with the waters just below where light is dimmer and less photosynthesis is possible. Slower growth of the plankton relative to the rate of grazing by zooplankton may mean less competitive displacement, allowing more species to survive side by side. The effect seems to carry on up the food chain to the invertebrate grazers, which are also richer in the less productive lower layers.

In the ice-clogged seas around Antarctica, the food supply is concentrated on the underside of the ice. The sea ice that floats at the surface lets through enough sunlight that algae can live in transparent channels formed by the ice melting from below, and also floating as plankton in the seawater just beneath. Grazers on the algae (e.g., small shrimps) congregate here but the supply of food is not great. By contrast open areas where the ice has melted back have much faster growth and abundance of phytoplankton, supporting dense populations of grazing shrimps. Despite the richer living to be made in the open areas, the diversity of phytoplankton and the invertebrates that feed off it is much lower than under the ice, and compared with less productive warmer waters in other climates. It seems that the dense populations and predictable resources lead to a pushing out of species (Huston, 1993). Tilman (1982) would explain this in terms of less diversity of nutrient concentrations, while Connell (1972) or Huston (1993) would see it in terms of reduced competition (relative to grazing disturbance) in tougher environments. Higher up the food chain, though, the effect is different. Seabirds are both more diverse and more species-rich in the most productive areas of sea that support more fish for them to catch. Again this may be an example of the initial rise in diversity from the low part of the humpback curve, among creatures whose food supply is less abundant overall.

1.19 SOME CONCLUSIONS ABOUT LOCAL-SCALE PATTERNS IN SPECIES RICHNESS

We do not even know for sure what enables large numbers of species to survive side by side in the most diverse communities. Still less do we know what causes much of the local-scale variation in species richness.

As the state of our knowledge stands at present, some local-scale differences in species richness are very easy to account for. For instance, it is obvious that a greater range of soils or local climates will add to the species diversity in a particular place.

Many other patterns are less easy to explain, but we have some good possible explanations. For instance, there is the humpback curve of species richness along nutrient gradients, among plants and other organisms that live fixed to the spot. This might be something to do with the balance between an environment which is too harsh, and another in which competition is too intense (Connell, 1970). On the other hand, it might have more to do with how easily species can arrange themselves among a range of combinations of nutrient concentrations (Tilman, 1982).

It is humbling to find that there is still so much mystery behind patterns in diversity that are just so strikingly obvious. To know with more certainty what is producing these local-scale patterns will need many more observations, and many more carefully designed experiments. But at least local-scale studies are relatively tractable—much easier than at the geographical and long-term scale, which is what the next three chapters (Chapters 2–4) are all about.

2

The Holy Grail of ecology: Latitudinal gradients

2.1 LATITUDINAL TRENDS

There are many patterns in the natural world, and most of them can quite readily be explained in scientific terms. For instance, there are differences in climate, which can be understood in terms of certain basic principles such as the amount of energy received from the Sun, and distance from the oceans. Yet there are some patterns in nature which are both striking and consistent, but so hard to explain that they almost seem to have been sent to mock us. They present us with a paradox: How can anything be so obvious, and yet so hard to explain?

The difference that latitude makes to diversity is plain to see. For example, the woods just outside my house in New Jersey contain about a dozen different species of trees—and I guess that I know all of them. But in Malaysia, in the tropical rainforest at Pasoh where I sometimes do fieldwork, I know for sure that I will never be able to name every type of tree that grows there. In an area of 50 hectares, about the same size as my local patch of woodland, there are several hundred different tree species. To learn them all might take years, a career change in itself. Why are there so many types of trees in Malaysia, and so few in New Jersey? The question is simple, but it is very hard to find any good answer to it.

The latitudinal trend is found in many different groups of animals and plants, and at least some microbes too. The diversity of species in most broad groups of life (e.g., the categories known as classes and orders) is low near the poles and increases fairly steadily towards the equator. It is not present in all groups, and finer subdivisions of life (such as families and genera) may not show this trend within them. But the trend is generally present, for most groups of organisms, at the higher levels of classification such as orders or classes. It is also generally present in ecological groups of organisms known as "guilds"—sets of species which make a living in a similar way even if they are not especially closely related. An example of this would be "mammalian herbivores", which can belong to any of several different orders.

2.2 THE DISCOVERY OF LATITUDINAL TRENDS

Latitudinal trends in species richness were so obvious that they had been already noted by European naturalists by the late 18th century. Alexander von Humboldt, who spent five years studying the Amazon Basin, was perhaps the very first to note the great diversity of form and appearance of organisms in the tropics. Charles Darwin, an avid fan of Humboldt, put it nicely: "In England any person fond of natural history enjoys in his walks a great advantage, by always having something to attract his attention; but in these fertile climates (i.e. Brazil), teeming with life, the attractions are so numerous, that he is scarcely able to walk at all" (Darwin, 1839).

It was clear to these early travelers that there were far more types of animals and plants in the tropics, even though just a fraction of the species there had been documented. A.R. Wallace, Darwin's contemporary and nearly as great a biologist, described and thought about this difference further, and came up with a possible explanation which we will consider below.

The latitudinal gradient shows up in almost every group of organisms that has been well enough studied to be able to talk meaningfully about patterns in its diversity (Table 2.1). There is an especially striking latitudinal trend in the species richness of trees, which rises exponentially towards the tropics (Figure 2.1). Similarly sized areas in southeastern Russia (at 45°N) and peninsular Malaysia (at 1°N) have 40 species and 1,400 species of trees, respectively (Adams, 1988). The trend is also present if all types of plants are taken together. Butterflies are another group that is consistently far more diverse in warmer climates, reaching a peak of richness in the tropical rainforests. For example, in the North American state of Maine there are 45 species of butterflies, whereas a far smaller forest reserve in Bolivia (in tropical South America) has 750 species (personal observations). When the species richness of butterflies down through the Americas is plotted on a graph, the consistency of the latitudinal gradient is clear—as is the fact that butterflies are most diverse in the tropical rainforests of the Amazon Basin. Ants also show a strong trend: on just one tree in the rainforest of Peru, the distinguished ecologist E.O. Wilson found 135 species of ants—more than in the whole of the U.K. which has only 48 ant species. In fact, insects generally, of all types lumped together, are far more diverse towards the equator. Birds also show the latitudinal trend very strongly (Figure 2.2). Mammals and reptiles have a weaker latitudinal trend, but they are nevertheless far more diverse towards the tropics (Figure 2.3). Freshwater fishes also increase strongly in species richness towards the tropics: Europe has only 250 species of freshwater fishes whereas South America with its large area of tropical climate has 2,200 species, and Africa has at least 1,800 (Groombridge, 1993). Also in the freshwater lakes and rivers, crayfish and amphipod shrimps are far more species-rich in the tropics that the higher latitudes (France, 1992). Generally even the species diversity within an individual genus tends to show a peak that is skewed towards the lower latitude end of its range. One such case is with the oaks, the genus *Quercus*: it is most species-rich around 25°N to 30°N in North and Central America, where there are some 160 species. In Asia the closely related *Lithocarpus* is found in temperate forests, but it reaches a peak of hundreds of species in the forests of the tropics.

Table 2.1. Some of the many examples of groups showing increasing species richness towards the low latitudes (the tropics) (after Huston, 1993).

Organism or guild	*Region*
Vertebrates	
Non-oceanic birds	New World
	New World
	New World
	Nearctic
	Palearctic
Mammals	New World
	New World
Fish	Nearctic
Reptiles	Nearctic
Anurans	Global
Lizards	Global
	Nearctic
Snakes	New World
	Global
Invertebrates	
Papilionid butterflies	Global
Sphingid moths	New World
Dragonflies	Global
Wood-boring Scolytidae and Platypodidae	Global
Planktonic foraminiferans	Nearctic
Permian brachiopods	Nearctic
Corals	Australian
	Global
Tunicates	Global
Calanid crustaceans	Global
Mollusks	Nearctic
Plants	
Trees	Palearctic
Orchids	New World

In the oceans, there are many examples of groups of organisms which become more diverse towards the tropics. In the shallow seas around the edges of continents, bivalve mollusks (commonly known as clams) and gastropods (snails) are more species-rich towards the equator (Fischer, 1960). Corals and fish are also a lot more diverse in the warmer seas close to the equator (Huston, 1993). In the surface waters of the world's oceans, the number of species of foraminifera (a group of protozoa) floating with the plankton increases towards the tropics (Stehli *et al.*, 1969). Recent work suggests that other groups of plankton, such as planktonic

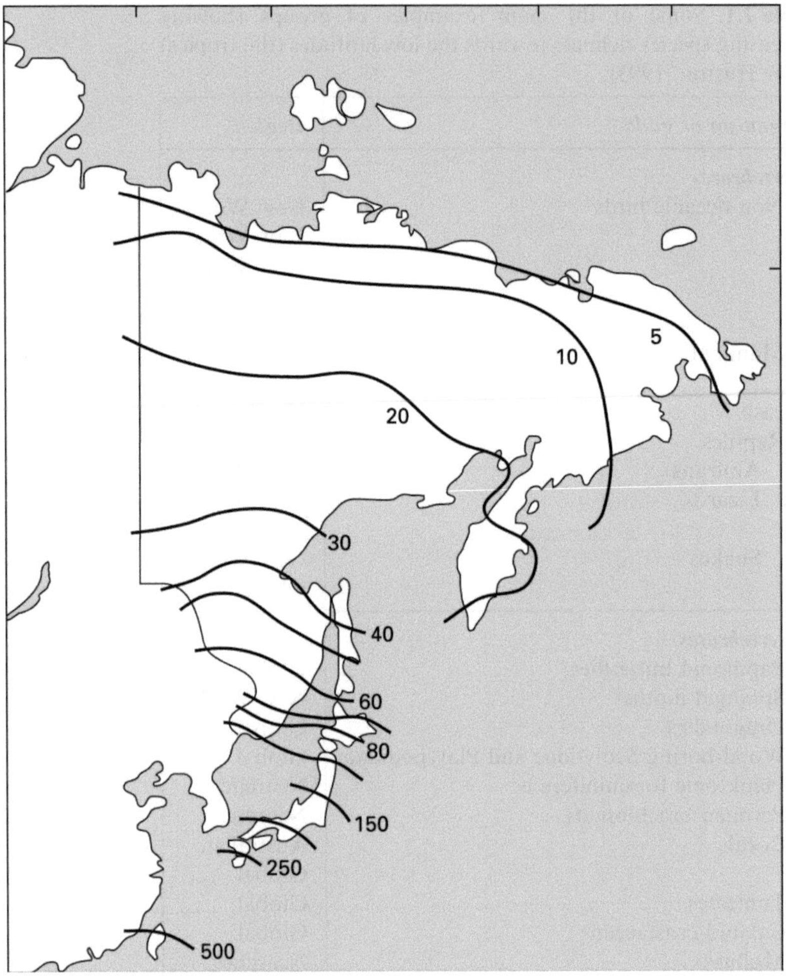

Figure 2.1. Wild tree species richness in eastern Asia (from Adams, 1988).

bacteria, may show the same trend towards being more species-rich in the tropics (Fuhrman *et al.*, 2008).

Latitudinal diversity trends like this have existed for many millions of years, and evolved various times independently. Even when we are dealing with lists of species that have all been extinct for tens of millions of years, the latitudinal diversity trend is often there, in cases where we have detailed enough data to look for it. For example, there was already the classic latitudinal gradient among the flowering plants in their early days during the Cretaceous, between about 90 and 65 million years ago (Crane and Lidgard, 1989). Even though the flowering plants became more diverse everywhere as time went on, the tropics continued to have the advantage, and this has continued into more recent times. In the seas, clams (bivalve mollusks) also showed greater diversity in the tropics well back into the Triassic about 200 million years ago. The little single-celled foraminifera that float in the surface ocean waters also showed this same latitudinal trend during the Cretaceous about 100 million years ago. Alroy *et al.* (2008) found that the sum total of animal life from shallow seas was already more diverse in the tropics way back during the Ordovician period some 480 million years ago. There was a similar latitudinal trend in shallow-sea faunas (though with different species and mostly different families) in the Paleogene some 50 million years ago.

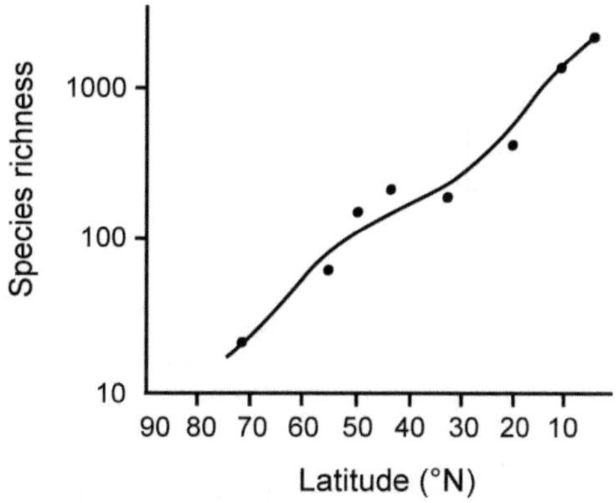

Figure 2.2. Breeding bird species richness in North and Central America is highest closest to the equator (after Huston, 1993).

Nevertheless, although an increase in species richness towards low latitudes is the general rule, there are exceptions. For certain groups, or in certain geographical regions, there may be no latitudinal diversity trend at all—or the trend may even be reversed, with fewer species in the tropics. Seabirds are most diverse in the high latitudes of the northern and southern hemisphere, including 17 species of penguins in the southern seas around Antarctica and 22 species of auks (alcids) in the sub-Arctic northern waters. Invertebrates such as snails and barnacles living on rocky shores seem to be most species-rich in the mid- and high latitudes, rather than the tropics. Lichens also seem to be more diverse in the high Arctic than they are in equatorial regions. Soil nematode worms appear to have their highest diversity in the mid-latitudes rather than the tropics, although this could be because they have not yet been studied thoroughly enough in the tropics. Parasitoid wasps, which lay their eggs in caterpillars and other insect larvae, also seem to be more species-rich in the mid-latitudes, as are certain plant-feeding insect groups such as sap suckers (aphids, mealybugs, and suchlike).

Even in those groups that do show an overall increase in diversity into the tropics, the latitudinal trend is usually not a simple steady rise with decreasing latitude, but instead a line that meanders upwards, complicated by highs and lows along the way (Figures 2.4 and 2.5). There are also longitudinal (east–west) trends in diversity that occur in addition to the latitudinal ones. Sometimes the trends that complicate or break with the latitudinal rule clearly relate to other environmental factors such as rainfall or topography—although the underlying reasons why they do so are a matter of debate (see below). For example, there is a decline in species richness of trees going westwards from the moist climate of the eastern U.S.A. into the drier Great Plains region of the central U.S.A., and going southwards through the deserts the tree species richness also decreases (Currie and Paquin, 1987). Mammal species richness does the opposite, increasing into the North American deserts, and is lower in places farther east even when these areas are more southerly.

We humans are land-living animals, and understandably we tend to forget that the oceans cover the majority of the Earth's surface: 71% of it in fact. The oceans are

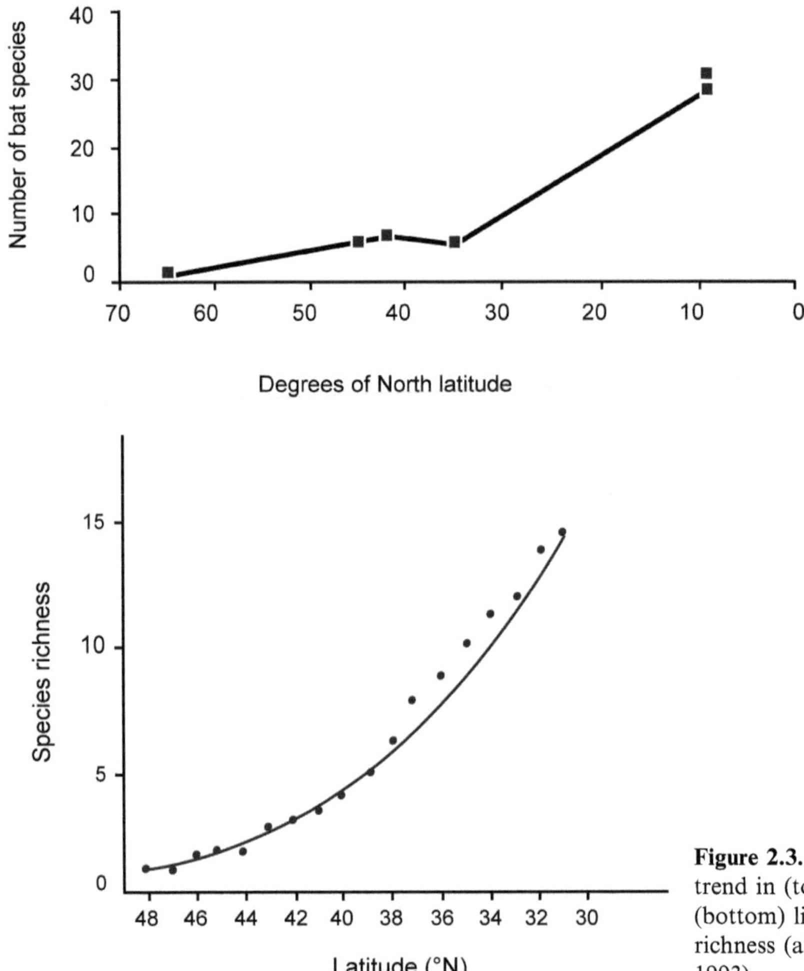

Figure 2.3. Latitudinal trend in (top) bat and (bottom) lizard species richness (after Huston, 1993).

unfairly neglected in discussions of species richness, considering that they are the predominant environment on our planet. And when it comes to deeper parts of the oceans (beyond about 4,000 m) there is not even very much data to talk about, although they cover more than 40% of the world. But there is another reason for this ignorance beyond just lack of attention: the ocean depths are very difficult and expensive for humans to visit and sample, and this has limited how much we know about the patterns of diversity down there.

From the sampling of the deep oceans that *has* been done, mostly from occasional scoops of ocean sediment taken on long ropes dangled off research ships at the surface, it is something of a moot point whether there is a latitudinal trend in species richness in the ocean depths (below about 300 m). If there is no trend, it would hardly be surprising, because a latitudinal gradient has to be caused by something in

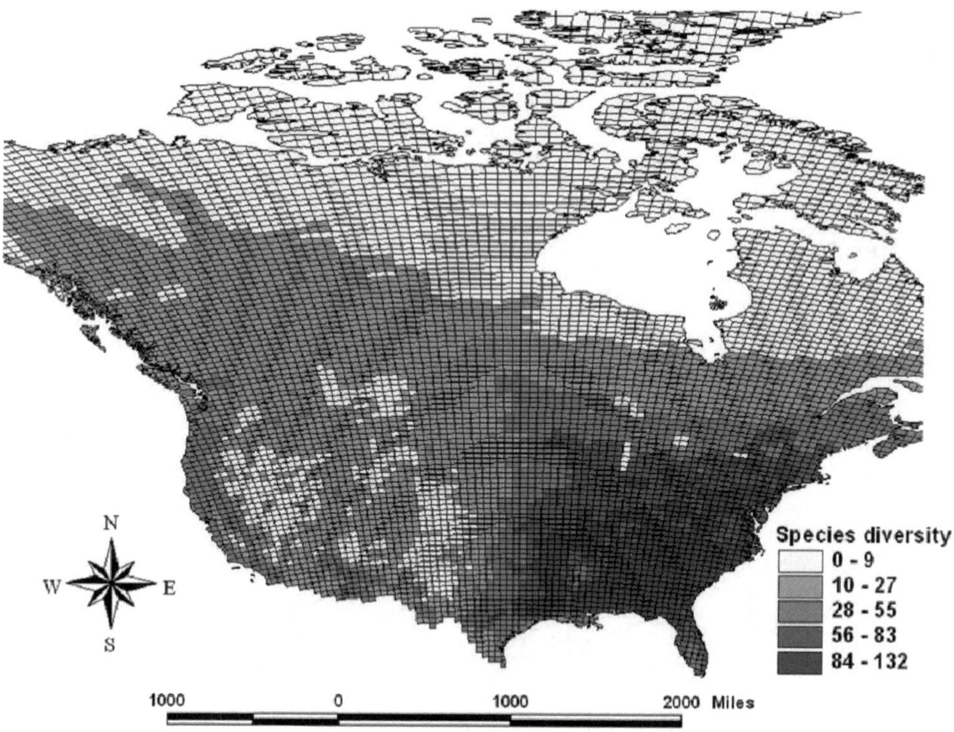

Figure 2.4. Tree species richness map of North America (Yangjian Zhang and J. Adams, unpublished). *See also* Color section.

the environment, and the deep ocean environment is uniformly dark and cold. No sunlight ever reaches there from the surface, and almost all of its waters have moved in from sinking regions of ocean surface near the poles which had first cooled to about the same temperature as a domestic refrigerator. Organisms on the land or in the sea do not generally come equipped with GPS devices, and if there is no variation in the environment they will not know what latitude they are at, and of course their ecology and species richness will not follow any latitudinal gradient.

However, despite our expectations, some studies over the last 10 years or so *have* suggested a latitudinal trend in the species richness of the deep oceans—in certain groups of organisms which have been studied in more detail than most. For example, Rex *et al.* (1993) and Culver and Buzas (2000) found latitudinal trends in the groups they studied: a group of deep-sea snails, and tiny protozoa called forams which live on the sea floor. Both groups of organisms turned out to be more species-rich towards the tropics. Assuming that these trends are real, what could be causing them? After all I just made the case that in the deep ocean, nowhere is any different from anywhere else. But there is actually one thing: the seasonal cycle of life in the top waters of the ocean. The waters at the surface that are reached and warmed by sunlight do follow the same general seasonal pattern of temperatures as the Earth's land surface. The

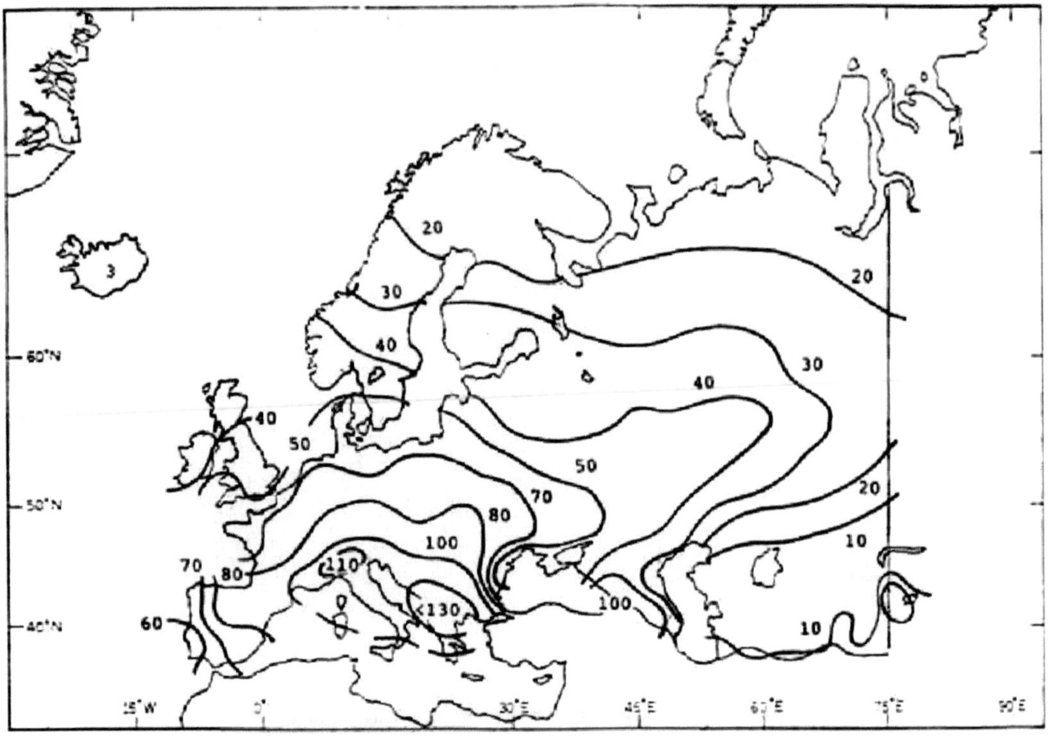

Figure 2.5. Tree species richness map of Europe. The east–west trends and more localized highs and lows of species richness are evident (author: unpublished data).

plankton and larger organisms within the surface seas grow and reproduce according to these changes. Near the equator, there is hardly any seasonal rhythm, because temperatures stay the same year-round. The seasonal rhythm becomes especially strong in far northern and far southern waters—becoming a big burst of productivity (known as the spring "bloom") as the sea surface warms up after the long winter. Nutrients stirred up during winter storms add to this burst of plankton growth. All this seasonal activity produces a pulse of material—organic waste—that slowly sinks down from the surface to the sea floor several thousand meters below, taking a few weeks to get there. So life on the deep-sea floor in the higher latitudes is not identical to that in the tropics, for there is a long period of starvation during the winter months, and then a burst of food during the summer. Even though we do not know why this would affect species richness, at least we can see that there is a clear environmental difference between the deep oceans in the high and low latitudes.

Though latitudinal gradients in species richness are plain to see in a whole range of environments and groups of organisms, they beg the question of what causes them. They must have something to do with the environmental gradients that follow latitude, but exactly *what*?

2.3 EXPLAINING LATITUDINAL GRADIENTS

With the discovery of latitudinal diversity gradients came the desire to explain them, and that is where the trouble started. Fundamentally, almost everyone who has thought about these gradients agrees that they have something to do with climate; mainly that temperatures get cooler and more seasonal as one travels from the equator to the poles. But despite more than a century of thought, it is still a mystery exactly why a trend in climate should translate into a trend in species richness. There are some 120 named hypotheses that have tried to explain the latitudinal diversity trend (Colwell and Lees, 2000). Explaining it is still one of the most fascinating and taxing questions in modern science: it has been called the "Holy Grail of ecology". A few years ago when the journal *Nature* came up with a list of the ten greatest scientific mysteries, this was one of them.

As a brief aside, one might wonder why in any case the climate gets cooler towards the poles. The reason is that less energy is received from the Sun at higher latitudes. It is all to do with the angle of Earth's surface relative to the Sun: on the equator the Sun climbs higher in the sky during the day, because there the surface of the Earth is directly facing the plane of the Sun in the solar system. The equator thus gets a full-on beam of the Sun, and this is important in two ways.

First, the Sun's rays do not need to travel so far through the atmosphere to reach the ground if they are shining straight down from above. Shining at an angle from lower in the sky, more of the Sun's light is absorbed and scattered by gases and tiny particles in the atmosphere. The farther towards the poles you go, the lower in the sky the Sun gets on average and the less energy reaches the surface (Figure 2.6, bottom). It is the same principle which explains why the Sun becomes very weak as it goes down towards the horizon at sunset: so weak that you can stare straight into it.

The second important principle is that as the Sun's beam becomes angled lower (as it is at high latitudes), the beam of energy spreads out across the surface. It is rather like shining a flashlight down onto a table; hold the flashlight pointing straight down and there is a strong beam onto the surface just underneath it. Hold the flashlight at an angle and the beam is splurged out across the table top, and is weaker. It is the same with the surface of the Earth when the Sun is lower in the sky (Figure 2.6, top). Each unit area of land or sea gets less energy compared with the lower latitudes, and so both it and the atmosphere just above it stay cooler.

In winter the angle of the Sun is at its lowest, because of the tilt in the Earth's axis. The farther away from the equator you go, the lower the Sun stays during the winter. As well as being lower during the days in winter, the Sun dips down below the horizon for a longer part of the day, so more time is spent in darkness receiving no energy from the Sun. Above a certain latitude (the Arctic Circle, and at the other end of the Earth the Antarctic Circle) it can even stay down below the horizon for 24 hours a day, so that weeks or even months of the winter are spent in complete darkness. Both these effects—low winter Sun angle and short day length—give the cold winters that become increasingly severe towards the poles.

Ideas about latitudinal trends in diversity have to build upon the differences in climate, and one can do this in various ways. It is possible to imagine how climate

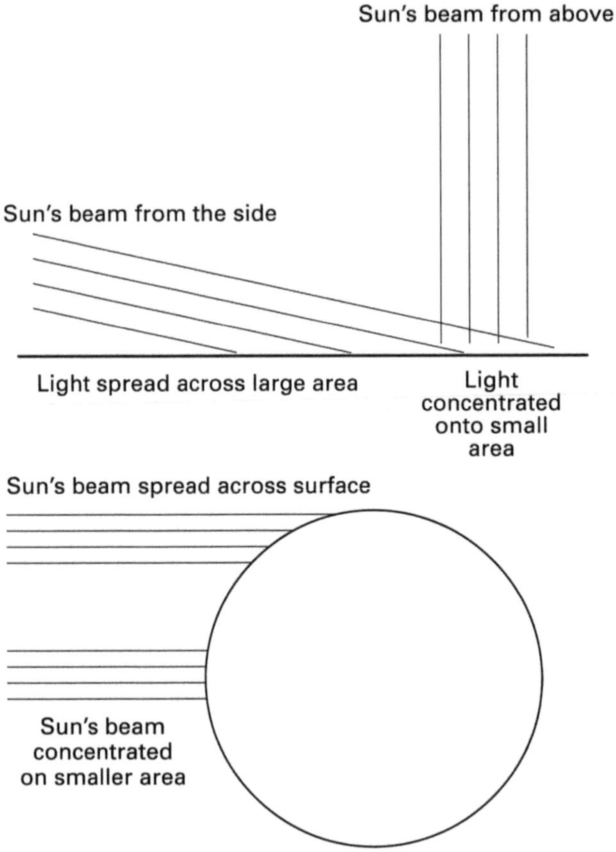

Figure 2.6. The tropics are hotter than the poles because: (top) a direct beam gives more energy (per unit area) than an angled beam; (bottom) passing through a greater depth of atmosphere absorbs more energy before sunlight can hit the Earth.

differences could affect the physiology of organisms, or the operation of whole communities and ecosystems, or the processes of species formation and extinction. Some of the theories invoke not only the present day temperature gradient, but also past changes in climate that may first have produced the gradient, or set things up so that a gradient would eventually appear in the present-day world.

Hypotheses that set out to explain latitudinal richness gradients basically fall into two groups. One set of ideas suggests that the species richness of the higher latitudes could potentially build up to be as great as it is in the tropics, if only species were given *enough time* to accumulate there. According to these theories, the reason that the cold high latitudes have fewer species is that they have not yet had enough millions of years to reach high levels of species richness. For some reason—these ideas suggest—there hasn't been enough time to form as many species as in the tropics, and that is the important limiting factor. The higher latitudes might have begun from a lower starting point in terms of total diversity, or new species might form less easily and less often at high latitudes. Or high-latitude environments might have suffered more extinctions to knock back the level of species richness, from which they are still recovering. Ideas which suggest that the low levels of species richness in high latitudes are not bumping up against any sort of lid—and have the potential to rise a lot higher—are called "non-equilibrium" (because species richness has not yet reached any sort of balance or equilibrium).

The other main group of ideas to explain latitudinal gradients suggests that there is actually a lid on how high diversity can rise in the colder regions of the world, and that the higher latitudes have already reached that limit. These are known as "equilibrium" hypotheses, because they assume that everything has reached a steady balance. So even if individual species sometimes go extinct and new species evolve and replace them, the overall number will stay about the same because of some process that regulates diversity. Equilibrium explanations suggest that even if they were given more time, and a fairly stable environment, the high latitudes would never be able to pack in many more species. These ideas assume that this is because there is something quite fundamental about the way that populations, communities, and ecosystems work within each latitudinal zone—something that sets a certain upper capacity for species richness in each place. What exactly is supposed to be different, and most important in setting this capacity, varies a lot from one hypothesis to another.

Going through the whole list of 120 hypotheses would be very tedious, and anyhow many of them are now dismissed as fairly implausible. Here I will just run through some of the most currently popular ideas, and see how they stand up to testing.

Box 2.1 Speciation: how are new species made?

The richness of species in the world fundamentally depends on the process known as speciation, where an ancestral species splits off one or more new "daughter" species. If there were no speciation, there would be no more than one species in the world, directly descended from whatever first ancient life form was the ancestor of everything else (presumably some sort of bacterium). Speciation is also essential to the maintenance of a healthy, diverse biosphere. If it were ever to stop happening, the loss of species through extinction (Chapters 3 and 5) would not be balanced. Like a population with only old people and no births, eventually the species richness of the world would decline away.

Frustratingly, given the importance of speciation, we are still not sure how new species are generally made. The favorite view among biologists is that most species form because of their ranges becoming split up, often by environmental change. This would allow the isolated populations to evolve in different directions, eventually becoming totally new species. The term given to this process of separation by distance followed by speciation is "allopatric speciation". The term allopatric comes from Latin words, meaning "separate homelands".

There has been a long-standing debate in biology about the importance of an alternative way of making species. This idea is that new species literally form *within* populations of the ancestral species, with novel forms arising and separating themselves off by refusing or being unable to mate with the original species. This is called sympatric, the suffix "sym-" meaning "together" in Latin (so sympatric literally means "having the same homeland").

And then there is an in-between scenario, where a semi-isolated population (though still in touch with the main population) changes and then becomes a separate species. This is called "parapatric" ("para-" meaning "sort of" in Latin). Most biologists who study this subject agree that it is much easier to make species by allopatric speciation, so this must surely be the way most species form.

What evidence do we have for each scenario? There are various classic case studies of animals or plants in which geographically distinct populations seem to be in the process of splitting into separate species before our eyes (e.g., various wild plants in California studied by Stebbins, 1950). It is assumed that this sort of situation, on various scales of both distance and time, has produced most of the world's species.

However, there are examples of groups of species in special situations where it is hard to explain how they could ever have originated allopatrically. One example is the rather unglamorous apple maggot fly (*Rhagoletis pomonella*), which has the annoying habit of ruining apples in the northeastern United States. This species seems to have arisen just in the last few hundred years, after settlers started planting apple trees in Hudson Valley of New York. It formed from a species that attacks the fruit of hawthorn (*Crataegus*) trees that grow in hedgerows and woodlands scattered through the countryside. Since both types of trees grow intermingled in the landscape, this certainly looks like a clear case of sympatric speciation (Via, 2001).

Another example of what really looks like sympatric speciation comes from sets of closely related species of cichlid fishes living in two small, closed crater lakes in West Africa (Schliewen *et al.*, 1994; Via, 2001). This is a fairly uniform habitat without any obvious barriers that could have separated off populations, and DNA comparisons show that all the species in each lake are indeed very closely related to one another. In these situations, it looks like sympatric speciation has somehow actually occurred.

There are also some intriguing experiments, done in laboratories, which have shown that in a situation of conflicting selection pressures ("you can be this or that, but not in-between"), a single population can actually split into two distinct populations and ultimately into two species, even though they remain in constant contact with one another (Via, 2001). Various mathematical scenarios also suggest other ways in which species can arise sympatrically; for example, if individuals who have one particular look about them always choose to mate with others who resemble them (this is called "assortive mating") (Doebeli and Dieckman, 1999).

Another important question about speciation is whether changes tend to arise gradually over millions of years, or pretty much all at once in a short period such as a few centuries. Charles Darwin believed that most of the differences that distinguish one species from another would have arisen very gradually, over many thousands or millions of years. But there is another view, which is now becoming more popular, that the changes associated with speciation often occur over just a few tens of generations, in small isolated populations. Isolation and small population size are thought to be the key, because for one thing a new mutation that changes the form of the species can rapidly become common just by luck in a small population. In a large population, it will always tend to be swamped out by the other more common genes, and so the new mutation cannot become a fixed feature—the species simply doesn't get to change. In fact, because it may make individuals that carry it temporarily less competitive or less attractive as mates, a newly appeared feature would soon be eliminated from a large population.

But in a small, isolated population, a process of chance variations in the abundance of genes (known as "drift") can allow successive mutations to accumulate and build on one another, until eventually a new, viable species is put together. This may be something capable of living in quite a different niche from its parent species. So, for example, imagine that there is only one male alive in the tiny population, and several females. He is likely to find a mate even if he does look fairly freaky to the females—for they have no choice. Once his genes get scattered through the small population, they become the standard of normality, and change goes on from there.

Nowadays in biology, the fossil record is generally taken as supporting the view that new species arise suddenly, not gradually, and then stay looking much the same through the rest of their existence—up until the time that they go extinct. This view, of sudden changes followed by long periods of calm, is known as "punctuated equilibrium" (Eldredge and Gould, 1972). It was the idea that first made the biologist and author Stephen Jay Gould well-known in academia, before he went into popular science writing and published his many excellent books. Eldredge and Gould based their theory on observations of trilobites (little marine creatures rather resembling pillbugs) which filled the seas before about 300 million years ago. Each species seemed to show no sign of changing gradually (gradual change is what Darwin would have predicted) during the time of its existence. Instead, new species appear in the fossil record "fully formed" with their characteristic features in place. Those features usually remain the same up to the point where they go extinct, or give rise to other daughter species with their own new characteristics. Once Eldredge and Gould pointed this out, biologists looked around at the fossil record, and had to admit that indeed there are rarely any signs of gradual changes accumulating within a species over time. Most change seems to occur when suddenly a population "breaks the mold" of whatever went before, and makes something quite new. So we almost never see anything in the fossil record that looks like big evolutionary changes accumulating one after another in a population—presumably because such change occurs in fleeting moments of geological time, and in small isolated populations.

I was once introduced to Gould, only a few weeks before he died as it turned out. I happened to be wearing a souvenir sweatshirt that had "Boston" written across it in large letters. Hearing my accent he exclaimed: "hey you're not from Boston!" Unfortunately that was the sum total of my dialogue with this great man. In fact, I may have met him twice. Some years earlier, on the bridge that connects Cambridge to Boston, I was almost run over by a bicycle ridden by Gould (or someone who looked remarkably like him). Fortunately, no words were exchanged on that occasion.

Yet another debate in the study of speciation is whether a new species first of all tends to become different from its parent species, and then only later becomes unable to mate with it after the two collide again. If two very different organisms, in two different niches, can still interbreed when they encounter one another again, this will tend to produce only "half-way adapted" offspring that will do no good in either niche. Imagine trying to merge the parts of a Volkswagen and a Honda: probably the result wouldn't work very well. To avoid such disastrous unions, evolving a way to prevent mating will suit both sides. In animals, the prevention methods that evolve might be based on behavior, so they each avoid one another and refuse to mate. Or they might be subtly chemical (e.g., in plants, where flowers of one species may refuse to let the pollen of the other grow on their stigma). This sort of process, accumulation of differences followed later by the adoption of a barrier to mating, seems to have been the main mechanism that split off species of fruit flies (*Drosophila*) (Coyne and Orr, 1989). But also there is another view, that most often things are done the other way around: that usually it is the barrier to mating that arises first by chance while the species are in isolation from one another. If this happens, once they are separated the two new species are free to evolve off in different directions and into different niches—even if they come back into contact and end up living in the same place.

The frequency with which speciation occurs is all-important in determining the species richness of any one group or region. While speciation tends to build up numbers of species, extinction takes them away. In a population of animals or plants, whether numbers stay steady, grow, or shrink depends on the balance between birth rate and death rate. It is much the same with species richness: whether it grows, shrinks, or stays steady depends on how often species are being "born" and how often they "die". If speciation slows down, but extinction stays steady, species richness will plummet.

2.4 NON-EQUILIBRIUM THEORIES: SPECIES RICHNESS CAN JUST KEEP ON RISING ...

2.4.1 The effects of ice ages

The first idea to explain latitudinal diversity trends was put forward by Alfred Russell Wallace, a great 19th century naturalist who ranks almost on a par with Darwin for his contributions to natural history. Wallace voyaged to the Amazon and the Far East and carefully observed and studied the rainforests there, coming up with the theory of evolution by natural selection quite independently of Darwin. Wallace was also a very good writer, producing several best-selling travelogues about his adventures in the tropics. Even nowadays they are a great read—full of observations and reflections on natural history and human cultures, mixed in with some right ripping yarns.

Writing in 1878 in a famous book of essays known as *Tropical Nature*, Wallace used the recent discovery that the Earth had been through great ice ages as the basis for his explanation for latitudinal richness gradients. He suggested that even though the numbers of species in any place (in the high or low latitudes) might tend to build up over time as life forms proliferated, an ice age every once in a while would obliterate much of the life in the higher latitudes. The severe cooling, combined with the great ice sheets that scraped down from the north, would have wiped out most species that lived in high-latitude environments. The tropics, by contrast, must have remained a reservoir of diversity—a benign, constant environment that accumulated species faster than it could lose them. Farther from the outer edges of the tropics, the effects of the ice ages would become increasingly strongly felt, giving a general gradient in species richness the farther away from the tropics one went. And each time the higher latitudes began to recover and form new species, they would be knocked back by another calamitous ice age.

As well as the complete extinction of many forms, some of the species that could potentially enrich the higher latitudes might still survive, but now are stuck at lower latitudes. Having been pushed there during ice ages, they might be unable to spread back to the places where they once lived before the ice ages began to hit, some two and a half million years ago. For example, many of the types of trees that still exist in the relatively warm climes in southern Europe or the southeastern U.S.A. may simply not have had time to move back north again to take up their full potential range, after being squeezed down into small refuge areas by the cold and aridity of the last ice age. The southeastern mountains of Europe have many species of trees that are unique to them; for example, the horse chestnut *Aeculus hippocastaneum* that survived only in the Balkans—even though it grows just fine when planted across northwestern Europe. Likewise in the Deep South of the U.S.A. a unique species of the plum yew, *Torreya taxifolia* (Figure 2.7), is confined to just a few localized wild populations around the Florida–Georgia border, but grows well when planted farther north. So even though the last ice age ended about 11,500 years ago, and there may simply not have been long enough for certain slow-moving organisms like trees to spread out to everywhere that they could live again in the modern climate. This inability of species to disperse out from the places where they have survived could be another factor that has accentuated the latitudinal gradients we see in the present-day world (Svenning and Skov, 2007).

The glacial legacy is a plausible idea to explain latitudinal trends, but in my view it has its weaknesses. If it were all-important, we would expect to see different regions that suffered glaciations having very different levels of diversity, because of the chance effects of extinction and survival. How many species get stomped out from each region is surely going to vary a lot with the local and regional circumstances. Some regions would happen to have more accessible areas of warm or moist climates to shelter species during ice ages, giving a better chance of survival—and this should show itself in the survival of more species. After the climate warmed again, their slow spread back out from refuge areas where they survived would also likely give arbitrary patterns of highs and lows in species richness—some areas species-rich and others species-poor despite having a very similar climate. After all, if the slow rate of

Figure 2.7. The Florida plum yew, *Torreya taxifolia*, has a very restricted distribution in the southeastern U.S.A. This young specimen is from the Torreya State Park in Florida (photo: Author). *See also* Color section.

spread of species can limit how far they move, variations in their diversity should depend very much on geographical barriers or opportunities, not climate. Yet the cool temperate climates of different regions—Europe, North America, eastern Asia, and New Zealand—tend to follow an almost identical pattern of tree species richness all across a wide range of temperatures (e.g., Adams and Woodward, 1989). This suggests to me that simple luck of survival and lack of spread of species has not been such an important factor (Adams and Woodward, 1989), and that something broader and older is controlling species richness—maybe one of the equilibrium factors that I will discuss below. By the way, tree floras can also show some very close relationships between climate and other characteristics apart from species richness, which suggests that they are often already well-adjusted to their potential climate ranges. For example, leaf shape correlates beautifully with climate in North America (Adams *et al.*, 2008, also see below), which is not what would be expected for a mixed-up rag-tag assemblage of species that is out of balance with climate.

Also as evidence against the idea of a glacial legacy in migration rates, we see plenty of signs from the past of the ability of trees to spread themselves very effectively across a region when the climate changed back towards favoring them.

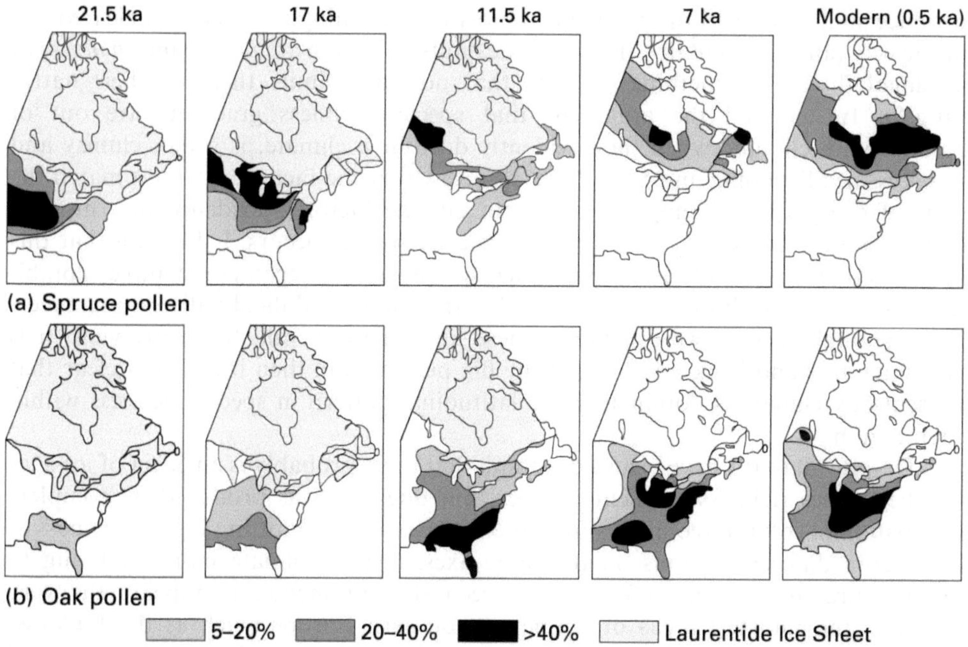

Figure 2.8. Maps of migration of two groups of trees—(a) spruce and (b) oak in the pollen record—as the world thawed out from the last ice age. Ka = thousands of years ago (after Adams, 1988).

For example, take what happened as the last ice age ended around 11,500 years ago. As the climate across the northern hemisphere warmed and moistened, there was a great burst of migration of tree species northwards. This shows up in the pollen records of lakes and backswamps scattered across the landscapes of Europe, America, and northeastern Asia (Figure 2.8). For many tree species, the rate of spread was even so great that it is hard to explain how they could ever have managed it given their long generation times. And then, after several thousand years of spreading north they stopped moving, as if they had each reached the limits of their potential range, back in balance again with climate. Not only that, since then some species have even retreated south a little, apparently as a result of a slight cooling of climate after about 5,000 years ago. This doesn't exactly look like what we would expect if many species are almost unable to migrate—instead we ought to see a lot of trees still slowly spreading out from their ice age refuges over the past several thousand years, and that is certainly not the case! Now, this is not to say that every species is growing everywhere that it could. Some species such as the horse chestnut might have stayed stubbornly put in their ice age refuges, at least until humans helpfully gathered their seeds and planted them elsewhere. But I would argue that these are the exceptions, not the rule. Even in the case of the horse chestnut, we cannot really be sure that it could survive long-term in Britain if the land was allowed to go back to forest, and competition began to bite. It is easy for a tree to survive and produce some scattered offspring when most of everything else has been cleared out of the way (and, incidentally, the horse chestnuts in Britain are now dying *en masse* of a fungal disease—something perhaps that normally limits them to their native range?). In the end the argument that species richness gradients are out of balance with climate, when they evidently do follow climate, is rather clumsy and unnecessary. Svenning and Skov (2007) suggest that the fact they cannot match the edge of every species' range to a present climate factor is evidence that modern climate is not important in limiting where each species occurs. I think all that this means is that the boundaries of individual species ranges depend on complex combinations of factors—both climate and soils—that we have difficulty discerning. Averaging enough species, we can factor these idiosyncrasies out. So in my view, it is necessary to stand back and look at the big picture, and then it becomes clear that present-day climate is dominating the latitudinal pattern in species richness within each region.

Despite all that I have just said, I think that there probably *is* an effect of ice ages on the species richness gradient in trees, but one that was produced by complete extirpation of certain tree species from their home regions. Plotting species richness of trees from different regions on the same axes, against climate indices relating to temperature, reveals the difference between the continents. It appears the very warmest and moistest areas of temperate southern Europe (such as the Balkans) and North America (the Deep South of southern Georgia, South Carolina, and northern Florida) have far fewer tree species than similar places in Asia (such as southern Japan and Taiwan) (Adams and Woodward, 1989; Figure 2.9). But this is only for the most southerly climates, for species that belong at the warmest extreme of the northern temperate zone, which would have had most trouble finding warm and moist conditions to survive in during the ice ages. It does not have any bearing on the question of whether tree species that are still present are out of equilibrium with climate *within* a region, to give latitudinal species richness gradients. If a species is completely gone, then it is not around to be out of equilibrium with climate within its home range!

I have concentrated on the example of trees here, but much the same goes for many other groups too, if we look at them in relation to climate, geography, and their migration patterns since the end of the last ice age. Most damning of all for the idea that the ice ages produced latitudinal gradients in species richness is the fact that such gradients formed themselves many times in the past, way back in time *before* the present ice ages. In many instances, they formed at times when the planet had not seen anything like an ice age for a hundred million years or more (Chapter 3).

2.4.2 "The tropics are more benign"

To the layman, it seems only natural that more species would want to live in the tropics, with their year-round warmth, rain, and lush vegetation. When I mention the mystery of latitudinal gradients to students, or taxi drivers, or my relatives, their immediate response is that this is the obvious answer, end of story. If the tropics are a

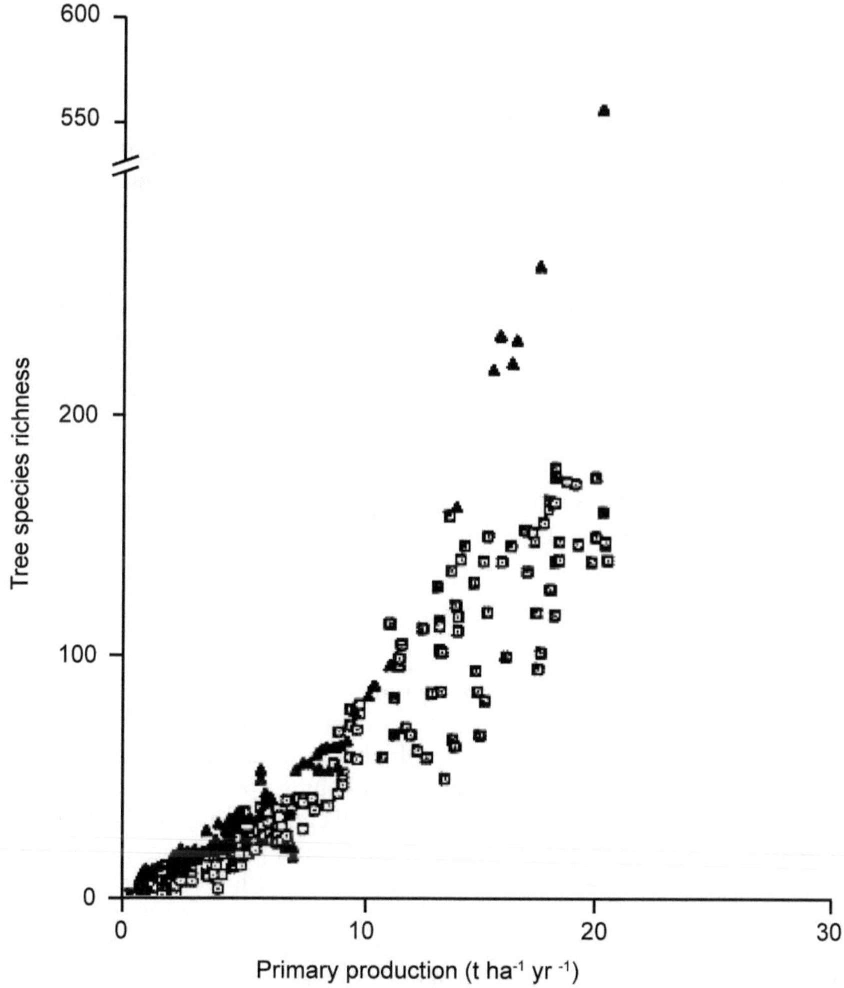

Figure 2.9. Within a region of temperate-zone forest, tree species richness follows a climate index relating to net primary productivity—mostly dependent on temperature. When the tree species richness of different regions is plotted, they follow the same trend in the colder, more species-poor areas. But in Asia, the areas with warmer climates are disproportionately rich in species compared with North America—possibly because species survived the ice ages better in Asia. ▲ = data points from eastern Asia (corresponds to map in Figure 2.1). □ = data points from eastern North America (corresponds to map in Figure 2.4) (from Adams, 1988).

better place to live, then surely they will accumulate species, rather like warm places such as Florida and Spain have accumulated people from more northerly climes over the last few decades.

However, this sort of explanation for species richness trends is based on vague anthropomorphic thinking, and the more closely we examine it, the less clear it seems that nature should work this way. For instance, the tropics certainly are not favorable for all forms of life. Many types of animals and plants thrive in cooler climates, and they will die if they are placed in an environment that is too warm, or lacking in cold or dry seasons. Antarctic penguins readily suffer heat stress in a warm summer, and their enclosures must be kept chilled for them to survive. Siberian larch (*Larix sibirica*) will die early if it is placed in a temperate climate with milder winters such as in England, and temperate trees cannot survive long in the tropics without the benefit of winter chilling. And furthermore, if some species of trees can survive the harsh winters of the northeastern U.S.A., then why not 20,000 species of trees, as there are in the rainforests of Southeast Asia? If a few types of butterflies can survive in Alaska, then why not several hundred species as one might find in a few square kilometers of tropical rainforest in the Amazon? All that these extra species would need to do is acquire the same adaptations found in species that already exist in colder climates, and the environment will then be wide open to them. And since some species have already made it into the colder places very successfully, why don't these groups now diversify into many thousands more species living only in cold climates—just like in the warm climates of the tropics?

2.4.2.1 *Are the tropics only more "benign" because most groups evolved in a past warmer world?*

To get anywhere in explaining latitudinal gradients we have to think carefully and precisely. Whether or not a climate seems "harsh" may actually depend on where you come from. Perhaps there is a sense in which higher latitude climates only really seem harsher because it was in warm climates that all things began in the past? If each group of animals and plants evolved first in tropical climates, in a world that was a lot warmer than it is now, it is not surprising that it would take a long time for their descendents to acquire the adaptations needed to cope with cold. Over time though, the higher latitudes should be tending to accumulate new species that have spilt off from the original groups and managed to make the leap into cooler climates.

For instance, consider the flowering plants (the angiosperms) that now dominate most of the world's land surfaces. There is a strong gradient in angiosperm species richness towards the tropics, where they are much more diverse. What if the angiosperms originated in the tropics? If they started out as a tropical group then they would most likely have diversified into many species there first of all. This would have given a head start to the tropics in terms of the ancestors of today's families and genera of flowering plants, that has only been built upon by splitting into more and more different lineages. Even though they eventually made it out from the tropics by evolving the means to tolerate colder and drier climates, the flowering plants can be expected to be much poorer in species within these more recently colonized climates. Over time then, various groups of angiosperms may have spread out polewards, but always the tropical latitudes were way ahead in terms of species richness.

I will talk a lot about flowering plants in this chapter. This may seem like unfair bias against all the species of animals, but whatever has happened with the plants is also likely to be important to the story of species richness of many other land-living groups, because they are the main habitat and food source for so many insects, birds, and mammals. More species of plants can mean more food types, nesting sites, hunting sites, etc. for animal life, allowing more different specialized forms of animals. In this way, the plants might provide the basis for other latitudinal gradients that build upon them. Through their interactions with other groups such as insects they might also tend to get caught in a spiraling increase in species richness; but more on this aspect below.

A lot of geological evidence suggests that the world used to be much warmer than it is now. This warmth could in a sense have provided the standard of "normality" on Earth from which life has had to adapt to increasingly colder environments. For example, around 55 million years ago, subtropical vegetation extended well north of the Arctic circle, with alligators and palms being found on the now frigid Spitzbergen Island. Generally speaking, between around 70 million and 40 million years ago the world was far warmer than it is now, with tropical or subtropical temperatures extending to the mid-latitudes above 40° north or south of the equator. This was a time when many groups of animals and plants were taking on broadly their present-day form. For example, many of the most species-rich families of flowering plants appear during this phase, shaped in part by the recovery of the world from the mass extinction that occurred at the end of the Cretaceous 65 million years ago (see Chapter 3). Likewise most of the modern mammal and bird orders and families appeared during this interval.

If we look even further back in time, it seems that some of the broader groups of organisms really did first appear and diversify in the warm equatorial regions. For example, the fossil record suggests that the flowering plants originated in the tropics. The earliest accepted angiosperm pollen turns up in rocks from the tropics where Africa and South America were splitting apart some 120 million years ago (Crane and Lidgard, 1989). After their first appearance, the diversity and abundance of the flowering plants increased much more rapidly in the low latitudes, judging by their fossil record. So indeed, diversity started to build up earlier near to the equator, and this may mean that they got a head start—so now the flowering plants in the higher latitudes have a lot of catching up to do.

David Currie and his colleagues (2004) asked what patterns we would now see in nature if "harshness" (i.e., difficulty of adapting to high-latitude climates) is really so important? One thing to expect is that the range limits of most species will often follow climate boundaries—since it is climate that is all-important and is supposed to be preventing things from spreading out any further. Currie focuses on birds, partly because these have been so keenly observed by birders for so long, that their distribution records are very accurate. And indeed, the northern winter limits of 62 bird species in North America do show up as being associated closely with a particular isotherm (line of equal temperature). However, another 191 bird species do not show any close association—which seems to disagree with the idea that "harshness" plays a role. Currie *et al.* (2004) also mentioned various other contexts in which species range limits seem not especially closely related to climate, and suggests that this too means that the difficulties of adapting to climate cannot be what is limiting their spread, and overall species richness. For example, they wonder why more bird species have not shifted their ranges in response to recent climate warming, if climate is what is limiting ranges.

My own feeling is that Currie and colleagues are putting a little too much faith in modeling species range boundaries in relation to climate. This is a fashionable tool at present, but I suspect that its users expect too much to come out of their results. It seems plausible to me that the computer software that finds matches between range boundaries and climate cannot take into account all the interacting subtleties of species interactions with climate. Several environmental factors acting in combination might determine exactly where a species range boundary will sit in any one region, yet this would confuse any simple model that tries to correlate boundaries with climate. It could depend partly on species interactions on a local scale, and yet on a broader scale of long-term averages climate could be a major factor pushing a species south or pulling it northwards. With all these interacting factors, plus a small element of luck, the position where a species boundary ends up could just be a matter of probability. In this case the overall pattern will only show up clearly in a statistical sense in large samples of species—in fact the latitudinal gradients that we are talking about here. Likewise, the many complex aspects of population biology and species interactions could limit how fast a species can spread in relation to climate warming—yet climate might still be important, it is just that everything takes a while to adjust.

If you need evidence that current climate *is* all-important in determining what grows where, one way to convince yourself is to look at characteristics unrelated to species richness—and how these pan out along latitudinal gradients. When my colleagues and I (Adams, Green, and Zhang, 2008; Figure 2.10) plotted the percentage of species of trees with toothed margins in the North American tree flora against latitude, we found to our amazement one of the tightest curves I have ever seen in ecology. Evidently the trees have arranged themselves precisely in relation to something in the environment, which is determining what species occur where. Whatever one makes of this curve, however one explains it, it certainly looks like North America at least has flora that (averaged across many species) is in close equilibrium with climate.

2.4.2.2 Do species "fall into" the tropics?

Earlier, I mentioned the layman's simple idea that the living is easier in the tropics. On further thought, maybe there could actually be something to this. Perhaps there is a way in which the tropics accumulate species, because they are indeed just easier to survive in? Warm, moist environments may require few special adaptations, unlike the seasonal higher latitudes with their hard winters, or the dry outer tropics with their long seasonal droughts. To resist frost may require protective antifreezes, and a special dormancy mechanism to cope with the worst of the winter—features which might be unlikely to evolve very often. And the colder the winters, the less often it should happen, because more and more complex adaptations are needed to cope with increasingly extreme conditions. Yet even if a particular species of tree or butterfly is

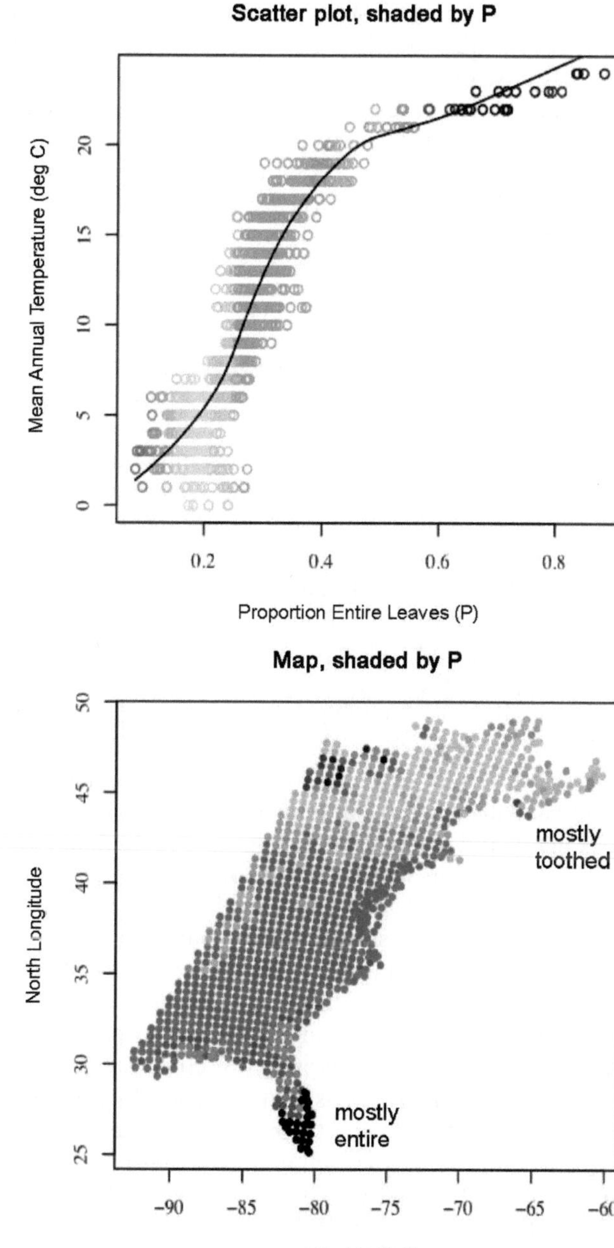

Figure 2.10. Proportion of species of trees with "entire" non-toothed leaves (bottom axis) depends closely on the mean annual temperature of the climate (vertical axis). This very close relationship implies a state of equilibrium between species distributions and climate (from Adams *et al.* 2008).

adapted to cope with cold climates in terms of its life cycle and physiology, all that it has to do is lose some of its special adaptations (and perhaps acquire a few other simple modifications) and it can now live in the tropics. It might evolve gradually, with the species extending its range into the tropics, or it might bud off a new species that is adapted to live in the warmer climate. Another thing to bear in mind is that most species that live in temperate latitudes are already adapted to a more or less tropical environment during part of the year, in the summer when conditions are warm and moist. In the world's semi-arid seasonal environments there is often also a warm and moist wet season during part of year. All that a species from one of these other climates has to do is lose its additional special adaptations to cold or drought, and it is set up to grow in the tropical climate.

Thus it may be easy for an evolving lineage to "fall into" the tropics, and the tropics will always be accumulating new species from the high latitudes as well as from within. It will be much harder, though, for tropical lineages to get out again.

The result, according to this idea, is a one-way flow of species from high to low latitudes. The ease of adaptation to warm climates will constantly be pushing more new species into the tropics. The higher latitudes, by contrast, rarely get new species from the tropics and must mostly rely only on locally generated species from lineages that long ago adapted to the cold. This could explain why the species richness of many groups of organisms, such as trees, mammals, amphibians, and birds, tends to follow climate gradients towards warmer, moister environments (a pattern found, for example, by Currie and Paquin, 1997 and Adams and Woodward, 1989)—paralleling the growth rate of vegetation.

My friend and colleague Axel Kleidon has developed this general idea about the benign-ness of the tropics a bit further, in the case of plants with the JeDi Model (which is short for Jena Diversity Model; Kleidon and Mooney, 2000; Kleidon *et al.*, 2009), in some work in which I have also played some small part. The "Kleidon vision" is to think of all the members of each plant species in a particular place as an imaginary plant, with random combinations of characteristics such as shape, size of root mass, and seasonal cycle. It is as if each plant species is made from a "kit" like a child's toy doll in which a range of combinations of clothes and accessories can be clipped on from an assortment available.

Any randomly assembled imaginary plant is far more likely to be viable in the year-round warm temperatures and moistness of the tropics than in the temperate zone. Axel suggests that in the tropics, idiosyncratic differences in the growth characteristics and form of plants do not lead to most of them getting pushed out or failing to grow: they can happily just coexist, each of them doing nearly as well as the others. The result of this is that there are lots of species from many different backgrounds coexisting in the tropics. In the temperate zone, by contrast, the stringent requirements set by the climate mean that few forms can do well. One can imagine that in this situation, only the very toughest and most exquisitely adapted species survive the cold climate, and all others are pushed out by competition or by simply being unable to survive and grow. In the models of Kleidon and Mooney (2000) and the JeDi model of Kleidon *et al.* (2008), species with various combinations of traits are randomly generated across the Earth's land surface, rather like mud thrown at a wall. In these models, only some combinations work, and most often in the tropics. Towards cooler and drier climates, the combinations "stick there" less often because of the rigors of those environments. What this seems to show is that in terms of

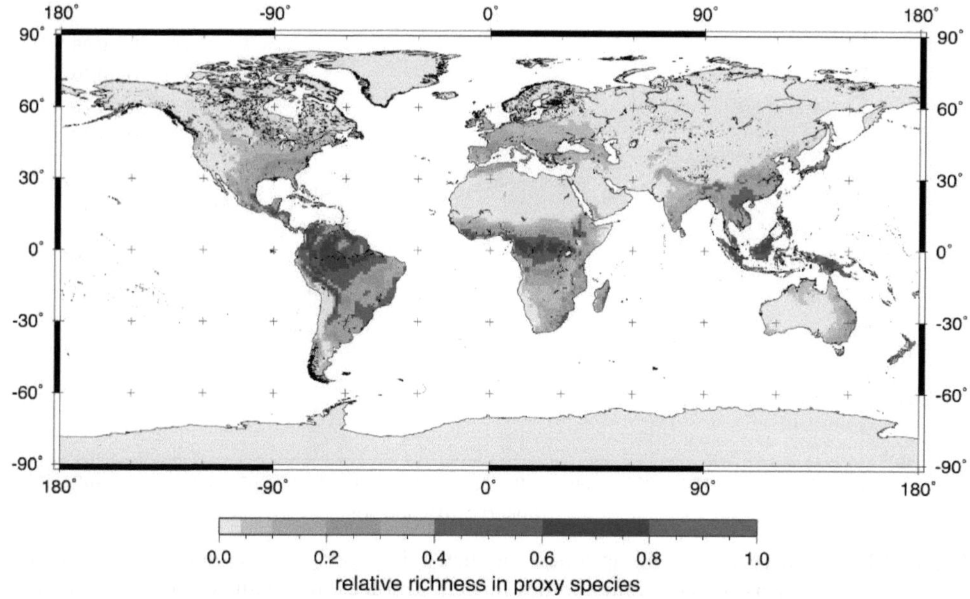

Figure 2.11. Predicted global map of plant species richness of the JeDi model. The patterns of species richness closely follow what is known for modern-day geographical patterns with latitude, and also manages to predict much of the east–west variation in species richness within regions (from Kleidon *et al.*, 2009). *See also* Color section.

reasonable first principles, the tropics really may be a less demanding place in which to survive as a plant.

In the real world, new species would be spreading out from the range of their ancestral species, and along a climate gradient. A new species is likely to have better luck in being viable if it moves in towards the warmer, moister tropics than out towards colder or drier climates—hence new species will accumulate towards the tropics, but tend to die if they move towards higher latitudes.

The JeDi model (Kleidon *et al.* 2009) predicts a certain general geographical distribution of woody plant species richness—and this can be compared with actual maps of species richness (Figure 2.11). Generally speaking, the predictions match nature very closely across the Earth's land surfaces: areas of relatively high species richness in the real world nearly all correspond to areas of high plant species richness in the model. This is an encouraging sign that there may really be something accurate about the way that this model depicts the workings of nature.

The JeDi model also comes up with a very interesting incidental prediction, that species should become more evenly abundant in the tropics, with a moderately sized slice of the pie for many different species. As I mentioned in Chapter 1, this evenness aspect is part of the conventional definition of "diversity" in ecology. In the temperate zone (e.g., at 45°N or 50°N), forests are typically dominated by one or two extremely abundant tree species with the other species pushed down to much lower levels of abundance. By contrast, in tropical rainforests the many tree species are

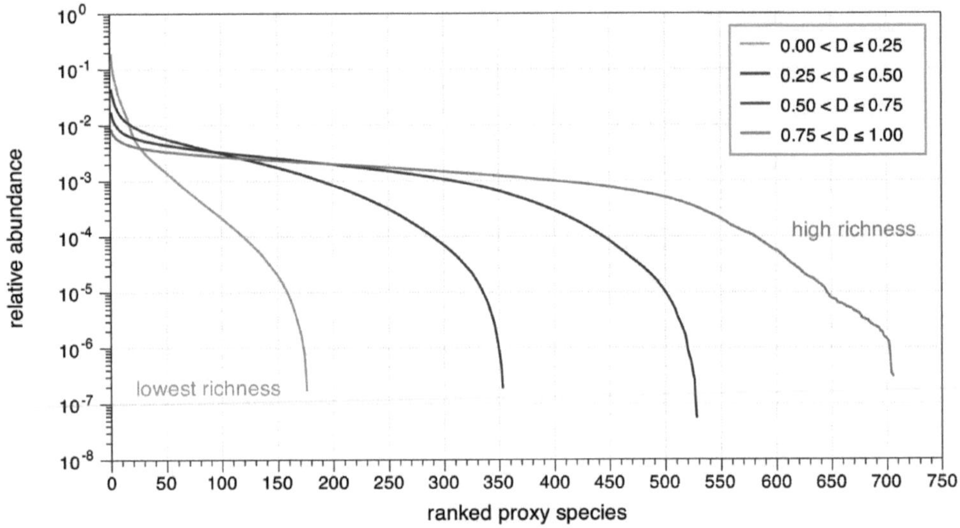

Figure 2.12. Relative abundance/evenness predicted in woody plant communities by the JeDi model. The least diverse communities in high latitudes are predicted to be dominated by a small number of species, with a small proportion of those being very abundant (green line). The most diverse low-latitude communities (red line) are predicted to have most of the species about equally abundant as one another (from Kleidon *et al.*, 2009). *See also* Color section.

mostly at low densities but fairly evenly abundant—normally there is no particular species whose abundance is far greater than all the others.

A remarkable thing about the JeDi model is that it predicts the relatively uneven distribution of species abundance within forests of colder climates—a few species are very abundant, with hardly any room left for anything else (Figure 2.12). The farther north one goes (e.g., into the conifer forests of Canada and Siberia), the more uneven the distribution becomes. This prediction came out incidentally, accidentally in fact as a result of the basic assumptions about the relative easiness of the tropics for the survival of plant forms. It was not a result of tweaking the model until it was able to predict both geographical patterns in richness and also the latitudinal gradient in the evenness of forests.

The trend in evenness is a big plus in favor of the JeDi model: it correctly predicts one of the most universal ecological patterns, which goes along with latitudinal species richness gradients in plants at least. Most of the other ideas to explain latitudinal gradients in plant species richness do not also produce such a prediction of greater evenness in the tropics, although the Janzen–Connell hypothesis is another which does (more about that one later).

Although I have primarily described the JeDi model as a disequilibrium model, one that sends species richness ever-higher, there might still exist a "lid" on species richness for both high-latitude and low-latitude environments, in a way that will favor greater species richness at the lower latitudes. This follows on from the evenness aspect of the model. Take, for example, the high latitudes as the Jedi model envisages them: with such an uneven abundance distribution and only small opportunities left over for the less successful species in the temperate zone, one should expect that there are fewer species there. This is because hardly any species can become abundant enough to form a "viable" population (see Chapter 8), in the face of competition by the few really successful growth forms. Kept at too low a density, the population of a species will simply fizzle out—unable to exchange pollen to reproduce and unable to survive the buffeting of random bad-luck events. This sort of uneven abundance distribution may thus provide an upper limit on the species richness that can be reached in any given climate. In effect there are only a limited number of thin slices of the pie that species can live off in the higher latitudes, because a few highly successful growth forms take most of that pie. It may explain why species diversity over evolutionary time tends to climb and then reach a stable plateau, as we see in the fossil record for both tropical and temperate zones of the world (Chapter 3).

Taken a bit further, tweaked to allow migration of species across the surface of the globe, the same model should also predict that plant species richness climbs faster to its plateau level in the tropics than in the temperate zone—and this also agrees with the fossil record (Chapter 3). This at least is my expectation, although it exists only in my intuition as this more advanced version of the model has not yet been written and run. What I imagine is that in effect lineages of species can easily "fall into" the tropics, by producing daughter species that survive in a warmer, moister, less seasonal environment. These species will simply have lost the features that allowed them to survive the rigors of cold or dry climates. But then once they have lost such features, those species cannot as easily produce daughter species that can move back out to drier and more seasonal climates, because such climates are a lot more demanding in terms of adaptations. It is a lot easier to lose a special adaptation than get it back again. Even an evolutionary group that originates in the higher latitudes, once it sends daughter species into the tropics, should tend to diversify there faster and contribute to the rapid tropical build-up of species richness.

And if it can predict plant species richness, a model like this can indirectly predict animal species richness in at least some groups, such as insects. This is because insects are often fairly specialized on particular host plants. So the more plant species there are, the more insect species (and perhaps bats, and frogs, and mammals) we can expect.

I am particularly partial to the JeDi model because it seems to predict so much about global species richness patterns, and it does it so simply and effortlessly. However, I have to admit that part of the reason that I am so fond of it could be that I have been involved in some aspects of its development! Its way of looking at plant communities is very different from what we are used to in ecology, which usually involves thinking at the level of the population. Normally, when we ecologists model populations they consist of little indivisible units, like atoms in physics: in ecology these units are individual organisms. The JeDi model takes a very different approach: looking at all the individuals of a particular species in a particular area as a big mass of living tissue that either grows or shrinks according to how successful it is.

There are no absolute winners or losers as there would be in the classical theory of niches—just some of these agglomerated species doing rather better than others, so that they grow to occupy more of the space and the living mass of the community. Even if they are all living in a totally uniform environment and in exactly the same niche, the result is the same—species are not extinguished by competition but instead take up an orderly, predictable and stable level of abundance, that tails off at some point to almost nothing in the weaker species. Being someone brought up in the standard tradition of population dynamics, this is a radical departure and one which I struggle with. The JeDi model should not work in predicting nature, and yet apparently it does. JeDi is not the only model out there that takes this approach; various other groups seem to be grappling with ideas in the same direction (Shipley *et al.* 2006). How can we reconcile a model like JeDi with the fact that there are populations that we know exist as discrete individuals, not just a merged mass of green slime. I can only assume that how well suited a plant is physiologically to its environment affects the probability of it getting established in a particular spot and then reproducing, even though there are no all-out winners and losers. It is not a predictable win or lose contest; instead it is a lottery, one that is weighted according to how well suited a plant is to its environment. But if a plant is very poorly suited to its environment, it will be pushed down the abundance scale so low that eventually chance fluctuations in its numbers allow its population to go extinct. Rather like a poor card player, by luck it may not lose every time, but it may end up starving anyway.

How would we ever test if the idea that "easier living in the tropics" has really allowed species to fall into the tropics, out there in nature rather than just in a computer? One way I suppose might be to look at the DNA "family trees" that trace the paths of evolution of organisms. If species can fall into the tropics more easily than making it out of the tropics, we might expect to see that the family trees have many branches that lead from higher to lower latitudes, but few that go out in the opposite direction. However, I do not yet know of any datasets that could be used to do this test.

There are actually some small-scale observations of nature that outwardly agree with what models like JeDi predict on the geographical scale; that plants are selected for relative abundance and size in the community in terms of how closely they approach functional perfection for the circumstances they are growing in—the soil, the climate, the successional stage, or whatever. Plants that are further from this functional perfection still stand a chance of growing there, but they are smaller and less abundant. Shipley *et al.* (2006) looked at the plant communities in a dozen old abandoned vineyards in southern France. Comparing one old vineyard site with another, they found that plants arranged their relative abundances according to how closely they approached the average property of the community. A plant community can be averaged out as a single generalized "plant" with a certain height, shape, etc., and the closer a species matches this average, the more abundant it is, and the more living mass is composed of it—for the individuals of this most successful plant (even if it really is several species of plants) are both common and grow relatively large.

2.4.2.3 The mid-domain effect: Is the latitudinal gradient just the sum of chance overlap of ranges?

One hypothesis to explain latitudinal trends in species richness focuses on the fact that each species has its own distribution range, strung out somewhere along the temperature gradient between the tropics and the poles (Colwell and Hurtt, 1994; Rosenzweig, 1995; Colwell and Lees, 2000). If different species have their own ranges randomly arranged along this gradient, extending north and south of the equator, by chance many will tend to overlap in "the middle" which is roughly where the equator is (Figure 2.13). This will tend to give a peak in species richness around the equator, declining gently outwards towards higher latitudes. This is known as the mid-domain hypothesis. It assumes then that there is nothing special about the tropics in terms of promoting diversity, except that it sits in the middle geographically between the poles.

How might we test the mid-domain hypothesis? One way would be to look at the shape of the peak of tropical species richness. Randomly overlapping species may be expected to give a bell-shaped curve with its peak near the equator, but one would not expect this peak to be especially high, having a gentle curve upwards to its high point.

In at least some cases, it has been claimed that the peak in tropical diversity has about the right shape for the mid-domain hypothesis (Colwall and Lees, 2000). For comparison, one can make an imaginary curve of species richness against latitude by overlapping a random set of latitudinal ranges. Willig and Lyons (1998) did this for marsupials (possums, mostly) of the New World, and found that there was a very good fit between what one would expect for a randomly placed set of ranges, and the actual diversity peak of each group. In the case of the marsupials, the peak was 20°S of the equator in the Americas—as would be expected by this hypothesis but not by other ones, given the shape of the landmasses. Another case is in Madagascar, where the overlapped ranges of 122 species of insects and vertebrates resulted in a diversity peak approximately where one would expect from random overlapping ranges scattered along the length of the island (Colwall and Lees, 2000).

However, if we look at the data on species richness in certain other groups, the peak in diversity is often many times higher in the tropics compared with that in the high latitudes. This is hard to explain by this mechanism involving random overlap, because chance overlap only produces a gentle curve and peak. For instance, take the exponentially higher levels of tree species richness found in the tropics compared with the temperate zone. As I mentioned above, there are 1,400 species of trees and shrubs in a 50,000 km^2 grid square of the Malay Peninsula, yet only around 40 in a similar sized area in southeastern Russia (Adams, 1988). There is simply no way that random overlap alone could explain such a high tropical peak in species richness. It is a similar story with other groups such as butterflies and birds, which also rise in diversity exponentially into the tropics—though the peak they show is actually a bit lower than for trees. Also, the geographical location of peak tree species richness in Amazonia does not agree with what would be expected for the mid-domain effect, given the shape of the continent (Silman, 2005). Overall, it looks like chance overlap of ranges might provide part of the explanation for latitudinal gradients, but it cannot be the main reason for the trend in many groups that show a latitudinal trend.

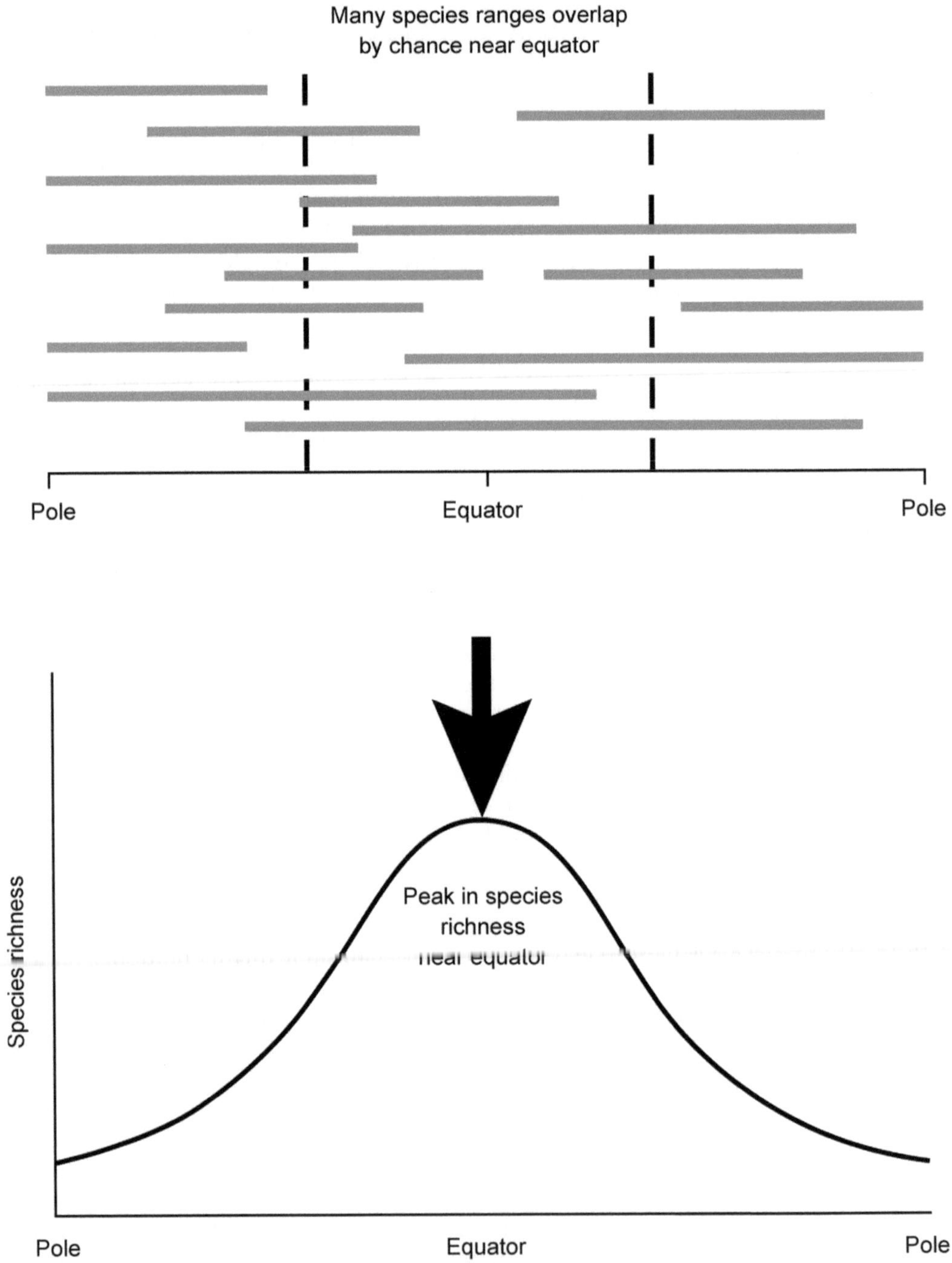

Figure 2.13. The mid-domain effect. If species have their ranges arranged across random spans of latitude between the equator and both poles, they can be expected to overlap by chance more often over the equator. This should give a rise in species richness close to the equator.

2.4.2.4 Are species originating more easily and more often at lower latitudes?

Another leading idea is that there is more diversity towards the tropical end of latitudinal gradients because the "splitting" process that gives new species can occur more easily there. This might occur because the processes that fragment and isolate populations—starting the process of divergence towards separate species—happen more often in the tropics for some reason.

So if speciation occurs more easily in low latitudes, this might tend to build up greater numbers of species in the tropics and warm temperate zones (Cardillo *et al.*, 2005). One can hypothesize that populations become fragmented more easily in the lower latitudes, allowing the numerous separate populations to evolve in different directions until they end up as quite distinct species. More splitting might occur because tropical animals don't move around as much, so different populations are less likely to keep in touch with one another through organisms migrating between them. In the case of plants, their seeds or the pollen might not move so far, maybe because the animals they rely upon to disperse their fruits or pollen are "lazier" in terms of traveling across the landscape. In each case, the result of lack of movement and exchange of genetic information will be more divergence between populations, and ultimately more new species forming. Or it could be that in low-latitude areas, once populations are initially separated they change more rapidly (perhaps because mutation rates in their DNA are higher)—to the point where they are no longer just different populations but fully fledged species. It has been suggested that the higher temperatures or UV light levels of the tropics give faster mutation rates.

In fact, for any latitudinal gradient to be formed due to whatever cause, there are probably going to have to be more species actually generated in the tropics. The question is, does the rate at which new species are formed provide the real limit on the number of species that are found in each part of a latitudinal gradient? It could be, for instance, that whenever there becomes more room available for extra species in the tropics, due to one of the density-dependent mechanisms mentioned below, life more or less automatically diversifies into more forms to fill up the available capacity.

Another leading idea is that there are only more species towards the tropical end of latitudinal gradients because the "splitting" (speciation) process that gives new species can occur more easily there. This might occur because the processes that fragment and isolate populations—starting the process of divergence towards separate species—happen more often in the tropics for some reason. Hypotheses that emphasize the rate of speciation as a primary cause of latitudinal diversity gradients instead suggest that both the tropics and higher latitudes are not yet "full" of species, and that the only reason that the tropics have more species is that species are formed more easily there.

The only way to figure out whether speciation rate is the real cause of tropical high diversity is to look and compare tropical and higher latitude ecosystems. Is there evidence that species are being formed more frequently at low latitudes? One thing we might expect—if speciation does occur more often in the tropics—is that species will on average be younger there. But how might we tell if they are younger?

There are essentially two ways to look at evolutionary family trees and tell how old species are (Currie *et al.*, 2004). One is to use the fossil record, which could in theory provide a faithful and detailed record of the history of each lineage. In practice, however, the fossil record is rarely as detailed as we would like; the preservation of remains and our finds of them are very patchy. One instance where we do have a detailed record over millions of years is with fossil seashells—bivalve mollusks—along the east coast of the U.S.A. and down into the Caribbean. It does seem that the lower latitude bivalves (clams) produce new species more frequently than higher latitude ones, given an equal-sized sample of species at the start. Hecht and Agan (1972) also found that tropical marine bivalve species are on average younger than those of higher latitudes. It also looks like evolutionary innovation in general might be greater in the tropical shallow seas. For instance, Jablonski (1993) noted that orders of marine invertebrates tend to originate more often in the tropics, because they tend to be found first in tropical seas before they spread out.

Another way to look for differences in the speed of evolution and rate of speciation is to look at living organisms in the present-day world. For instance, one can use molecular evidence to look for signs of more recent and more frequent origination of species in the tropics. This can be done using DNA sequencing: since DNA seems to accumulate random mutations at a constant rate over time, it can be used as a "clock" to estimate rates of divergence. We might expect to find evidence that related and similar-looking pairs of species have only diverged relatively recently in the tropics, compared with pairs of species in the temperate zone having split a long way further back in time, if evolution and speciation is more "sluggish" there. In a study comparing the DNA sequences of bats, Stevens (2006) found that indeed, the most closely related pairs of bat species split further back in time in the temperate zone than the tropics—certainly giving the impression that there is more rapid, frequent speciation in the tropics. Other studies have disagreed with the idea that speciation has occurred more frequently and recently in the tropics. Bromham and Cardillo (2003) found that in very similar-looking pairs of bird species ("sister species"), some pairs in the tropics and other pairs in the temperate zone, there was no difference in the time since the two diverged, as estimated from comparing the number of differences that had accumulated in DNA from the mitochondrion. Bear in mind, however, that it is not completely certain that the "clock" of mutation rates is really accurate—for instance, one hypothesis to explain latitudinal gradients is even based around the idea that mutation rates are higher in the tropics!

Novotny *et al.* (2006) also produced a DNA "family tree" of all the woody plants in a temperate forest plot, and did the same for the woody plants within a tropical forest plot. They had other purposes in mind for their study, but if we compare these two family trees it looks (to me) that the branches of the tropical lineages are older—suggesting actually that speciation and diversification is more recent in the colder temperate zone. Perhaps there ought to be more work done like this, to verify whether temperate tree lineages have generally split more recently than in the tropics. In any case, more recent splitting in the temperate zone looks like the opposite to what we would expect if the tropics are still a great engine of speciation.

A different approach to looking at present-day organisms is simply to compare species that fall in the same general group, for the overall degree of difference in form. This approach doesn't look at time since divergence, as in the bat study—it just asks "if two closely related groups find themselves in the tropics and temperate zone respectively, does the tropical one proliferate into lots more species?" That is what we would expect if speciation happens more easily in the tropics. So, for instance, one can look at bunches of species that form a family tree straddling both the low and high latitudes—with some species in the tropics, some in the temperate zone. Then we can compare the branches of species (known as "sister clades", a clade being a biologist's term for a family tree) from this family tree that exist in the tropical and temperate zone respectively. Given the degree of overall similarity in the group, we need to ask "Have the tropical parts of the family tree more species within them, reflecting a tendency to split into many species within the tropics?" Cardillo (1999) compared 48 tropical–temperate pairs of clusters of species, in passerine (perching) birds and in swallowtail butterflies. He found that, indeed as we might expect, just being in the tropics somehow promotes speciation, because the tropical branches of each clade had more species on average. However, in studies such as this the lack of any objective standard by which to compare such family trees (such as a fossil record or DNA dating) makes me rather reluctant to take the results very seriously.

Overall then, there seems to be some evidence for more rapid speciation in the tropics as the generator of greater tropical diversity—but this is not always the case (e.g., the molecular study on sister species of birds by Bromham and Cardillo, 2003, above). In any case, it is difficult to know from the few case studies that have been done whether they are representative of most of the vast range of life forms. Also, we cannot really tell whether more frequent species formation is really the driving force of tropical diversity, or merely something that responds to extra capacity for species in the tropics.

2.4.3 A general test of the disequilibrium theories: Has the build-up of species richness continued over time?

Disequilibrium mechanisms for latitudinal richness gradients invoke processes that have continued over differing periods of time, or at different rates, at different latitudes. If the all-important thing determining the species richness of a region is how long diversity has been building up, we should expect to see signs that diversity in the temperate zone did not just stop rising in the early days (the "early days" being, for example, in the first 10 or 20 million years following a mass extinction, see Chapter 3). It should just have continued to increase over time (e.g., throughout the 65 million years or so since the last great mass extinction to hit the Earth). With each million years that passes, there are more opportunities for groups of tropical organisms to evolve the adaptations to make it out into the temperate zone, and more chances for the groups that are already present in cooler climates to split off new species there. At the same time, diversity should also have continued to rise in the tropics—just as it did the when the angiosperms had recently evolved back in the Cretaceous around 120 million to 70 million years ago.

So we have a general test of the idea that time is all-important in bringing about latitudinal diversity differences: all we need to do is look at the fossil record to see if diversity has continued to increase steadily during the last several tens of millions of years. The best picture we have is for the flowering plants. Counting up the number of types of angiosperm fossils over the last 50 million years or so, it turns out things are not as we would expect from this theory. In fact, the overall diversity of angiosperm genera in temperate climates has not increased (see Chapter 3) for tens of millions of years. The heyday of the temperate tree flora had already been reached by 40 million years ago, when a wonderful array of different types of trees filled forests that stretched from Europe all the way across Siberia, eastern Asian, to North America and Greenland. Why new genera stopped appearing in this ancient Tertiary flora is a mystery. Since reaching a plateau 40 million years ago, the temperate tree flora has actually lost much of its diversity (see Chapter 3)!

There is a broadly similar picture in the fossil record from the tropics too. The diversity of plants in tropical rainforests in the Americas actually seems to have reached a peak in the early Tertiary, around 50 million to 40 million years ago, not so long after the first tropical rainforests formed (Chapter 3). Since that high point, diversity has declined, not increased. What it was that gave the maximum diversity back then is unclear, but this was a time when the world went through an exceptionally warm phase. Tropical rainforest seems to have covered the majority of the Earth's land surface, which may have led to many opportunities to accumulate new species, with few things going extinct in the very large area of forests. It seems that tropical tree diversity reached as high as it could go early on, when the world was warmer. As it has cooled off since then, the capacity of the tropics to contain species has decreased.

It looks as if there is some invisible lid which has stopped tree species diversity in both the temperate and tropical forests from increasing any further. Each climate zone appears to have reached its own capacity for species many millions of years ago: and that capacity is for some reason different between the high and low latitudes (and in both regions, it has since declined over time too). This might be due to one or more of the "equilibrium" mechanisms that I outline below.

So it really does *not* look like tropical origins, plus lack of time to diversify out from the tropics, are the reason why there is a latitudinal gradient in flowering plant (angiosperm) species richness. Perhaps, in the early days between about 120 million and 65 million years ago when this newly evolved group of plants was spreading out to higher latitudes, the legacy of their origin in the tropics really did limit their diversity in colder climates. Maybe back then, nature was not yet "full", and the disequilibrium mechanisms were all-important in determining how fast diversity built up? But it seems that stage has long passed, and the angiosperms have now reached some sort of upper limit of diversity that happens to vary with latitude. How true is this of other groups of animals and plants? We cannot be sure. But we know that such abundant plants as the angiosperms, which form the base of the food chain and the habitat itself in so many places, are likely to be important in determining what happens with other groups of land organisms too.

2.5 EQUILIBRIUM THEORIES: THERE IS A LID ON SPECIES RICHNESS THAT IS HIGHER IN THE LOW LATITUDES ...

In contrast, a second group of ideas suggests that the reason richness increases towards the low latitudes is that warmer climes are able to "contain" more species. These theories propose that both the high and low latitudes are essentially already full of species, about as full as they could ever get (unlike the scenario in the other group of theories mentioned above). So each latitudinal band is actually in balance, in "equilibrium", as regards species richness—nothing is really changing over time, just some species going extinct and roughly equal numbers replacing them. And it just so happens that more species can exist side by side in a tropical warm climate than a cold higher latitude one, because they have a higher capacity, due to something about the way ecosystems work. This explanation sounds compelling, but it is vague, and makes a lot of assumptions. So equilibrium ideas each have to have their own detailed explanation for why warmer low-latitude climates might allow ecosystems to contain more species.

2.5.1 The tropics are just bigger

If we look at a flat "Mercator" style map of the world, it is easy to forget that tropical climates occupy a larger area than the colder mid-latitude temperate zones of the world, or the chilly Arctic zone. This goes for both land and sea, because the Earth is fatter around its equator, giving a lot more area than is present near to the poles. Terborough (1973) suggested that since tropical lands are more extensive, there is simply more room in these places, and this allows more species—each existing in its own corner of the tropics. Terborough suggested that this has allowed the numbers of species to build up to higher levels in the tropics, until they became "full" and stopped accumulating extra species.

While this is a reasonable idea, it is hard to understand why there should be quite *so many* more species in the tropics. Comparing the number of types of trees, or butterflies, or birds, for instance, there is far more than the several-fold increase in species richness in the tropics, that would be expected from their greater area alone. And if we compare broad habitat zones, the idea looks shakier still. There is actually a smaller area of tropical rainforest in the world than northern conifer forest, because most of the tropics are either ocean or lands too dry for rainforest. So by this token, we should expect the conifer forests of Siberia and Canada to be as diverse as tropical rainforests—and they clearly are not!

2.5.2 More energy, more food in warmer climates: the "species–energy hypothesis"

One leading idea to explain large-scale species richness trends is that resources such as food are more abundant in warmer climates. This may be because in climates that are both warm and moist, plants grow faster.

A very interesting observation is that in many groups of animals and plants, species richness along a latitudinal trend actually follows a climatic index called "actual evapotranspiration", more closely than it follows latitude as such. Actual evapotranspiration, or AET for short, is derived from a combination of the warmth of the climate and the amount of rainfall. It generally seems a very good predictor of the growth rate of wild vegetation; plants tend to photosynthesize and grow fastest in warm temperatures with plenty of rainfall. The growth rate of vegetation is also known as "net primary productivity", shortened to NPP, a term often used by ecosystem ecologists. NPP is interesting because it represents the food supply for the whole ecosystem; including the rate at which the plants have been making food for themselves, the rate at which plant material is available for herbivores, and thus the number of herbivores available to be eaten by carnivores. It also provides a rough approximation to the rate at which dead material and waste becomes available for decomposers to break it down.

Somehow it seems intuitively reasonable that more food around could feed more different types of organisms, but we need to explain why it should go into feeding a greater diversity of types rather than all being gobbled up by one species that becomes super-abundant. An important point to bear in mind is that in an ecosystem, each species of organism can take a "slice" of what is available. That is its ecological "niche'—its specialty—a job which it is better at than most of the surrounding species. When it is in its *forte*, no other species can push it out. Similarly, in the traditional picture of ecology at least, every other species around it will have its own niche, doing what it is good at.

An important principle here is that if there is an abundance of food available, there will be enough available for each species to evolve towards being good at occupying a very narrow niche. More food from plants can allow each type of herbivorous insect, bird, or mammal to specialize in being especially good at gathering and using a narrow range of plant food materials. For example, if fruit is very abundant, birds might evolve to eating only particular types of fruit of a certain shape, size, or composition. There is an advantage in specialization, because it enables a species to push out competitors around it and secure its own stable food supply. In nature there is a lot of meaning to the phrase "jack of all trades, master of none"; yet specialization has its costs in that it cuts off other potential sources of food if the supply gets scarce. If a species became too specialized on a narrow, rare food resource, its population has to become very sparse and small. Being so few in numbers even at the best of times, it risks going extinct if its population happens to dip down a bit below normal—so this occasional extinction process should prevent species becoming too specialized where food is less abundant.

In the tropics then, in our hypothetical scenario with each bird species specializing in on only certain types of fruits, there can be more species of birds around in total. In cooler climates with less vigorous growth of vegetation and a sparse and intermittent supply of fruit, there are fewer opportunities to specialize. It will be the same with insects which might eat leaves or seeds: with more rapid growth of vegetation and thus new leaves or seeds being supplied frequently and abundantly, these herbivores might each specialize in on only one particular type of tree. In cooler higher latitude climates, slower vegetation growth means less plant material produced per unit of time, and this sparser food supply means that there can be less specialization. The same principle might be true of any type of organism that depends directly or indirectly on the resources provided by plants. Carnivorous insects and birds might specialize more if there are more herbivorous insects around in general to eat, evolving to become very good at catching and digesting only a particular subgroup of insects (e.g., hairy caterpillars to the exclusion of all other insects). Adding to this may be the fact that there are simply more different species of insects around (due to the effect of the high growth rate of vegetation) which means additional opportunities to specialize. We can imagine that decomposers might be able to specialize too: perhaps, for example, on the waste products or dead bodies of only certain particular types of large mammal herbivores (which are more abundant because of the increased growth rate of edible vegetation) again enabling more species of decomposers to exist side by side.

Plants too might respond to an abundance of resources that allows a higher growth rate in the tropics, by themselves becoming greater in numbers. More trees can be packed into a given area of forest, or more wild flowers into an area of grassland. Each species of plant might tend to split new species off that can very effectively grab a part of the diverse resources that they must feed off. These extra plant species might themselves each tend to support their own specialized set of insect herbivores and microbes. The whole interacting system, with more resources and more types of resources, could then spiral upwards until the community is absolutely packed with species.

Eventually what will stop this upwards spiral of species richness at a stable high point will be the limit of how much more finely the food supply can be subdivided. The degree of specialization in nature may be limited by this balance, between on one hand the tendency of each group of organisms to specialize—splitting into narrowly adapted types—and on the other the likelihood of their going extinct if they become too specialized and their populations are too sparse and small to maintain themselves. We might imagine the total amount of a resource as a cake, from which each species can take a slice (Figure 2.14). If the cake is small, the slices must be wide to provide the resources to "feed" each species. If a species takes too narrow a slice, it is likely to "starve"—or in actual fact go extinct.

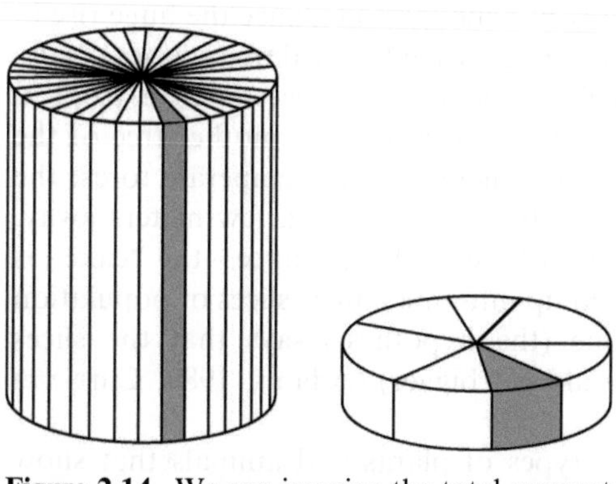

Figure 2.14. We can imagine the total amount of resources available to similar species in a community as a cake, from which each species can take a piece (its "niche"). If the cake is large (left), a species can afford to take a narrower slice and still get enough to survive. If the cake is smaller (right), the slices must be wider. In the former, there can be more different niches and more species packed into the community.

So the species–energy hypothesis suggests that more species can exist at lower latitudes because they can each specialize more in the warmer climates, and this is because there is generally more food around to divide up. We might expect that over time, each species will keep on narrowing down its diet (its niche) until it is stopped by having populations that are too sparse to keep themselves going. A prediction of this hypothesis is that we should expect to see in warmer lower latitude environments that there is about the same population density of *each individual species* of herbivorous insect or tree or whatever, compared with cooler high-latitude environments (Adams, 1989). But the total density of individuals of *all species taken together* should be much higher in warmer and more diverse environments. This is because in the tropics the "cake" is predicted to be bigger, but the "slices" remain the same size per species. So here then is a fairly straightforward test of the energy–specialization hypothesis. If there are more species present in a forest or coral reef, or whatever ecosystem we are studying across different regions, there should be roughly the same number of individuals per species all the way down latitudinal gradients.

How do real biological communities stand up to this test? A classic study that provides a wealth of data on latitudinal trends in tree species richness was conducted by Alwyn Gentry (1988). Around the world Gentry and his colleagues looked at 220 plots, in all sorts of different natural forests, and recorded how many tree species were present and at what level of abundance, in small sample areas a tenth of a hectare in size.

As well as showing that the latitudinal gradient in tree species richness holds true at a smaller sampling scale, what the data of Gentry and his colleagues also show is that there are slightly more trees—of all types taken together—in a hectare of tropical forest than in a hectare of temperate or boreal forest. This is because in a tropical forest their trunks are packed more densely into the tropical forest (with a lot of fairly skinny tall trees). However, although the number is greater, the total of trees per hectare is really not that many more: certainly not enough to balance the huge rise in total numbers of species (Adams, 1989). On average, each tropical tree species occurs at far lower population densities than each temperate tree species. In a tropical rainforest, it will usually take quite a determined search to find another tree of the same species as the one you have just sampled, whereas in the temperate forest the next tree of that particular species is likely to be standing just a few meters away. This means that the hypothesis fails the test, at least in this instance—the "cake" is not really much bigger and is actually divided up into much finer slices of population in the tropics than in the temperate zone (the hypothesis says that the slices should stay the same size but the cake should get bigger) (Adams, 1989; Currie *et al.*, 2004).

It is more or less the same with other types of plants and animals that show strong latitudinal diversity gradients. For example, two extensive, regular surveys by amateur naturalists in North America provide a good standardized set of data to compare both species richness and total abundance across a wide latitudinal gradient (Currie *et al.*, 2004). The North American Breeding Bird Survey involves thousands of car loads of birding enthusiasts careering around the countryside, each making 50 stops on a 40 km route, and counting all the birds they can spot during each stop which lasts only three minutes (probably this is a good day to stay off the country roads in North America). Total abundance of all birds taken together increases slightly towards lower latitudes, and in parallel with the best climatic factor that predicts richness (actual evapotranspiration: the moistness and warmness of climate, thought to be the key to higher ecosystem productivity and energy supply). However, even this weak trend flattens out into the subtropics and stops increasing. And most importantly, the population density of individuals of *each species* of bird does not increase towards lower latitudes: instead they thin out in numbers quite a lot.

Another similar study conducted each year is the Fourth of July Butterfly Count, a glorious occasion for getting out and spotting butterflies on a warm summer's day that is also a public holiday in the U.S.A. In this case, butterfly enthusiasts identify and count all the butterflies that they see in 25 km circles at 514 sites in the U.S.A. and Canada. From the results of the Butterfly Count, it seems there is no relation between *total* number of butterfly individuals (taken across all species) and latitude, or the climate factor that richness best correlates with—actual evapotranspiration (Currie *et al.*, 2004). And yet, the number of individuals per species decreases towards lower latitudes, rather than increases as the species-energy hypothesis would expect.

So when ecologists census the total numbers of individuals of both broad types of organisms and of particular species, what they generally find is that there is not any real latitudinal trend in total abundance that parallels richness. Overall then, none of the three studies—on three very different types of organisms—supports the general idea that species can get to be more specialized in the tropics simply by dividing up similar-sized slices of a bigger cake. There might be more total weight of mammals, in a given area, in tropical ecosystems (Huston, 1993), but unfortunately no-one has yet worked out if that means just as many individuals per species in the tropics, or whether populations thin out like trees, birds, and butterflies do.

To make the species–energy mechanism "work" despite all that we find, it is necessary to make extra assumptions. For example, one might assume that species in the tropics are rather better able to survive at low population densities, perhaps because the warmer environment is more stable with its lack of the winters and hard frosts that might knock down a sparse population every once in a while. It is possible to go on endlessly making excuses and special assumptions, if one really wants to defend the species–energy hypothesis. There comes a point at which the whole exercise becomes so unscientific that without any really good evidence that the rules are different in the tropics, it is not worth arguing about. As things stand, in its simplest most straightforward form the species–energy hypothesis obviously fails the test, at least in those cases where it has been tested.

2.5.3 More strongly seasonal environments mean less chance of occupying a narrow, specialized niche

Species richness tends to be higher in warmer climates close to the equator, and also in wetter climates regardless of what latitude they are at. It might be something like NPP (see above) that is important in this, or it may be something else. Generally speaking, environments that are warmer and wetter on average also happen to have less seasonal fluctuation. There may be less of a temperature dip during the winter, or less of a dry season. Also moister environments tend to have less year-to-year fluctuation in rainfall. So average climate parallels something else, seasonality, and that might actually be what is more important in determining species richness. Equatorial rainforest environments have little seasonal fluctuation in either temperature or rainfall, compared with most other parts of the world, so this might be the reason why their species richness gets to be so high. It is interesting that the areas of tropical rainforest which have the highest tree species richness are those which are not only wetter on average, but also have the least seasonal and year-to-year variation in rainfall (Chapter 4).

Why would seasonality be important? Seasonal variation may constrain how precisely a species can narrow down its niche. For example, in the tropics where trees fruit all year round, there are many types of birds, and bats too, that are specialized only in eating fruit. In the higher latitudes trees stop fruiting during the winter, and anything that was precisely adapted to eat only fruit would starve. Hence in the temperate and boreal zones, the eating of fruit is only ever an opportunistic thing, done by more generalized species of mammals and birds that also eat other things during much of the year. The wider, less specialized niches of the high-latitude creatures might then mean that fewer niches can be packed in side by side there: this translates then into fewer species. Or imagine a plant that is adapted to a very precise range of soil conditions, perhaps a precise percentage of soil water content that depends on the local hydrology. In a more variable climate, with occasional droughts or even a regular strong dry season, water content in the soil fluctuates a lot. Hence there is no advantage in becoming specialized towards exploiting soil of a particular water content, because everywhere it fluctuates so much. A narrow specialist would lose out because of this, and be unable to survive. This then might help to limit how many specialized niches there are for plants in the forest.

Another example would be the possibilities of living as an epiphyte, a small plant sitting up on the branches of big trees. Epiphytes are very diverse in tropical rainforests, and are far less species rich or completely absent elsewhere. Surviving as an epiphyte is only possible in a very reliable, almost non-seasonal moist environment because these plants have almost no reserve of water available to them (sitting as they do clinging to a tree's bark with hardly any roots). Where there are frequent droughts or a regular dry season, any plant that tried to live this way would not last long. This is another way in which a very predictable environment provides more possibilities for different niches.

While the seasonal environment idea has been discussed a fair amount by ecologists, it is not always clear how it differs from the physiological harshness idea. Whether an epiphyte can survive can also be said to be an aspect of how physiologically "benign" the climate is (below), rather than simply a quite separate matter of whether the environment fluctuates. Also the supply of fruit is perhaps partly an aspect of the size of the overall "cake" of productivity (see above). Sometimes, different hypotheses seem to blur into one another!

In a general way, the idea that the stable annual environment of the tropics is important in allowing large numbers of narrow niches does make a lot of sense. However, I doubt that this factor alone could produce the huge rise in species richness towards the tropics—although it could be just one more thing that tends to promote high species richness in the tropics, acting in parallel with others. I am also not sure how exactly one would test the idea.

2.5.4 More specialized enemies of plants mean more species can exist side by side at lower latitudes (the Janzen–Connell hypothesis)

Possibly the most elegant hypothesis to explain latitudinal gradients in species richness depends on the idea that in tropical environments, a species that becomes too abundant is quickly attacked and eaten back by specialized herbivores, predators, or diseases that will go for only this and no other species. Certainly, my experiences trying to grow vegetables for the family kitchen in a small garden in Malaysia seem to bear out the idea that insect attack in the tropics is relentless: most of my carefully tended bitter gourds, beans, and ochra were devoured by insects, with astonishing rapidity.

Mostly this idea of more precisely honed and intensive attacks in the tropics is used to explain latitudinal trends in species richness of plants and other sedentary organisms, such as corals. The idea was originated independently in the same year by Daniel Janzen (1970) and Joseph Connell (1970), just another example of the synchronicity that often seems to happen with good ideas in science. Janzen was mostly thinking about tropical forests, while Connell (1970) was also thinking more about it operating in shoreline and coral reef communities. In fact it could probably operate in some form in almost any type of ecosystem.

Most attention though has been given to the idea that it operates in forests, giving the dramatic latitudinal trends in tree species richness that we see around the world. Because of this, I will mostly concentrate here on the evidence for and against it operating in forest ecosystems.

In the Janzen–Connell hypothesis, as it is called, it could be the big adult trees that are important, or it may be just the seeds and young seedlings. If the plant cannot reproduce, the effect is just the same anyhow regardless of the stage in life at which the attack occurs. This population-regulating process might prevent a few tree species that are strong in competition from doing well and pushing everything else out. The result: many species are able to coexist in the tropics, side by side. Over time, plant species generated by the process of speciation fill up the available capacity in the tropical ecosystem to exploit this opportunity.

In the temperate zone, by contrast, it is suggested that herbivores cannot become as specialized due to a less predictable, unstable environment. Their total numbers are often knocked back by frosts and harsh winters, and thus they are not locked into such an intense battle with their plant hosts. It is really the intensity of this battle which is key to the story—in the tropics plants are in an arms race with the insects that eat them, always evolving new poisons to keep them under control. Each insect species is under pressure then to evolve resistance to the new poisons in the plants that eat it, but that requires specialization.

As well as being less specialized in the temperate zone, insects should often be getting knocked back by harsh winters so they are also less abundant on average: the result is that temperate zone plants often can build up their population and yet still escape being eaten. Overall, with fewer specialized attackers and those attackers that are there are being less abundant, the "pest pressure" on the strongest and most abundant species of plants is less. This means that they can push out the weaker species around them unimpeded: and the effect of this is that fewer species of plant can fit into the temperate ecosystem.

Not only does this sort of mechanism potentially allow more new species to live side by side in the tropics; it may also in itself generate more species too. According to this "species-generating" aspect of the idea, new plant species in the tropics are continually forming in response to attack by very selective pests. Every once in a while, a population of a particular plant evolves a new poison that can fight off its own pests, and bursts forth as a new and separate species that thrives for a while with less of a burden of attack. Eventually, though, a new pest finds a way to get through the defenses and itself forms a new species that is adapted to attack this one host plant. The cycle goes on *ad infinitum*, generating new plant species and insect or fungal species that attack them—at least until something else begins to constrain the number of tree species that are packed in together (perhaps the minimum population density at which each species can exist and still get to reproduce reliably).

The very high insect species richness of the tropics would follow if each type of insect is specialized on one of the many plant species there. According to the Janzen–Connell hypothesis, in the temperate zone the lid on species richness is low, because the lack of selective herbivore and disease pressure allows some plant species to build up in abundance and push the others out (because the plants have overlapping niches). In the tropics, the lid on diversity is very high, and it may not even have been reached yet. The ultimate limit may be set by how low the population density of a tree species can go in the tropics without the population actually dying out.

Janzen and Connell's hypothesis is widely talked about, and is seen by many ecologists as the most elegant and persuasive mechanism to explain latitudinal richness trends. So to summarize it again briefly, in its original form the mechanism essentially operates through insects cutting back the population densities of forest trees, allowing more tree species in total to coexist. Under relentless selection to escape their attackers, trees also diversify into more species, allowing proliferation in the number of specialized insect species that attack them.

How would we put this idea to the test, to see whether it is important in nature? There are some fairly clearcut ways to do this, looking at several different aspects of the hypothesis. Unfortunately, not very much testing has actually been done yet, partly because it is time-consuming and expensive to gather the sort of meticulous data that are needed. But there are at least a few studies that we can discuss.

Fundamentally, the Janzen–Connell hypothesis relies on the idea that population densities of each individual species of tree (or whatever else is being attacked, such as algae, corals, or meadow plants) are thinned out by damage from herbivores, predators, or diseases when they get too abundant. Whenever individuals are packed too close together, they will tend to suffer more damage and mortality—either to the adults or the offspring. This is what tends to keep them spaced out from one another. If it turns out that in nature they do not show this density-dependent type of damage and death, the whole underpinning mechanism falls apart. Janzen and Connell's idea also suggests that tree populations in the tropics should be more strongly controlled by this mechanism than trees at higher latitudes, and their being "thinned out" more intensively in the tropics is basically what allows more species to exist there.

Looking for evidence of strong density-dependent control on tree populations, has been the focus of various studies. Signs that this thinning mechanism has been at work should show up if we plot the locations of all the different types of trees in a patch of forest onto a little map. If each species is totally randomly scattered through the forest, it will often by chance form clumps—just as if you were to throw a bunch of pennies on the floor, some would end up landing close by one another. Furthermore, the large number of seeds produced by each mother tree should be able to increase the chance that the offspring will grow nearby, and instead of a random pattern we should expect to see a still more "clumped" distribution. So if most trees in the forest are randomly scattered—or actually tend to grow clumped together because of offspring being next to mother trees—the clustering we will see is useful evidence. It shows that the population density does not really matter and is not controlled by selective damage from insects or whatever. If there really *is* close control on density and spacing in tree populations, we should find that clumps are not present when we look on the map. Trees will instead be spaced out more evenly than expected by chance, as if they are avoiding one another. In fact, the trees that have tried growing too close to other trees of the same species will have died under the onslaught of pests, leaving only the more widely spaced ones alive. The term that is used to describe this spaced pattern is "over-dispersed" (i.e., over-dispersed relative to what you would expect by chance).

What do the studies show as regards the spacing of trees? It seems that in some cases species of trees in the rainforest are actually spaced out relatively evenly—which supports the Janzen–Connell mechanism being at work in the tropics at least. But many other species in the rainforest seem to be spaced randomly, and many are actually more clumped than would be expected by chance (Hill and Hill, 2001). Overall then, the evidence for this hypothesis is disappointingly weak. Another possibly relevant observation is that tropical forests (such as the research forest at FRIM, near Kuala Lumpur) that have recently started off poor in species, often because they were plantations or because they were heavily logged, become increasingly diverse over time. Importantly, rarer species became more common in the community at the expense of more common ones, which suggests that some sort of control on population densities was buoying up the rarer ones. It is as if the main forest trees at the start are being suppressed by pests and diseases that allow other tree species in.

The big trees that one plots on a map of an area of forest will just be the survivors, from among all the seedlings that did not make it. Looking at seedlings might give a more direct clue to what is going on in regulating where trees grow. It is thought that the youngest life stages of trees are key in terms of where and how often they establish in the forest. Most seedlings die, but if a young tree survives this stage then beyond a certain point its long-term survival is almost assured (Grubb, 1977). Hence it will be at the seedling stage that the selective thinning of populations is most likely to occur, adjusting the population density according to the amount of attack by pests.

There are many studies that show greater mortality among seedlings of the same species that start growing at high densities, compared with others in sparser populations (Hill and Hill, 2001). Likewise, seedlings closer to a mother tree often suffer greater mortality than those farther away (Hill and Hill, 2001). But perhaps the key question is whether this selective mortality on dense populations occurs with any more intensity in the tropics than the temperate zone. Observations on the mortality of naturally occurring seedlings that occur at patches of varying densities within forests (Ris Lambers *et al.*, 2002) have shown that this pest pressure effect tends to thin out seedlings of each species in both the tropics (at Pasoh in Malaysia, and BCI in Panama) and in the temperate zone (on seedlings of a range of North American tree species such as red maple, *Acer rubrum*, and tulip tree, *Liriodendron*). In fact, mortality that depends on density was found just as commonly in the temperate trees as in the tropics, which is not really what one expects to see when trying to explain why there is a latitudinal gradient in tree species richness.

Such studies are based only on observations of trees or seedlings already growing in forests; but it should be possible to do experiments which show that if you deliberately clump together trees from a given species, they will die more often due to attack by pests. Easiest to work with, and in fact probably most relevant, are the seedling stages of trees.

One can even simulate this in a greenhouse, by using soil taken at different distances away from parent trees, and this has been done in the temperate zone at least (where admittedly the mechanism is thought to be weaker). Packer and Clay (2003) performed a rather elegant greenhouse experiment with seedlings of a single temperate tree species, the cherry *Prunus serotina*. They took soil from directly underneath adult trees, and also soil from varying distances away from them. The seedlings grown in soil taken from well away from parent trees did much better, suffering fewer deaths due to a fungus (*Pythium* sp.). So it looks like the fungus is usually present close to the parent trees but absent farther away. The fact that a temperate tree shows this pattern perhaps gives some clues to how a mid-latitude forest maintains its diversity—but it weakens the case that pest pressure is something that operates only in the tropics to allow many more species to exist there. It is useful to know that such a mechanism *can* work, but to produce a latitudinal gradient in species richness it would have to operate more strongly in the tropics than in the temperate zone, and this has not yet been shown.

Another fundamental component of the Janzen–Connell mechanism (at least in terms of it being responsible for latitudinal gradients) is that insect herbivores should more often confine their attack to a single species of tree in the tropics. In the higher latitudes, insects are suggested as being less specialized, so they attack several or even many different species of trees within a forest community—and this will mean that their effect in selectively thinning out each species will be much weaker. In these cooler climate forests, if there is less control on the density of each type of tree, fewer tree species can coexist because "stronger" ones can now build up in numbers and push out nearly all the others. We ought to be able to see if this is true by comparing how specialized insects are with their host plants at different latitudes. So far, only one really thorough study has tackled this. This work compared a tropical rainforest in New Guinea with a temperate forest in Slovakia, in central Europe (Novotny *et al.*, 2002, 2006). Hundreds of species of insects were looked at, and the results suggested in fact there is no greater host specificity in New Guinea than in Europe. However, another more limited study (Dyer *et al.*, 2007) just on butterflies and moths did suggest that in Central America's rainforests caterpillars are more specialized to fewer host plant species than they are in North American temperate forests, and the farther north you go within North America the less specialized they become.

Novotny *et al.*'s findings do not look good for the Janzen–Connell hypothesis, while Dyer's observations do support it. Clearly, these observations on host specificity need to be backed up at more locations before the hypothesis can be declared either dead or alive, or partially alive. Another study on host specificity that seems to give the theory a knock is a study in Guyana in South America where Bassett (1999) found that most of the insect herbivores on seedlings of five sampled tree species were generalized, attacking a range of different species. It is important to realize, however, that slightly different variants of this mechanism could operate in different ecosystem types, and involving different groups of organisms in these forest ecosystems. Even if insects are not host-specific enough in the tropics to explain why tropical forests stay so much more diverse, fungi or nematodes might turn out to be. Dealing with such complex ecosystems, it is hard to eliminate all of the many possibilities. And even if the insects are not specific to particular species of plants, they might still be specific to *groups* of species, maybe in the same genus. This might tend to disguise the effect if one looks only at individual species, yet the effect would be present if one were to look together at spacing of members of a tree genus or family rather than a species.

Overall, although the Janzen–Connell hypothesis is intuitively compelling, there is not much solid evidence in its favor, and a fair amount of evidence against it. As the observations show, there clearly is some tendency for density-dependent thinning of tree populations to operate in forests; the unanswered question is whether it works strongly enough to make any real difference to species diversity, and whether it is any stronger in the tropics. If the mechanism turns out to operate about equally strongly at high and low latitudes, then it cannot be the underlying cause of latitudinal gradients. However, by weakening a few tree species that would otherwise turn out to be stronger in competition, it might be the precondition that enables large numbers of species to exist as long as there is some other mechanism that works in the low latitudes, to provide more species through evolution and speciation. This is the important difference between a mechanism being *necessary* as a precondition for high species richness to evolve, and *sufficient* in itself to actually produce so many species.

2.5.5 The latitudinal gradient is produced by a balance between growth and disturbance

This idea is often known as the "intermediate disturbance hypothesis", and it appears in several variants (Chapter 1). Essentially the idea is that in almost any community of plants or animals, many species are too similar to one another in their ways of living (their niche) to coexist in a situation that is purely based on competition. Because so many of the tree species in a forest prefer the same soil conditions as one another, or the birds eat the same types of insects as one another, they will eventually suffer as a result of this overlap. One species of tree, or bird or whatever, will gradually push out all the others that overlap with its niche because it is the strongest competitor around. The only way that so many types can manage to continue to coexist is if (every once in a while) some random disturbance event wipes out a large part of the population of every species involved, arbitrarily killing individuals regardless of species. Such a crash in populations would effectively reset the clock for the pushing-out process (which is known as competitive exclusion). After the disaster, the survivors of all the local species of trees, or birds, have a chance to start building up their populations again, and with plenty of resources to go around now, so competition is not so much of an issue. If these population crashes occur often enough, the stronger types will never, ever manage to push out the weaker ones, and the community will maintain a long list of species.

Many different types of events could cause the crashes that may be necessary to allow high levels of species richness. Fires are one obvious example; they can devastate populations of almost any organism. Droughts, floods, storm waves, or extreme winter freezes are other examples of disturbance events that knock back populations. Occasional catastrophes such as landslides, tsunamis, or volcanic explosions might be frequent enough to have effects on populations of long-lived organisms such as forest trees. However, it is quite likely that for many types of organisms, the sort of disturbance event that is enough to reset the clock of competition can be very localized, such as a big old tree falling over in the wind, or even the hoof of a grazing animal scuffing the surface of the turf in a grassland. This is the scale that can affect the seedlings of forest trees or meadow flowers, or fungi and invertebrates that live in the soil and just above it.

So, disturbance may be necessary for maintaining species richness. However, there can always be too much of a good thing. If disturbance events are too frequent or too devastating, many species will not be able to keep their populations going because they cannot recover their numbers before they are knocked back again. If they cannot keep up their numbers they will die out locally. Hence the level of species richness in the ecosystem may be a balance between the competition-relieving and population-destroying effects of disturbance.

Michael Huston (1993) suggested that latitudinal gradients in richness are produced by a gradual shift in the balance between the growth rate of populations, and the intensity of the knock-back in populations that disturbance causes (Figure 2.15). This is just a scaled-up version of the idea that is used to explain humpback diversity curves on the local scale (Connell, 1974; Huston, 1993)

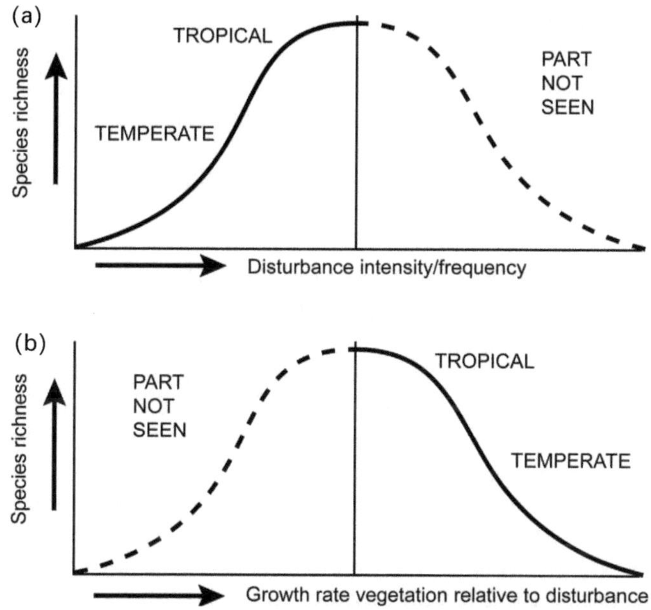

Figure 2.15. Ideas on how disturbance intensity/growth rate may differ between temperate and tropical forests.

(Chapter 1). Thinking on a geographical scale, imagine, for example, that storms are more frequent or more severe at low latitudes; this might keep knocking back populations more effectively at low latitudes and allowing in more species of trees, butterflies, birds, or whatever (Figure 2.15a). On the other hand, it could be that there is *too much* disturbance at high latitudes, which batters down populations so often and so severely that very few types can adapt to cope with it.

Huston suggests that in the case of forest ecosystems, there is not actually much difference in disturbance frequencies between tropical and temperate zones. The more important thing, he proposes, is a lower growth rate of trees in the very nutrient-poor soils of the tropics (Figure 2.15b). This will mean that tree populations in the tropics cannot recover from disturbance as fast, and that this limits the rate at which species are pushed out by competition. Less pushing out of competitors, then, means that many more species are able to coexist in overlapping niches in the tropics.

Is there any evidence for the latitudinal difference in growth rates that Huston proposes? It is hard to pin down how often disturbances actually occur, partly because they vary a lot between different parts of the landscape—and often because they are rare events that could occur anything between once in a few decades to once in every several thousand years. What about the other half of it, the growth rate of trees? Huston (1993) presents some figures compiled from around the world which do seem to show that growth rates of tropical trees are no greater than temperate trees, and often less. But there does not seem to be very much slower growth in the tropics, as one might expect if this mechanism were true. Huston also suggests that disturbances only really matter if they affect a tree during its growing season—and since the temperate forest trees grow faster during their (brief) growing season than tropical trees do during the whole year, that means effectively less disturbance for tropical forests. I am not sure about this—for isn't a landslide or a wind storm just as severe if it affects the trees when they are dormant? Overall, the idea that a shift in the balance between growth rates nd disturbance produces latitudinal gradients does not seem quite so convincing to me as the widely held belief of the importance of intermediate disturbance in producing more local-scale patterns in species richness.

2.6 THE PROS AND CONS OF THE VARIOUS THEORIES FOR LATITUDINAL GRADIENTS

Science works by subjecting ideas to rigorous examination, and the best way to do this is devise "tests" that could potentially disprove an idea. If a hypothesis can make it over the hurdles that these tests provide, it looks a lot more convincing than some idea that has never been tested.

Unfortunately, it is hard to test ideas about species richness trends that extend across thousands of kilometers. Not all of the various hypotheses provide much to discuss in terms of finding evidence for or against them. This does not necessarily mean that these hypotheses are wrong, but for the present they can only be judged on their own innate plausibility (or lack of plausibility). Bear in mind also that even if a particular hypothesis "fails" a test in its original form, this does not necessarily mean that the idea should be thrown out completely. Nature is complex, and it is still possible that similar but slightly different mechanisms might be at work producing gradients in species richness. And because organisms are so varied, different processes might be important to different extents in different places or with different biological groups—so if we test a theory in the wrong situation, we might unfairly dismiss it.

Ultimately, one has to admit that the causes of latitudinal diversity trends remain largely mysterious. Looking among the ideas that it has been possible to test, all we can conclude for now is that some hypotheses have survived tests rather better than others, but none of them seems to do especially well.

For instance, the species–energy hypothesis has not stood up well to the instances where it has been possible to test it. But then again there are so many different groups of animals and plants showing latitudinal trends in so many different places, that it is possible that it might turn out to work in certain other cases. It is also possible that the species–energy mechanism also works in some subtly different way—somewhat similar to the original species–energy hypothesis, but then not quite the same—in some way we have not anticipated. Frustratingly, nature is so complex that it could well be like that, even if philosophers of science would cry "foul" if we kept trying to alter our hypotheses as soon as they failed the tests we tried out on them.

Because broad-scale species richness trends like the latitudinal gradients are so hard to test, often we have to look around quite purposefully for evidence that either supports or seems to deny the explanation that we are interested in. This is rather like a court of law may aim to gather evidence for and against a defendant. To me, given the difficulties we are faced with in explaining species richness patterns, this seems like a perfectly reasonable way of approaching the problem. Philosophers of science may also regard this as a sin, but they are not the ones who actually face the challenge of understanding the natural world!

Generally, we have so little good evidence and so few good tests, that it is hard to choose between the leading ideas to explain the latitudinally related gradients in species richness. For the present, my personal favorites to explain broad latitudinally trending gradients are the predation pressure hypothesis of Janzen and Connell, and the "harshness" hypothesis of the JeDi model (Kleidon *et al.*, 2009). Yet despite its elegance, the Janzen–Connell idea does not seem to have stood up too well to testing, though certainly more testing is needed before we can completely throw it out.

The JeDi "harshness" explanation is strange and disconcertingly different as a way of looking at communities, but empirically it seems to work. Exactly why it should work so well in a world that also consists of populations (not blurred heaps of biomass, as it implicitly assumes), is something I am not sure how to explain. The next stage I suppose is to find more ways of testing this harshness idea.

As things stand at present, latitudinal gradients in species richness remain a great mystery. It is going to take a lot more effort before we can arrive at any firm conclusions about which processes are or are not at work here. Given how varied the world's organisms and their ways of living are, we should probably not expect to find a single unified cause for latitudinal trends. Instead the cause is likely to vary from one group of organisms to another, and from one place to another. In some situations there may even be several quite different factors acting in parallel, all contributing to the same latitudinal gradient. So we might actually find that all along there were several different holy grails, waiting to be discovered.

3

Deep time and mass extinctions

3.1 THE DEPTH OF TIME

Diversity patterns are as basic a geographical feature of our world as mountains, seas, and deserts. However, that does not mean they are static. There is plenty of evidence that, like all geographical features, diversity can undergo changes. Often, these changes are on about the same slow time scales that mountains build up and wear down, or that new seas open up or vanish. But diversity can also change over just a few thousand years or less, when the environment changes rapidly enough to cause this.

At times during Earth's history, biological richness proliferated into a sort of baroque splendor, with all sorts of exquisite flourishes and gilding. Occasionally it suddenly faded, leaving a world that seems like a monotonous grey concrete façade by comparison. Some of these changes seem to have been well explained scientifically, but others remain a mystery. The stories which we do know fairly well include events so terrible that they would be scarcely imaginable, if we did not have such good evidence for them. Some of these disasters may almost have ended life on this planet altogether.

Biologists and geologists are fascinated by past changes in diversity in themselves. However, there are also broader reasons for studying them. First, we fundamentally need to understand how diversity changes over time in order to explain present-day patterns in species richness. The mechanisms that have been used to explain present latitudinal gradients in species richness (Chapter 2) depend upon past processes which have allowed richness to build up, or left it depleted. Likewise, explanations for species richness hotspots (Chapter 4) often invoke species generation acting over many millions of years, plus the legacy of ancient climate changes. Second, it may also be that there are practical lessons for species conservation in the future (Chapter 8), from the study of past ups and downs of biological diversity.

3.2 SPECIES RICHNESS CAN CHANGE ON A RANGE OF TIME SCALES

The number of species occurring in a particular area can vary on almost any time scale. To an ecologist out in the field counting the bird species in a patch of forest, it will fluctuate over just a few minutes as birds fly in and out of the trees. On the time scale of a few months, the lists of species of birds or butterflies in a region can vary with the seasons as some migrate. Over centuries and thousands of years, species can change their natural ranges to shift with global climate cycles such as glaciations. And on time scales of more than a few hundred thousand years, the evolution and extinction of species becomes most important in shaping diversity.

3.3 SAMPLING THE PAST: THE FOSSIL RECORD AND SPECIES RICHNESS

Humans have always been keen observers of nature, and for most of our existence survival depended upon this. But mostly this knowledge existed only as spoken memories, and was easily lost. Going back 2,500 years, there are a few written observations of wildlife, beginning in ancient China and Greece. The oldest cave paintings and carvings of animals date to around 30,000 years before present. Before then, we have no direct human observations to show us what animals or plants occurred where.

Instead, to get any sort of record of diversity in the distant past we must rely on the buried, preserved organisms (or at least parts of organisms) that form the fossil record. Although fossils are the best clue to past species richness, the picture that they give is far from perfect. A fair amount of chance is involved in what happens to get preserved in the sediments of a lake, river, or a shallow coastal sea. Sheer luck—from the point of view of the paleontologists—is especially important in determining the fossil record of land-living animals, which only rarely end up falling into water. Generally, by far the best fossil record is for marine animals that secrete a shell and live in shallow shelf seas (those less than about 200 meters deep) along the edges of the continents. The next most complete fossil record is for land plants, especially trees, whose pollen gets carried by the wind and ends up in lakes and rivers, or carried down to the sea.

The amount of information available from the fossil record also decreases further back in time, a phenomenon that geologists call "the pull of the recent". The most complete record of past life forms is for the most recent phase of geological history known as the Quaternary, which spans the last 2.4 million years. The most recent 100,000 years of this is especially good, but only the last 11,000 years or so (comprising the present warm phase) has the level of detail that we would ideally want for the entire span of Earth's history. The reason that the fossil record becomes poorer further back in time is that a greater part of the sediments that were originally laid down has either been eroded away, or altered by heat and pressure to the point where the fossils they contained have been lost. Hence before a few million years ago there are many gaps in the fossil record, even for the types of organisms such as clams and other shallow marine invertebrates that get preserved most often—and these gaps in the record tend to become bigger the further we look back in time. Every so often though, we find a stunning exception, where the right combination of conditions has preserved the creatures that lived there during one brief interval of time in near-perfect detail, and the rocks have come down to us almost unaltered. In these special places and moments in geological history, we have a sudden, lucid glimpse straight into the past, like a shaft of light that penetrates down into a deep dark cave. Paleontologists speak the names of these places with almost sacred reverence: the Burgess Shale, the Rhynie Chert, and Solnhofen Lithographic Limestone, to name just three of them. But they are so outstanding because they are unusual. In the fossil record, there are so many intervals when the picture is not clear that it is often difficult to distinguish a "real" rise or fall in diversity, from the effects of changes in the chances of preservation. Only with a lot of careful piecing together of the incomplete record from rocks around the world, can we begin to reach firm conclusions about the fate of diversity in such distant times.

Most of the organisms that occur in the fossil record are now extinct, and this in itself presents problems for trying to understand how species richness has changed. For one thing, it is often hard to figure out what exactly corresponds to a "species" in the distant past. In the present-day world, the biologist's definition of a species is partly based around whether it can interbreed and merge with the most similar forms around it. If two rather similar-looking types of organisms do not appear to interbreed and do not intergrade in nature, they are usually counted as separate species. Even though we often lack the observations to be certain that any two similar types in the present-day world really *are* separate species by this definition, we could in theory study them in more detail if we wanted to check on this. In the fossil record, we certainly do not have this luxury. Because the organisms we are studying as fossils are nearly all extinct, we cannot be sure about what they could interbreed with. The fossil record is also usually too patchy to tell if some of the distinct forms we see back in the past were really just different-looking populations that often intergraded, or whether they formed discrete species that never interbred. We know from organisms that are alive in the present-day world, that sometimes species that are obviously distinct can be indistinguishable from their remains. Lions and tigers do not interbreed in nature, and their appearance and behavior are very different, yet their skeletons are essentially identical. How would we ever know from the fossil record alone, that these are separate species?

The fossil record selects only certain items for preservation. What we tend to find as fossils are only the toughest parts of organisms, such as shells, wood, teeth, and bones. Yet, as the example of lions and tigers illustrates, it is often instead the soft tissues which biologists use to identify species in the present-day world. This is true not only of animals, but plants too. For example, closely related plant species tend to be distinguished from one another by the anatomy of their flowers and fruits which are the most complex and distinctive features of plants (and that is the reason why they tend to be used to draw distinctions). Yet flowers and fruits are not often preserved as fossils, and instead as fossils we usually get only leaves, stems, or roots.

The crushing and chemical alteration of the remains of dead organisms after burial also often means that even the toughest parts of animals and plants that do turn up only retain a rough outline of their original shape, which makes it even more difficult to distinguish individual species among them.

For such reasons, when paleontologists discuss diversity in the fossil record, the unit they often prefer to talk about is the genus—a broader cluster of similar species—rather than the species itself. Numbers of genera will tend to be paralleled by numbers of species, as each genus usually contains between one and several species. So when we talk of genera, we are in effect talking of a measure of species richness by proxy. In some groups such as mammals, the majority of genera contain only one or two species, so talking in terms of numbers of genera is almost the same as talking numbers of species. In some other groups such as flowering plants, genera tend to contain a larger number of species. In these groups with lots of species per genus, the relationship to species richness is far from one-to-one—but still likely good enough to be a general guide to species richness. In the earliest days of animal life, before about 550 million years (Myr) ago, even genera might not readily be distinguished because the first animals were anatomically so simple that there is not much we can use to distinguish between different types.

3.4 THE BROADEST SCALE PICTURE OF BIOLOGICAL RICHNESS, SINCE THE BEGINNING OF LIFE ON EARTH

The Earth's history is thought to extend back some 4.6 billion years. Chemical signs in the rocks of microbial life extend back at least 3.8 billion years, tentatively even 4.2 billion years. Calcium carbonate mounds known as stromatolites that were formed by blue-green bacteria (cyanobacteria) date back to 3.5 billion years ago. At this time, before any complex unicellular or multicellular life existed, diversity could only have manifested itself in a wide range of biochemical forms, just as we can find nowadays in living samples of microbial life in soil and lake waters (Chapter 6). A very wide separation between modern day groups of bacteria shows up from comparing their DNA sequences, indicating that they diverged and then accumulated changes starting billions of years ago (Figure 3.1). However, to talk of distinct lines of ancestry is not always particularly meaningful, because many of these groups have been exchanging small fragments of DNA with one another ever since.

A range of "chemical fossils", distinctive markers for particular groups of bacteria, also show up way back in time. These chemical fossils indicate that already by around the time in the record when the first physical fossil traces of bacteria appear, several major branches of microbial life were already present in the world, floating in the waters and working away within its sediments. So we can reasonably guess that this purely bacterial world already had a great diversity in terms of cell chemistry and ways of living, even though in terms of morphology it was fairly monotonous.

A step-up in life's diversity of both form and function occurred by sometime around 2 billion years ago (maybe earlier, according to some), when the larger,

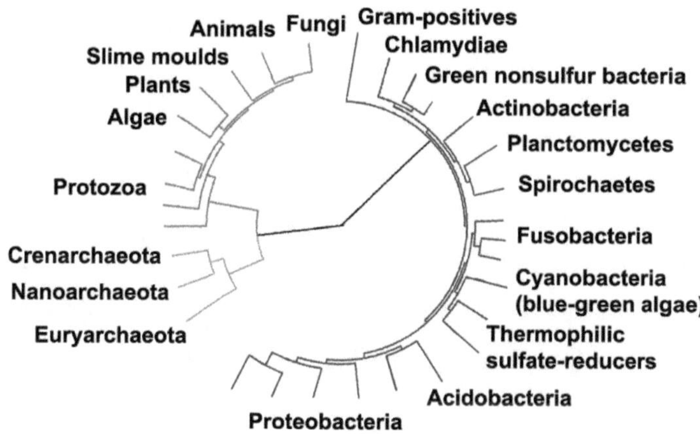

Figure 3.1. Family tree of all life on Earth, from comparing DNA sequences. Most of the broad- scale diversity of life is in the form of different groups of bacteria (lighter gray) (source: Tim Vickers, Wikipedia Commons). *See also* Color section.

compartmentalized eukaryotic cells—equipped with nuclei, mitochondria, and other components—start to appear in the fossil record. The origin of the eukaryotes may have been fueled by the build-up of atmospheric oxygen, which provided enough energy to maintain their intricate structure. Back at this early stage in Earth history, such fossil cells are still too rare and poorly preserved for us to say much about levels of richness, although once again there are molecular traces about the same time that point to a variety of types. The time depth indicated by the differences in modern-day DNA of eukaryotic lineages (Figure 3.2) also suggests that several major groups of single-celled organisms had already diverged from one another by then.

Another jump upwards in biological diversity occurred with the appearance of multicellular organisms. The earliest definite fossil animal, the little seed-like *Vernanimalicula* that burrowed in soft sea floor sediments, is first found around 600 Myr ago (Chen *et al.*, 2004). If simple animals existed for a long time period before then, we have not yet found them as fossils. Over the next couple of hundred million years, the range of animal life forms known from the fossil record increased enormously. A first little burst of diversification occurred after around 540 Myr ago, the time known as the Cambrian Explosion (Figure 3.3). Starting from just a few genera around 580 Myr ago, the overall richness of animal life climbs broadly on upwards (Alroy *et al.*, 2008). Adding to this rise has been the explosion of life onto land during the last 450 million years, which has given a whole range of exclusively land-living groups.

However, there is a flaw built into the fossil record that can easily give us the wrong impression if we are not aware of it. This is the effect of better preservation towards the more recent past, that misleadingly adds to the impression of increasing diversity over geological time. This is "the pull of the recent", which I mentioned above. With more rocks and sediments to preserve them, and with less alteration of their remains, more recent life forms simply stand a better chance of being found as fossils. However, it is thought that in a very general way at least, the upward trend in diversity for the first several hundred million years of animal life is something real, not just a misleading by-product of preservation (Alroy *et al.*, 2008).

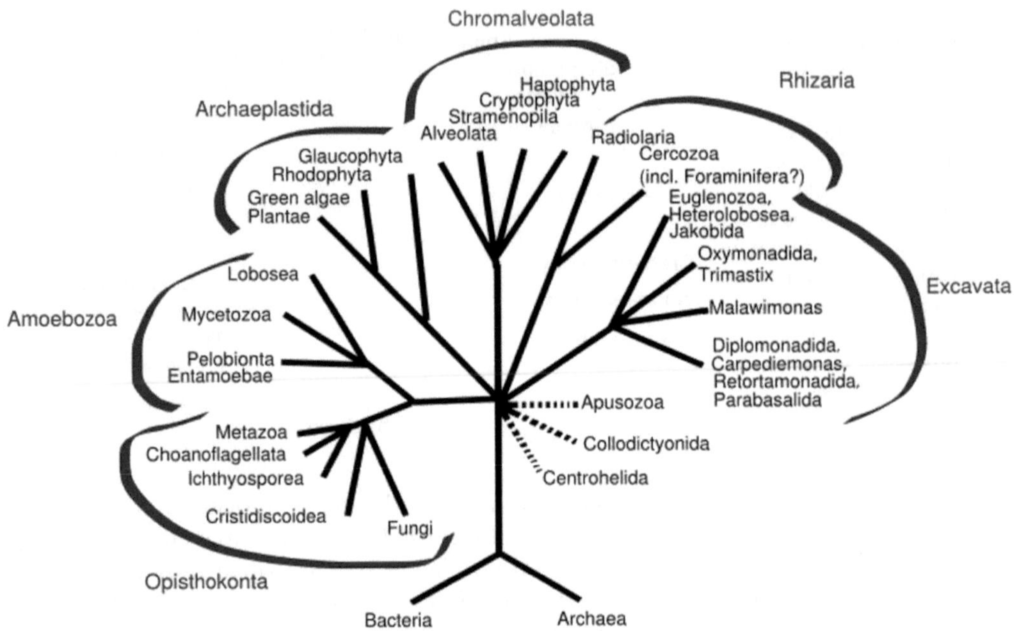

Figure 3.2. A family tree of the various groups of eukaryotes (complex cells). Fungi and animals (Metazoa) are crammed together in one small branch of this tree along with other groups, under the "Opisthokonta". Plants and seaweeds end up in another branch, the "Archaeplastida". The other branches are mostly single-celled "protozoa" (source: Tim Vickers, Wikipedia Commons).

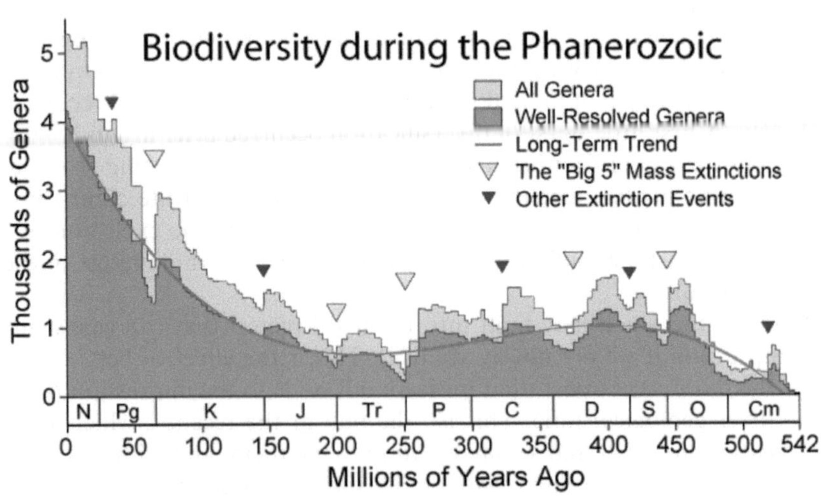

Figure 3.3. Build-up of richness at the level of genera known from the fossil record, since 540 million years ago (based on Alroy *et al.*, 2008, from Wikipedia Commons, diagram by DragonsFlight. *See also* Color section.

Box 3.1 The highest level divisions of biological diversity

Biologists categorize life according to a nested hierarchy of categories, just as librarians would classify the book you are reading here according to "non-fiction", then "biology", then by the category of "ecology", and at the lowest level by the name of the author. It may be no coincidence in fact that librarianship and biology are so similar—for it is suspected that the man who started it all, Carl Linneaus, based the system on one that librarians were using back in the 1700s (Chapter 6).

In biology, one of the highest-level categories is the phylum, which comes from an ancient Greek word meaning a "clan". Each phylum represents one common architectural plan, a fundamental way of building an organism and arranging its tissues and organs. So, for example, life forms that look so very different from one another as an octopus, a clam shell, and a snail are all thrown together in a single phylum known as Mollusca, or mollusks. We vertebrates—with all our pretensions—end up in the same phylum as the unimpressive-looking sea squirts, which as adults have no eyes, no limbs, no brain, and often live cemented to the sea floor. There are something like 30 phyla in all (biologists disagree about some of these groupings, hence the approximate number), many of them fairly obscure organisms found only on the ocean floor or in other inaccessible places. Beneath their superficial differences, members of a phylum turn out to have a lot in common in the way their bodies are built, which is why they have ended up being thrown in together. These basic common plans were noted way back in the early 1800s, two centuries ago, and they have been confirmed by DNA evidence which shows that members of each phylum are really quite closely related.

One pattern which is evident from the present-day world is the way that diversity is skewed mostly towards a few groups of organisms that are very, very rich in species. And there are a lot of other broad groups—including both phyla, or lower-level subdivisions such as classes—that seem to have hardly any species in them.

So, for example, there are several animal phyla for which there seem to be less than 30 species in the present-day world, and one phylum (the weird, amoeba-like but still multicellular Placozoa) which apparently contains only a single species, *Trichoplax adhaerens*. At the other extreme, the phylum Arthropoda (arthropods) contains the vast majority of the 1.7 million species so far discovered in the world, most of them being insects, mites, and spiders (the arthropods contain nearly 1.2 million species!).

For any group of animals or plants to become distinctive enough to form a phylum, it would probably have split off from all other life forms at least several hundred million years ago. Many of the more obscure phyla must barely have continued as a thread of life down throughout those eons to the present, perhaps only ever existing as just a few species. Yet in the same time span, other phyla have exploded into tens of thousands or even millions of species. For reasons that we cannot entirely understand, nature has favored some body plans far more than others.

Below the very broad categories of phyla are finer subdivisions based on the body plan. There are *classes*, and many phyla contain several classes. So, for example, in the mollusk phylum, the snails, clams, and octopuses all end up in different classes. Fortunately, we mammals at least end up in a separate class from our relatives the sea squirts.

Zooming in below the level of classes, and focusing on ever-finer details of body plans, there is a complex nested hierarchy of categories. For instance, a few steps down below classes are orders: for example, monkeys, lemurs, and we humans are grouped into one order known as primates. Below orders, a couple of steps further down, are families—another much-used level for describing where organisms belong in the scheme of things. An example of a family would be the great apes: chimps, gorillas, and orangutans and also ourselves. This family is known as Hominidae. Another very useful and much-used level (again skipping a few steps on the way down) is the genus—a little cluster of very similar species. We modern humans all belong to a single species of the genus *Homo*. In prehistory in our same genus there used to be other species of humans—but all of these are now extinct (see Chapter 5).

In all, about 250,000 species have so far been identified in the fossil record, according to the paleontologist David Raup. The number is still rising: it increases each year as new fossil forms are found and identified. Strictly speaking, it is more accurate to say that 250,000 "species names" have been handed out to past life forms known from fossils—because it is a moot point how many of these were really distinct species from one another when they were alive. We can reasonably guess that some fossil "species", that we identify as a single species from the fossil record, were really clusters of similar species that we would have been able to distinguish from one another if we had seen them when they were alive. In other cases, fossil forms described as being distinct species would have turned out to be just slightly different populations of the same species if we had been able to study them in more detail. Some of these "species" might even have been just males and females of the same species. It has been suggested, for example, that some of the fossil "species" of our extinct human-like relative *Australopithecus* were actually males and females of one species, with the females being smaller and more lightly built than their males (Rob Foley, Cambridge University, pers. commun.). In the case of fossil plants, each fossilized part of a plant (root, leaves, seed, pollen, etc.) that turns up separately in the rocks tends to get its own species name—and in fact its own genus name. One example is the weird scaly tree *Lepidodendron* that lived about 320 millon years ago when the Carboniferous coal measures were being laid down. Originally, only the bark of the tree was called *Lepidodendron*. Its roots, with a distinctive criss-cross pattern of holes on them, were called *Stigmaria*. The leaves went by another name, and the cones too. When someone finally figured out they were all parts of the same plant, several genera went "extinct", at least from the lists of fossils. It is probably the same for some of the other plant fossil genera that only identify parts of plants—if we knew what they were really part of we would have fewer species. So, only when the different plant parts eventually get found connected to one another is it possible to say that they are actually just one species, and then cross several species names off the list.

In addition to such problems, there are undoubtedly many, many species that existed but have not been found in the fossil record, because they lived in environments that do next tend to preserve many fossils (e.g., tropical rainforests, or deserts, or the very deep sea floors). Even in the "best" environments such as shallow seas, preservation is a rare and chancy event and it is likely that many species did not have the luck to fossilize in sufficient numbers to get discovered by geologists.

3.5 THE SUDDEN INITIAL INCREASE IN DIVERSITY 540 MYR AGO

The spectacular richness of animal and plant life that we are used to seeing in the present day has only appeared during the last several hundred million years. Though this seems an immense span of time in itself, it is just a small fraction of the total age of life on Earth, going back something like 3.8 billion years. While the first animal life has been found dating to around 600 Myr ago, it apparently waited tens of millions of years to really start diversifying. Our modern, diverse world began with a first jump of diversification starting around 540 Myr ago during the Cambrian period, and lasting about 25 Myr. While it did not give rise to many genera (Figure 3.1), this event gave rise to a wide range of distinct animal and plant body plans. Biologists are so awed by the pace at which new forms appeared, seemingly out of nowhere, that they call this the "Cambrian Explosion". Most of what has happened in evolution since then is just variation on the basic themes that appeared during the Cambrian Explosion.

All at once it seems, many of the familiar phyla of the modern world sprang into view: annelids (the group that nowadays includes earthworms and leeches), mollusks (snails and squid), arthropods (insects, shrimps, and spiders), chordates (vertebrates), echinoderms (starfish and sea urchins), and coelenterates (jellyfish and corals), and so on. Alongside these were other phyla that have now vanished. These latter just seem to have been evolutionary experiments, thriving for a time but eventually losing out in the tough world of ecological competition. They include, for example, enigmatic quilt-like creatures known as vendobionts—as far as we can tell, these are unrelated to anything alive today.

Though it was a drastic contrast with what had gone before, the world during the Cambrian Explosion lacked the sort of lower-level diversity that we are used to seeing nowadays. Each broad phylum or class seems to have been represented by just a few genera, though that could just be because most types of animals around at the time lacked shells and did not fossilize to be found by us. Also, there are few signs of the geographical differences in faunas that are normal nowadays. Fairly much the same forms of animal life occurred everywhere in the world, in the shallow seas around continents that may have been located thousands of kilometers away from one another at the time. For example, an early set of life forms known as the Ediacaran

Fauna (around 580 Myr to 540 Myr old, named after the Ediacara Hills in South Australia), mostly squishy soft-bodied creatures, turns up in Australia, Canada, and Namibia. The somewhat later Burgess Shale Fauna (around 505 Myr old) is a different set of organisms that also seems to have been very widespread in its day—being present in North America, China, and Greenland. However, some reconstructions of the world's geography at that time have most of the continents grouped together, which could explain the geographical uniformity, because it would have been easy for things to spread around. But also it may be just that change was happening slowly in those days. If new variants on each broad theme were not evolving often, then there would also not be much likelihood of local types to evolving, from ancestors which had by chance managed to spread across oceans to shallow seas around other continents. They would have stayed about the same, giving the same look to creatures everywhere.

The basic groups such as phyla that first appeared during the Cambrian Explosion set the stage for the spectacular proliferation of orders, families, genera, and species that followed over the last several hundred million years. Yet, essentially all that has happened since then has been variation on the same basic themes which were set down back in the early days.

3.5.1 What could have caused the explosion of animal life after 600 Myr ago?

The appearance of so many animal forms between about 600 Myr and 540 Myr ago (including as its final stage the Cambrian Explosion) was an initial step in the expansion of diversity, which ultimately gave rise to the full richness of animal life in our modern world. But what caused it to happen, and why did it not occur earlier in the long history of life on Earth?

Many different explanations have been put forward, some of them more widely believed by the scientific community than others. One view is that it simply took time, and luck, for the right combination of characteristics to build up in single-celled protozoan ancestors that in time gave rise to simple colonies of cells. When some of these colonies accumulated the necessary set of genetic innovations, they could begin to produce the complex tissues and organs that animals are made of. Comparison of DNA evidence from a variety of present-day phyla suggests that all animal life only arose once, from one common ancestor—for there are no single-celled protozoans mixed in among the branches in the family tree of animal life.

Another view is that rather than innovation within the genes being the limiting factor, it was a change in the environment that actually lit the fuse for the Cambrian Explosion. The diversification that produced the Ediacaran Fauna after 600 Myr ago began at the end of a long series of extreme global ice ages, which had brought sea ice and glaciers even down as far as the tropics. When thick ice covered most of the world's oceans, there could have been little photosynthesis and oxygen production by plankton, and the ocean water would also have been cut off from oxygen supply in the atmosphere above. As would be expected, some rocks which formed from layers of sea floor sediments at the time of the great Precambrian ice ages have a chemical composition indicating that the ocean waters were starved of oxygen. But after the ice melted back, the same rock sequences show an abundance of oxygen seeping into the sea floor sediments, because the sea above them was now well aerated. It has been suggested that the increase in oxygen in seawater was necessary for complex multicellular life to evolve (Canfield *et al.*, 2007; Shen *et al.*, 2008). Without the extra energy supply from plenty of oxygen for respiration, complex organisms with tissues and organs could not survive. The ice ages could then have delayed the appearance of animal life, with their end providing the trigger for it to finally evolve and diversify, culminating in the big burst of appearance of phyla about 540 Myr ago in the Cambrian Explosion.

Another variant on the idea of an environmental trigger essentially emphasizes the opposite—that the Earth had been too hot for animal life for most of its history, up until the time when the great global ice ages hit from around 900 Myr to 700 Myr ago. It would then have been the coolness of the ice ages that allowed diversification of complex life forms to begin, culminating in the appearance of animals after about 600 Myr ago (Gaucher *et al.*, 2008).

3.5.2 Is the Cambrian Explosion just an effect of better preservation?

The most dramatic part of the early build-up of animal diversity was the Cambrian Explosion itself at about 540 Myr ago, when so many phyla suddenly appeared in such a short space of time. It seems strange that it happened in such a concentrated burst, and not spread out more evenly over the period after about 600 Myr ago.

It could be that the Cambrian Explosion was in a sense an illusion, caused by a different sort of change in ocean chemistry. This would have made it easier for the many groups of organisms which *already* existed in the tens of millions of years before the explosion to form hard mineral shells and get preserved.

Having a hard shell greatly increases the chances of an organism leaving some sort of trace in the fossil record. Before that time, it is possible that many small, soft-bodied forms of animals had existed in the oceans, but were not getting preserved because soft tissues rot and fall apart easily. These early oceans may have been too acidic for any animal to form a hard shell, because they were packed full of carbon dioxide from a CO_2-rich atmosphere. Under these conditions, the minerals in the shell would simply dissolve as quickly as they formed. It is quite striking that up until the Cambrian Explosion, all we see is a limited range of soft-bodied "Ediacaran" forms and nothing at all that has a shell. The Ediacaran Fauna could in fact be just the tail end of a true diversity of soft-bodied animals that existed much further back, but by chance there were none of the right sort of fine-grained undisturbed sediments that could preserve them. Then about 540 Myr ago the acidity in seawater decreased, and it became possible for organisms to form hard carbonate or phosphate shells and skeletons. The empty shells could also now survive sitting on the sea floor (without rapidly dissolving) and eventually end up getting fossilized. It is interesting that the first shelled organisms had phosphate, not carbonate shells, and this fits into the general scenario that it was progressively decreasing acidity which first allowed hard shells. This is because phosphate can exist in more acidic conditions than carbonate.

Another interesting thing is that the first tiny single-celled protozoa to use mineral shells (e.g., radiolarians) also appear during the Cambrian Explosion. This seems quite a coincidence, since they are not at all closely related to animals. It tends to suggest that there really was a fundamental change in the environment around that time.

There are estimates of how far back in time the main groups of animals diverged from one another in evolutionary terms, from comparing present-day DNA sequences (Bromham *et al.*, 1998). These suggest that the ancestral lines of different phyla actually separated at widely different times, spanning several hundred million years before the Cambrian Explosion when they eventually appeared in the fossil record. If we accept the reliability of this dating based only on DNA (and its reliability is not certain), it raises the question of why the initial true burst of animal life in the fossil record has not been found so early on, and what the divergence of lines of ancestry really means in terms of diversity. For example, we do not know whether the real anatomical differences between the main animal lineages (the "phyla") all arose way back at this earlier time. Instead it could be that separate lineages of organisms stayed looking more or less the same as one another, until they suddenly morphed into many different-looking forms at the time of the Cambrian Explosion.

3.6 HOW MANY SPECIES HAVE EVER EXISTED?

There is continual turnover in life, at every level from the molecules within a cell, to the cells in the body, to the birth and death of individuals in a population. From the fossil record we see that species are also continuously turning over: forming through speciation, and later going extinct. So the groups of organisms that were common and varied during one slice of Earth history may be quite different from those 50 million or 100 million years later. Given the rates at which species come and go in the groups that we can observe fairly closely in the fossil record, it is generally agreed that most of the species that have ever existed on Earth are now extinct.

Exactly how many species and genera have ever existed? We can never know for sure, but Jackson and Johnson (2001) have tried indirectly estimating how many genera—most of them nameless, unseen and now extinct—might have occurred in the seas during the last 540 million years or so. The estimate depends very much on certain assumptions. One such assumption is the shape of the graph of build-up of richness of genera globally in the fossil record, starting from just a few genera 540 Myr ago in the days of the Cambrian Explosion. For example, perhaps global diversity built up to a stable plateau level and then already leveled off (see Figure 3.4a), hundreds of millions of years ago? If it did, the number of invertebrate genera we have in the marine realm in the present-day world (which Jackson and Johnson estimate at 67,000 genera, assuming 5 species on average per genus and 1 million marine invertebrate *species* in the present-day world) may be quite a good indicator of how many species existed at any one time in the past. Now assuming turnover of these marine genera, with them each lasting some 28 million years on average before

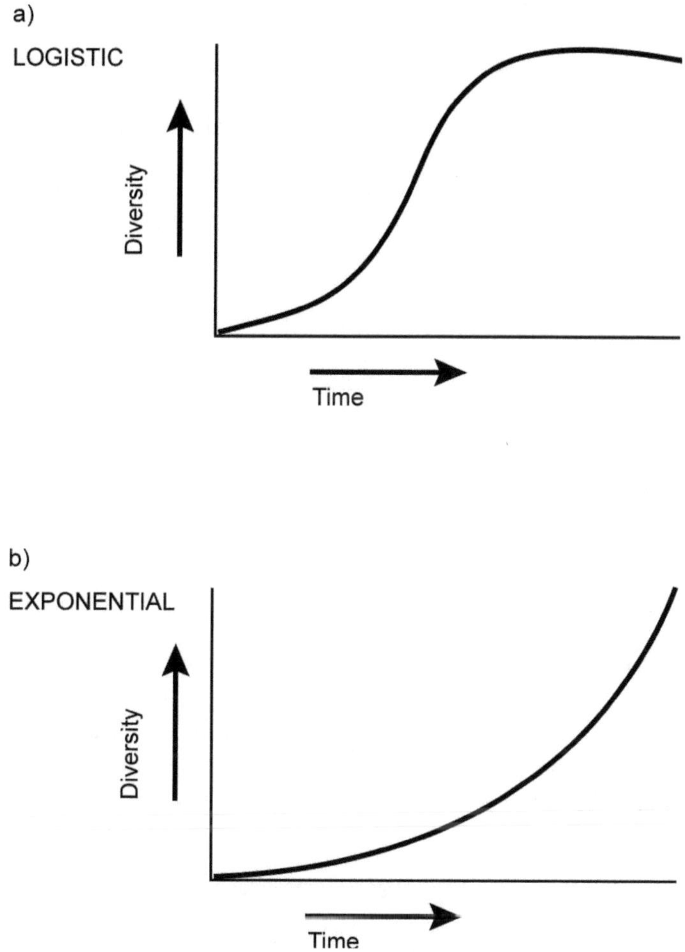

Figure 3.4. Hypothetical curve of rising species diversity through geological time: (a) logistic, (b) exponential.

they go extinct, this makes for an estimate of *one million genera* having existed over the past several hundred million years (I am not sure how reliable this estimate of species lifetime is: Ellen Thomas of Yale University tells me that actually the lifespan of species is a lot shorter in the group she works on, single-celled forams). Assuming 5 species per genus, that equals then about 5 million species, just in the marine realm: so four out of every five species that ever existed there are now extinct.

But diversity could have grown along a different path. If instead one assumes a more drawn-out exponential growth (see Figure 3.4b) in numbers of genera over time, then disproportionately more of the genera that ever existed are alive today—since the curve for numbers of genera was much lower for most of the past few hundred million years. With this assumption, 200,000 now-extinct marine invertebrate genera have existed at some point or other in geological time—which itself makes for "only" 1 million vanished marine species. Intuitively, it seems likely that the actual curve of build-up of diversity was somewhere between these extremes, so that the number of genera that existed in the past was more than the lowermost estimate, but less than the uppermost. Sepkoski *et al.* (1981) compiled a graph that showed how the numbers of marine genera that we do know of accumulated over the last 600 million years. The same general picture comes from a somewhat larger dataset and more advanced set of analyses by Alroy *et al.* (2008). The trend for diversity does look intermediate between the plateau and exponential curves, although the poorer fossil record further back in time may be tending to push down what would really have been a plateau which extended further back towards the early days of animal life.

Some more recent data in favor of the "plateau" view of build-up of diversity come from the good-quality fossil record of Caribbean mollusks over the last 10 million years. In the Caribbean, there has been turnover of species and genera, but on average about as many new types have replaced the ones that went extinct, keeping diversity fairly constant. On the other hand, from the fossil record of North America and Europe, the growth of diversity over time seems to have been exponential. Either way, it all adds up to a huge number of genera—and species—that have evolved and been lost. On land, the number of extinct genera could be far greater than this, and surely includes many millions more unknown vanished species of insects and rainforest plants, if we think of the present diversity of tropical rainforests (Chapter 6).

Whatever the real curve for diversity growth over time, only 18,700 genera have ever been found in the marine fossil record (Alroy *et al.*, 2008), and that is only about 3% to 16% of those which actually existed by Jackson and Johnson's estimation. Even among these organisms which live in the environment that makes their preservation the most likely, most of the diversity that ever existed has come and gone unseen by fossil hunters and biologists.

3.7 BACKGROUND EXTINCTION

Most of the turnover of species that are found in the fossil record seems to happen individually, species by species, not noticeably associated with other groups originating or disappearing. Eventually, when all the species in a genus or family have blinked out of existence one after another over millions of years, the whole genus or family is also lost. Loss of diversity in this individualistic way—one species at a time—is part of the run of the mill of life on Earth, and is known to geologists as "background extinction".

3.7.1 The causes of background extinction

Why do species come and go in the fossil record? After all, if they are well suited to their way of living, shouldn't they last forever? Though we do not know much about the biology of most extinct life forms, many of the species that we find in the fossil record do look like they were rather well adapted to their ecological niche. For example, over the past 60 million years or so there have been many different species and genera of herbivorous mammals, each with suitable adaptations to its way of living such as the slicing and grinding teeth, and heavy muscular jaws that are needed to deal with tough vegetation.

Some of these herbivores belonged to familiar groups that we would broadly recognize, such as horses or deer (though they were each different from present-day forms). Other types of herbivores were quite different from anything that exists nowadays. For example the brontotheres were a very diverse group of herbivores (related to horses, though only distantly) that ranged from about the size of a pig to a rhinoceros. They lived by browsing vegetation—at least 43 genera are known, and many of these must surely have had more than one species. Outwardly the brontotheres seem well adapted to their role, and the larger ones must have been able to defend themselves against any predator. In their time they were evidently very successful, being diverse and widespread. Yet one by one over time, all these genera went extinct—the last one dying out about 34 Myr ago. Even in the other groups of herbivores that are still with us (e.g., deer, elephants, pigs, and horses) many different genera and species have come and gone—their place often eventually taken by other ecologically similar species that moved into the area long afterwards. So the picture we get on the time scale of the fossil record is of constant and aimless turnover. Why if species filled an ecological role quite adequately for millions of years, did they have to go extinct, only to be replaced by others which each in their turn also went extinct?

There are two possible explanations for why so many species eventually went extinct. One is that in the end, they simply suffered some bad luck. For example, imagine a species of bird that is common and widespread across a continent and well-suited to surviving in the environments there. By chance, a drastic shift in climate reduces the area which it can live in. This causes a decline in the numbers of this species and breaks up its natural range into just a couple of small, surviving populations. Perhaps other suitable areas of climate have now appeared elsewhere, but our bird species might be unable to reach them because a mountain range or sea is now blocking the way. Then one of the two remaining populations is wiped out by a volcanic eruption. The last population gets hit by a big drought or severe winter, and that is the end of the species. Thus a perfectly viable, well-adapted form of life has been hammered down to extinction, even in a world in which there are places it might survive, if only it could reach them.

Runs of bad luck rather like this have been seen in the recent historical past (e.g., the unfortunate tale of the heath hen, *Tympanuchus cupido cupido*, a subspecies of the prairie chicken of the eastern U.S.A.; see Chapter 5). The demise of each species within each genus of brontotheres, or each extinct form of deer or horse, may well have involved such combinations of events. Yet although each species extinction is an individual story of bad luck with its own twists and turns, certain broad underlying forces can also be seen at work pushing out certain groups of species that shared similar characteristics—even though these species or genera went extinct one by one, spread over millions of years. The demise of many of those groups of brontotheres and early forms of horses can partly be put down to a drying and cooling of the global climate over the last 40 million years. This led to less of the woody vegetation that they browsed, and its replacement by grassy vegetation that they were not adapted to eating. So background extinction can be both part of a broad-scale story, and yet very individual too.

A particular species might also blink out of existence it if it is outmoded and outcompeted by other life forms that happen to acquire new and better characteristics. For example, a predator that is the main enemy of a particular type of squirrel might hit upon a subtle modification to the shape of its body that makes it much better at catching squirrels, along with the other types of prey that it eats. The squirrel might then go extinct, with the predator continuing to make a living on all the other prey species that it can still catch. The squirrel species blinks out of existence in the fossil record, without us seeing what actually caused it to go extinct. The process could work the other way round if a squirrel evolves better characteristics for running along branches or down burrows and escaping its predator. If the predator relies heavily on that type of squirrel for its food supply, it may go extinct and disappear from the fossil record—quite inexplicably from our distant point of view.

Ecologists view nature as being full of "arms races", which resemble the arms races of military superpowers. In the world of human arms races, new tanks or warplanes are continuously developed, and then outmoded as the other side develops something more lethal. In nature, each species is continually under pressure from natural selection to evolve new features, to better escape its predators, catch its prey, or outperform its competitors. Diseases and their hosts are also in a continuous arms race, with new strains breaking out and then resistance arising in the host population that survives. Most of the time, these arms races stay in a sort of balance, with temporary slight gains and losses that merely perpetuate a stalemate. But every once in a while, one side in a biological arms race gains so conclusive an advantage that it uses it to wipe out its adversary. This then may be another reason why many species go extinct in what we see as background extinction.

Zooming out to see the bigger picture, we might find that each of these individual extinction events fits into a broader story. It may be that a group of predators, or prey, or a competitor, is spreading out across the world evolving more and more new species. In the case of the brontotheres, their ultimate demise might well be partly due to competition from other grazers that lived in the same way—but were better at doing it. Perhaps, for example, such newly evolving groups as rhinos, pigs, and horses helped push them out. Each extinction, while in a sense random, may also fit into a broader pattern of change in either the physical environment, or the biological environment of species that it has to interact with.

The fact that all species are continuously jostled by new and unpredictable challenges may thus be the reason that they mostly go extinct over the time span of a few million years. Whereas most of these challenges can be successfully overcome, eventually any species will meet hurdles that are too great for it to jump over. This explanation for background extinction in terms of an ever-shifting environment of climate, food supply, and enemies goes by the label of the "Red Queen Hypothesis". This peculiar name, coined by van Valen (1974), comes from a scene in *Alice in Wonderland*, where a character known as the Red Queen "had to keep running just to stand still". Like the Red Queen in the story, if a species does not continuously change in response to the challenges that arise, it will lose its place and vanish. Sooner

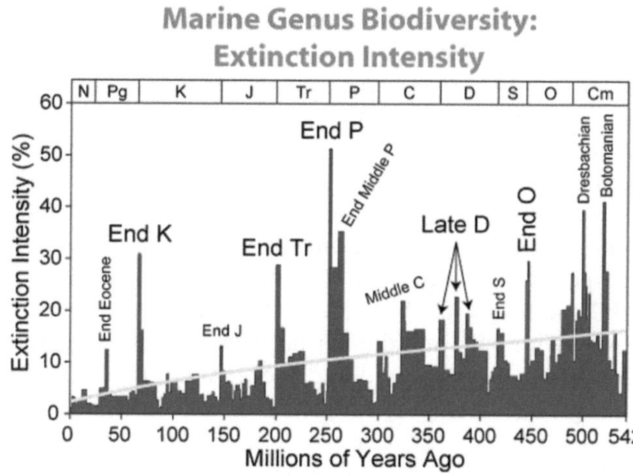

Figure 3.5. The mass extinctions: each peak is a rise in extinction rate. The Big Five mass extinctions are named in larger letters (source: DragonsFlight, Wikipedia Commons).

or later, any species will be thrown a series of challenges it cannot keep up with, and that will be the end of it.

If extinction was all that ever happened to species, then before long there would not be much biological diversity left in the world. But alongside background extinction, new species are continuously appearing in the fossil record, when preexisting species spilt and evolve in different directions (see Box 3.1). The level of diversity we see in the fossil record—and in the present-day world—is a balance overall between the rate of formation of species, and the rate of extinction.

3.8 MASS EXTINCTIONS

Seen against the usual gentle turnover of life forms in the fossil record, certain times stand out in their slaughter. On several occasions over the last several hundred million years (Figure 3.5) there were dramatic worldwide dips in the numbers of species, families, and genera, known as *mass extinctions*. Depending on exactly how one chooses to define them, there have been five or more mass extinctions in the last 600 million years. Each of these dips in biological richness was followed by a period of recovery, when new forms appeared at a faster rate. Eventually the total numbers of genera seem to have leveled out each time, as if the gaps in nature had once again been "filled up" with different types of organisms.

3.9 THE MAIN MASS EXTINCTIONS

The effects of mass extinctions on the history of life were so dramatic that they provided the first identifiable subdivisions in the Earth's geological history. In the early 1800s, even before ideas of evolution were widely accepted, the great Victorian geologist James Hutton noticed three quite distinct phases in the history of life. He called them "Paleozoic", "Mesozoic", and "Cenozoic", from Greek words meaning "ancient life", "middle life", and "new life", respectively. When each new phase started, many of the older forms that had been common and diverse before vanished, and were replaced by other groups.

The ancient life of the Paleozoic era was typified by the trilobites (flattened, segmented animals that seem to have been distant relatives of the horseshoe crabs and pill-bugs). The trilobites were very successful during much of the Paleozoic—abundant and diverse, although they had already undergone a major decline to just a few species before the last of them finally disappeared at the end of this era. Around 6,000 species of trilobites have been described for the Palaeozoic, but none of them survived into the middle period, the Mesozoic. In the Mesozoic a whole range of other life forms appeared and proliferated, most famously the dinosaurs and several other groups of large reptile-like animals of the land, air, and sea. Then at the start of the last period—the Cenozoic—these large beasts mostly disappeared and were replaced by mammals. Accompanying the big beasts into oblivion went such groups as the ammonites—shelled relatives of the nautilus which had been abundant and diverse in shallow ocean waters for several hundred million years.

As the field of geology developed, other big transitions in the history of life were identified. Though not quite as dramatic as the 'biggest two" (the end-Palaeozoic and end-Mesozoic) extinctions, these nevertheless swept away a large proportion of the life forms on Earth. These other "smaller" mass extinctions include ones that took place scattered through geological time, in geological periods known as the Cambrian, Ordovician, Devonian, Jurassic, and Eocene. Although estimates vary between authors, it is thought that the five biggest mass extinctions (those which swept away the greatest proportion of the genera and species on Earth) were the following: the end-Permian (251 Myr ago), the end-Cretaceous (66 Myr ago), the late-Devonian (364 Myr ago), the end-Triassic (201 Myr ago), and the end-Ordivician (about 425 Myr ago). These are often called the "Big Five".

How much of life's diversity was lost in these mass extinctions? The figure seems to vary from one event to another (Alroy *et al.*, 2008). Also if we take any one mass extinction, there is a fair amount of uncertainty in terms of what proportion of all the world's biological groups were lost, even for the better-studied ones such as the Cretaceous and Permian events (Alroy *et al.*, 2008). Certainly in the bigger mass extinctions, a large slice of the world's known species and genera disappeared. In fact, the decimation was so extensive that a large proportion of the families (see Box 3.1) were lost too—and to lose a family every last species within it must go extinct. In each of these, the majority of the world's species were wiped out: in the biggest ones probably more than 70% of all species on Earth.

3.10 THE END-PERMIAN MASS EXTINCTION

The greatest mass extinction of all seems to have been one of those that Hutton had first spotted, marking the end of the Palaeozoic. It brought to an end a subdivision of geological time known as the Permian, the final stage of the Palaeozoic, and for this reason the event is generally called the "end-Permian" mass extinction. The latest geological dating techniques put it quite precisely at 251 Myr ago.

Something like 60% of all the known families of animal life vanished at this point in time. It is important to bear in mind though that species losses at the end of Permian have not been counted up directly (because paleontologists do not quite trust themselves to identify species accurately in the fossil record this far back), and in any case those that are known from the fossil record would probably be a fairly small sample of all those in the world at the time. Instead of trying to count species losses directly, paleontologists have attempted to estimate how many species must have been lost on average to make around 60% of *families* vanish at the end of the Permian. To do this, they need to make a reasonable guess about how many species would have been present in each family, which is based on looking at how many species occur in each modern family. All of the species in a family would have to be wiped out to eliminate that whole family. Many other families survived, but they too most likely suffered very heavy losses. All it takes is just one species surviving, for that particular family to avoid extinction—so obviously a very high percentage of species in the world have to vanish in order to completely wipe out the majority of families. David Raup (Raup, 1979) estimated that an extinction that wiped out 60% of families would have killed off somewhere between 80% and 96% of all species of animal life, depending on one's guess of how many species there would have been per family on average.

Another striking feature of the larger mass extinctions is how unselective they were, affecting creatures of the land and sea alike. Among the major animal groups lost at the end of the Permian were the dominant large herbivores and predators of the land, all belonging to a lineage known colloquially as "mammal-like reptiles" (the synapsids). These rather peculiar-looking beasts included somewhere in among themselves our own ancestors. Except for a few small surviving forms that carried on into the Mesozoic (and eventually happened to give rise to us mammals), they were all swept away in this extinction. Also on land, several orders of insects were lost in the mass extinction. In the sea, several major groups of invertebrates vanished completely at the same time (e.g., the blastoids, distant relatives of sea urchins in the echinoderm phylum). The last few types of trilobites, which had been gradually declining from their heyday hundreds of millions of years before, were also snuffed out. Their loss marked the end of an entire class of animal life. But as on the land, it was not just groups already in decline that were hard hit by the extinctions—even the most abundant, successful types of marine animals were almost completely wiped out. For example, the mollusks, the broad category that includes snails, clams, and squids, had been thriving up until the end of the Permian but did badly in the mass extinction (even though it would later come back spectacularly). In the Permian seas before the mass extinction, one of the most abundant groups of mollusks in terms of abundance and diversity of types were the ammonites, which had coiled shells that resembled the present-day nautilus. Only a single genus of ammonites managed to squeak through into the Mesozoic: all the rest were lost. Almost all of the many types of snails (gastropods) in the seas disappeared, though they eventually managed to make a comeback millions of years later from some unknown and unseen survivor that must have hung on somewhere. Two important classes of corals—the rugose and tabulate corals—that had been the major builders of reefs through earlier parts of the Paleozoic, were not so lucky and vanished completely. Many genera of plants also went extinct at the end of the Permian, although none of the broadest groups such as orders or classes were lost at that time. Instead, most of the Permian groups of plants seem to have continued millions of years longer into the Triassic before finally disappearing and being replaced by other, more modern-looking types.

3.11 THE END-CRETACEOUS MASS EXTINCTION

On the scale of mass extinctions, the second most devastating event took place around 65.5 Myr ago, marking the end of the Mesozoic. The final phase of the Mesozoic is known as the Cretaceous period, named after the Latin name for chalk (because so much chalk rock was laid down in the seas of the Cretaceous across western Europe), so this tends to be called the end-Cretaceous extinction. It also rather confusingly goes by the name of the Cretaceous–Tertiary extinction, occurring as it does at the borderline between the Cretaceous and the succeeding geological era of the Cenozoic which is often known as the Tertiary. As if this were not enough, because "Cretaceous–Tertiary" is a bit of a mouthful to keep repeating, geologists like to abbreviate it by using initials: not C–T as one might expect, because the letter initial C is already reserved for an earlier phase in geological history, known as the Carboniferous. Instead this one gets called "K–T". And so any gathering of paleontologists—when overheard—seems to be constantly talking about someone called "Katy". Here for simplicity, we will just call it the end-Cretaceous extinction.

Adding even further to the complications, the nomenclature of geological time has recently been messed around with by committees who have declared that (a) the Tertiary no longer formally exists, and must be broken up into the Paleogene (its first part) and the Neogene (its later part) and (b) the Cretaceous–Tertiary boundary should be called the Cretaceous–Paleogene boundary or "K–Pg". This idea of having to call K–T the "K–Pg" has been the last straw for geologists who study this mass extinction event, inspiring a rebellion. So far, many are holding out and sticking stubbornly to their old "K–T".

The best estimates are that around 40% of families of animal life were lost in the end-Cretaceous extinction event. Scaling down to the species level, this would tend to suggest somewhere around 70% of species were lost at the time. Like the end-Permian extinction, this event occurred across the board, affecting most habitats and most groups of organisms.

The most famous victims of end-Cretaceous extinction were the dinosaurs. In the Mesozoic world they had diversified into scores of genera, occupying a whole range of ecological niches, and ranging in size from some forms about the size of a chicken to the mighty 60-tonne *Sauroposeidon*. Several other groups of large creatures that are often confused with dinosaurs—but were actually quite distinct from them—also vanished at the same time. These included the bat-like pterosaurs, with broad leathery wings. Real live sea serpents, the mosasaurs and plesiosaurs, were also important and diverse in the Cretaceous seas, and they too went extinct at the boundary with the Tertiary. The descendents of the dinosaurs, the birds, also did very badly in the end-Cretaceous extinction. The two most common bird groups, both of which had rows of teeth in their beak, perished leaving only a third and toothless group that gave rise to all the birds of today. Most genera of mammals, although they were at that time rather few and small in size, were also lost. And in this particular extinction the plants too suffered heavily. Several major groups of seed plants such as the cycadeoids (looking rather like the modern-day cycads, but apparently unrelated to them) vanished, and the flowering plants (angiosperms) that had been proliferating and spreading across the Cretaceous landmasses also suffered a severe knock-back. A large proportion of angiosperm species went extinct in some areas including an estimated 60% of the species of broadleaved trees in North America. From the reduction in the diversity of types of insect damage to fossil leaves across the boundary, Wilf *et al.* (2006) suggested that there was also a mass extinction of herbivorous insects, even among the ones that fed on the species of plants which survived.

The ammonites, which had survived the end-Permian extinction to come back and flourish in the Mesozoic seas, were not so lucky this second time. They vanished at the same time as the big reptiles, never to return. The major reef formers of the time, a strange group of clonally sprouting clams known as rudists, also disappeared completely—along with 98% of the reef-forming coral species. Along with these went most of the other marine invertebrates for which we have a fossil record, and many types of microscopic plankton. Even the organisms that gave the Cretaceous its name—tiny algae called coccolithophores whose countless shells make up the chalk rock of western Europe—mostly perished along with all the rest. More than 95% of their species vanished, putting a stop to chalk deposition. Although the survivors managed to make a comeback in the Tertiary seas, the coccolithophores have again never become as abundant as they were when the chalk was being laid down.

3.12 THE END-ORDOVICIAN MASS EXTINCTION

Third on the scale is a mass extinction that occurred in the middle of the Paleozoic, 444 Myr ago at the end of a period known as the Ordovician. This early one swept away many groups that had proliferated in the first burst of animal life in the seas, the Cambrian Explosion. The end-Ordovician mass extinction provides the natural boundary that geologists use to mark off the Ordovician from the period that followed it, the Silurian. What suffered badly on this occasion were the typically Paleozoic groups of marine animals such as trilobites, brachiopods, and tabulate and rugose corals: about half of the known genera of marine animals went extinct.

3.13 OTHER MASS EXTINCTIONS

Almost as dramatic were such events as the end-Devonian mass extinction 354 Myr ago, which swept away many marine families, hitting primitive groups of fishes especially hard. The end-Triassic mass extinction 205 Myr ago removed 30% of marine genera, 50% of the genera of land vertebrates, and more than 95% of the genera of vascular plants (gymnosperms, ferns, and suchlike) (McElwain *et al.*, 1999).

There are other, lower peaks in extinction rates in the geological record, and in the end it becomes difficult to draw a distinction between what is really a mass extinction and what is just an incidental rise in the rate of background extinction. While these other pulses of extinction were less impressive on a global scale, some of them were still quite devastating for certain groups and environments. For example, the end-Paleocene extinction about 55 Myr ago wiped out many of the world's deep-sea foraminifera, tiny protozoa that are important in marine food chains, though it apparently did not affect much else.

Raup and Sepkoski (1984) have suggested that looking back over the last 500 million years or so, there were peaks in extinction rate about every 26 million years—a regular cycle which, if true, demands explanation (below). However, as the geological dating has improved over the last few years and the timings of mass extinctions rearranged, the supposed 26 million year cycle has become less clear and less believable.

3.14 MASS EXTINCTIONS AFFECTED SPECIES INHABITING MANY DIFFERENT ENVIRONMENTS

An intriguing aspect of mass extinctions is that they affected a very wide range of life forms (Table 3.1). For example, the end-Permian extinction wiped out land-living and ocean-dwelling animals about equally, and as far as we can tell it hit invertebrates in the sea as hard as it hit marine vertebrates. Within the Permian seas, animals that floated or swam within the water (the plankton and nekton) were about as likely to go extinct in the mass extinction as those that burrowed in sediment or lived fixed to the sea floor (the benthics). The only pattern that stands out in this particular event is that land plants did not suffer as badly as animals, although they too were hit hard.

Table 3.1. The general characteristics of mass extinctions.

—A major proportion of families, genera, and species go extinct in a short span of geological time
—Mass extinctions affect many different groups of organisms living in a wide range of environments
—Mass extinctions occur all around the world affecting different landmasses and ocean basins

The end-Cretaceous extinction was similarly wide-ranging; indeed even more so in some ways in that we find clearer signs of whole broad groups of land plants becoming extinct.

Explaining such widespread extinctions must take into account the full range of environments that they affected, and some important clues can come from seeing which types of organisms or habitats were less affected. Although the Permian extinction was very much across the board in the way it affected animals, other mass extinctions did show some selectivity for or against animals living in particular types of niche. In these other instances, certain groups of animals or plants living in particular environments seem to have survived relatively unscathed. For example, in the end-Cretaceous extinction many lineages of terrapins, turtles, and crocodiles seem to have made it across the boundary into the succeeding Paleogene when all of the larger beasts went extinct. As we will see, the pattern of extinction and survival might give some useful clues to what went on during a really big mass extinction.

3.15 THE CAUSES OF MASS EXTINCTIONS

Looking at the present-day world around us, it is hard to envisage a natural catastrophe that would be capable of stamping out life on the scale that we see during mass extinctions (though we humans seem to be working hard at present on a mass extinction of our own making, see Chapter 5). Does something fundamentally different happen during mass extinctions, compared with the general background of life's history when the gradual coming and going of species predominates? Or could it be that what we think of as mass extinction is really just an uptick in the same familiar processes that normally lead to background extinction? Hypothetically, mass extinction might not be especially sudden, but something spread out over tens of millions of years—just a statistical increase in extinction rate that was not quite balanced by the formation of new species.

To really understand mass extinctions, we need to gather as much information on them as possible, and this has been a major preoccupation of geologists over the last three decades of so. Essentially we must ask ourselves various questions about mass extinction events, and then set out to find the answers.

3.16 WERE MASS EXTINCTIONS SUDDEN, OR GRADUAL?

If mass extinctions were basically just a chance upswing in the same processes that tend to lead to background extinction, we should expect to see each mass extinction taking place over millions of years. Species going extinct individually, due to many different and localized causes, would not all go extinct at exactly the same moment (unless perhaps they happened to play an unusually important role in an ecosystem, that many others around them depended upon). If on the other hand everything vanished almost simultaneously, we have to look for something that was far more dramatic in scale than the usual causes of background extinction.

Unfortunately, the imperfect patchy nature of the fossil record tends to make it difficult to tell what happened suddenly, and what happened gradually. This is because it is often hard to see from the fossil record exactly when things died out. No species of organism will always be getting buried and fossilized everywhere, and there may be gaps in its fossil record that last for millions of years. A species may disappear from the record for a long time, then reappear because the right sort of sediments were being laid down once again. We only know that it had been alive all along because it finally turns up again. If a species vanishes permanently from the fossil record, it is still possible that it survived somewhere else for quite a while after we last see it, and finally went extinct unseen at some unknown time later on.

Even if all the species lost in a mass extinction were killed off at exactly the same moment in time, we would expect that just by chance some of them would already have been absent from the fossil record during a span of several million years before the actual extinction event. This is because by chance the deposition and preservation of sediments was not happening where they lived. Such gaps in the record would tend to make it look like many species went extinct one by one over several million years, and it would be possible to reach the wrong conclusion about the speed at which the mass extinction took place. This "blurring" is known as the *Signor–Lipps effect*. Because of this uncertainty, there has been a fair amount of academic dispute about the suddenness of mass extinctions.

Some paleontologists look at the fossil record around the time of a mass extinction and see a gradual picture of extinctions, occurring one by one over several million years. Others looking at the same data suggest that there is really no reason—given the imperfections in the fossil record—to assume that the extinction was spaced over such a long span of time. They may be able to point to certain fossil localities where several species occur right up to a time that falls within the general interval of a mass extinction, then vanish simultaneously and are never found again. The most useful sorts of species for making the case for suddenness in a mass extinction, are those that were especially widespread and abundant, and also living in the right environments to get fossilized reliably. The best of all are microfossils—protozoa and other little life forms that were extremely abundant and widespread within the seas. They get fossilized reliably (depending on the species) in the shallow seas around the edges of continents, or the oozes of the deep ocean floor. When microfossil species go extinct, you know that they really *did* go extinct at that point in time and did not hang on anywhere else. So when extinctions of these most widespread and abundant species occur apparently simultaneously throughout the world, it certainly looks very likely that the mass extinction was a sudden cataclysm. If such key species go extinct suddenly and all at once, this strengthens the case that the many other less reliably fossilized species—which might seem to die out individually before this point in time—were actually still alive but unseen right up until that point. One can suggest that they too disappeared from the Earth at just the same time as the ones that there is a better record for.

3.17 THE PARADIGM SHIFT TOWARDS ACCEPTANCE OF SUDDEN MASS EXTINCTIONS

A paradigm is a way of looking at things, a fundamental viewpoint. Up until recently, most geologists and paleontologists held to the paradigm that all important events in the past occurred gradually over many thousands or over millions of years. Darwin, for instance, believed that there had been no catastrophic mass extinctions. But during the last few decades, a clear majority have become convinced that mass extinctions can occur suddenly, in the geological "instant" of a few thousand years or even just in a few months. A mass extinction might even occur during one especially bad afternoon.

What has brought about this paradigm shift is a combination of different types of evidence. The first is the careful study of the last occurrences of organisms that had seemed to disappear gradually (one species at a time) in the rocks leading up to the suggested sudden mass extinction event. In several instances of mass extinctions, closer investigation of rock sequences around the world has found that these species were actually alive right up until the same point in time when many other species vanished. For instance, when the geologist Peter Ward studied the ammonites of the Cretaceous, he at first thought that many species had gone extinct well before the Cretaceous–Tertiary boundary. Then with a lot more careful study of a key site in Spain he found fossils that filled in the gaps and showed that at least some ammonite species did go right up to the boundary, and then finally all went extinct.

There used to be a fair amount of disagreement about the speed of the end-Permian mass extinction. In the past few years, geologists have been fortunate enough to find some very detailed sequences of rocks that go right through the end-Permian mass extinction event, and reveal a lot about its rapidity. One such sequence is found in northern Italy, another in eastern Greenland, and a third in southern China. They all show a very diverse sea bed fauna that suddenly vanished, with most species failing to come back again in the layers above—presumably because they were now extinct (Benton and Twitchett, 2003). The most detailed of these sections, at Meishan in southern China, starts off with a very species-rich marine fauna typical of the late Permian. Then suddenly, a large proportion of the species disappear. There is a pause of several tens of thousands of years, and then another devastating pulse of extinctions that takes away most of the remaining species. The overall picture is of two discrete sudden pulses of extinction, both within the space of 60,000 years or so. It is now widely thought that it was in the intervening period between these two marine pulses of extinction that mass extinction on land took place, for that is the time when fungal spores start turning up in large quantities in the sequence.

Such reassessments of the fossil record have led to the conclusion that the end-Permian mass extinction 251 Myr ago and the end-Cretaceous mass extinction 65.5 Myr ago were both very sudden in geological terms—occurring over at most a few tens of thousands of years—and perhaps much less. It is important, however, to note that not all mass extinctions necessarily occurred in a geological instant. One study of the late-Devonian mass extinction suggests that it was really spread out over several million years. And what may have caused the dip in diversity was not an increase in probabilities of extinction as we might expect, but a decrease in the rate of speciation—the generation of new species in the fossil record. So old species dropped out at a constant rate due to background extinction, but they were not replaced at the same rate that they were going extinct. The result was that the overall numbers of species fell. However, there have been other studies of the late-Devonian extinction published since that paper, which reach the different conclusion that the mass extinction really was sudden and did in fact involve a big increase in extinction rates. Partly then, this might have been the Signor–Lipps effect causing confusion again.

3.18 THERE IS EVIDENCE FOR DRAMATIC ENVIRONMENTAL UPHEAVAL DURING MASS EXTINCTIONS

The second general line of evidence, helping to convince most paleontologists that mass extinctions can be sudden, is the general environmental record from around the times of mass extinction (see below). In some cases, there are signs of catastrophic events on a global scale—exactly the sort of thing that would be expected to wipe out the majority of life forms on Earth.

Partly the evidence comes from a range of indicators of the chemistry and temperature of air and waters, preserved in the composition of sediments laid down at the time. Additional clues come from studying in detail the patterns of what survived and what did not, and which organisms proliferated earliest in the aftermath of an extinction event.

3.19 GLOBAL COLLAPSE OF ECOSYSTEMS IS ASSOCIATED WITH THE "BIGGEST TWO" MASS EXTINCTIONS

Just what does it take to kill off most of the world's diversity? There is a great deal of evidence that the biggest two mass extinctions—the end-Cretaceous and end-Permian events—were times when normal ecological processes ground to a halt for many thousands of years. This overall picture comes partly from study of the layer of sediments which appears to mark the precise time of the mass extinction around the world: this is known as the "boundary layer". We cannot be sure what span of time each boundary layer represents, but generally in the Biggest Two events it is thought to represent at most a few tens of thousands of years, and possibly very much less.

Conclusions about what went on during mass extinctions are also based on studying the sediments *above* each boundary layer, representing the hundreds of thousands of years which followed on from the mass extinction itself—this often gives a picture of the lasting disruption to ecological processes that followed on after the main extinction event, and by extension this gives clues to what might have happened at the boundary itself.

Box 3.2 Boundary layers: the geological moment of mass extinctions

Both the Cretaceous and the Permian mass extinctions seem to take place at a clearly identifiable moment in geological time, marked by a "boundary layer". Boundary layers are points in the sequence of rocks where there is a sudden change in the texture and composition of the sediments—as well as the simultaneous disappearance of many life forms. For the end-Cretaceous, the most recent of these Biggest Two mass extinctions, the boundary layer turns up in many places scattered around the world, where rocks laid down at the time are exposed in cliffs or road cuttings, or pulled up in the drill cores taken by prospecting oil companies. By contrast, the end-Permian boundary layer is found in only a few localities where sediments were being deposited at the time, and where the rocks they formed have survived and been exposed at the surface.

For both the Permian and the Cretaceous, the boundary layer tends to be marked off by a sudden temporary change in the color and texture of the sediment, varying from a few centimeters to a meter or so in thickness. In some other places, it is difficult to tell just by looking at it where the boundary level is, except that most of the life forms from below it vanish, and there may also be subtle changes in the chemistry of the sediments. For example, the boundary layer at Meishan in China is like this.

The Cretaceous boundary layer is famously rich in iridium, at all of the localities where it has been sampled. This is used to make the case that a large meteorite impact must have caused this particular mass extinction. The boundary layers also yield a range of other clues about the environment at these times, such as combustion products and spores, and also oxygen, sulfur, and carbon isotopes that can reveal changes in the global cycles of the elements.

Something else interesting about boundary layers is that the sites where they are found outcropping at the surface seem less common than they ought to be, given the thick sequences of rocks that we have exposed in cliffs around the world. For several of the mass extinctions, including the end-Cretaceous and end-Permian, we have many locations with sediments from shortly before the mass extinction, but the boundary layer itself has been scoured away by erosion that followed on afterwards—and eventually another layer of rocks has formed from sediment that was dumped on top much later. This sort of gap in the record is known to geologists as an "uncomformity", and the widespread nature of unconformities at times of mass extinction seems too much to be just coincidence. What may have caused these uncomformities is not certain. Some geologists suggest it was a drop in sea level at around the time of the extinction. Whether the drop in sea level helped cause the extinction, or was just part of the effects of whatever really caused the mass extinction, is a moot point. Many of the boundary unconformities may actually have been caused by the dissolving away of layers of carbonate-rich sediments by an acidified ocean, and the formation of scouring deep currents that carried sediment away (Ellen Thomas, pers. commun.).

3.20 DEATH OF VEGETATION

At the time of the end-Cretaceous extinction event, there are signs of a dramatic die-off in vegetation around the world. Wherever any evidence that relates to plant life is found from across the boundary layer, we see that trees and other large plants simply vanish from the fossil record for tens of thousands of years above the boundary. Some areas seem to have taken longer to recover than others (North America was especially hard hit, perhaps because it was close to the thing that caused the extinction: see below). Obviously, something that happened during the mass extinction killed most of the vegetation on land. During this strange time, ferns were essentially the only common land plants in places thousands of miles apart. In North America, the fern that came to cover the continent was a species of *Stenochlaena*—a tropical fern that still occurs as a climber on trees in the rainforests of Southeast Asia (Figure 3.6). Their spores often turn up in abundance in the sediments just above the boundary layer—the peak in fern spores being known as the "fern spike" by paleontologists. A fern spike has been recorded in rocks just above the boundary across North America, Europe, Antarctica, and New Zealand (Vajda *et al.*, 2001): in fact, wherever a record of what was growing on land exists.

Figure 3.6. The tropical rainforest fern *Stenochlaena* blanketed North America for thousands of years after the end-Cretaceous mass extinction. In the present world, it only survives in Southeast Asia (photo: Author, northern Borneo). *See also* Color section.

Why ferns? Well, in the Cretaceous world before the mass extinction, ferns tended to fill up open, disturbed areas in forests before trees could get in and outcompete them. They still play this ecological role to some extent nowadays although their importance has lessened since the grasses (Graminae) and the daisy family (Compositae) evolved and took over from them. After a dramatic event such as a volcanic eruption in the present-day world, a "spike" of fern spores still occurs in lake sediments as the devastated landscape is colonized by ferns, before other types of plants return. For example, following on from when Mount St. Helens in the U.S.A. erupted in 1980, fern spikes occur in the local sediment record in lakes. So knowing this, what are we to make of the Cretaceous fern spike? It seems that during the end-Cretaceous boundary, tree populations were completely wiped out over vast areas, allowing ferns to thrive in the complete absence of competition. Just before the fern spike, an abundance of fungal spores has also been found at one site, lasting at most of few decades (Vajda and McLoughlin, 2004). It looks as if these fungi were decaying the dead plants which were all that was left of the vegetation, until the ferns came in and blanketed the landscape. A fern spike also apparently occurs after another earlier mass extinction event, at the Jurassic–Triassic boundary, in a couple of localities.

The picture of utter destruction of land vegetation during the end-Permian mass extinction is backed up by various sources of evidence. In some places, for thousands of years after the boundary layer at the end of the Permian, the only evidence of life washing off the land surface consists of fungal spores. These spore layers turn up in the boundary layer in northern Italy and in Israel, for example (Benton and Twitchett, 2003). It seems that all plant life had been wiped out, leaving only dead material that fungi could live off by decay. So it looks like at the end of the Permian the devastation was even more complete than at the end of the Cretaceous. However, this interpretation is not universally accepted: it could just be that the fungal spores were tougher than pollen washing off the land, so they were the only things preserved in the fossil record.

After the "fungal spike" at the end-Permian boundary, there is a sort of "fern spike" of fast-growing plants coming back in (although the groups of plants that played this role were not strictly speaking ferns). For hundreds of thousands of years, the only common land plants that left fossil remains behind were small shrubby or weedy species—plants that might thrive after a disturbance had swept away the forests.

At one site in Greenland, Cindy van Looy and colleagues (van Looy *et al.*, 2001) traced a detailed story of the devastation of the forest ecosystem as the Permian ended. In this particular place, most of the forest trees died at the boundary but some vegetation remained: mostly the sorts of weedy fast-growing plants that get in to exploit open ground and sunlight. Over thousands of years some of the forest trees began to come back. But then the trees were knocked back again (by what exactly, we do not know), and after that most of them never returned—they had gone extinct. For a long period, from this point on, the weedy plants had the land to themselves.

Backing up the general scenario of global devastation, Sephton *et al.* (2005) have found evidence of the organic components of soils being washed down in large quantities to the sea just after the end-Permian mass extinction, suggesting massive soil erosion from denuded landscapes. Study of the landscape processes revealed in ancient river sediments from this time also agrees with the picture of a land surface stripped of vegetation. In sequences of river sediments from southern Africa, at the extinction event there is a sudden change over from straight-flowing rivers normal in a forested landscape, to shifting braided streams full of gravel and sandbanks. Braided streams are typical of a landscape stripped of vegetation cover, so all the soil washes into rivers and chokes them.

Fitting in with the picture that the world's land vegetation was destroyed during certain of the mass extinctions is evidence of charcoal washed off the land surface around the world at the Cretaceous–Tertiary boundary. For example, in sediments from Spain there is more than a 100-fold increase in the chemical by-products of combustion in sediments across the boundary (Arinobu *et al.*, 1999). It looks as if vegetation was consumed either in one initial vast fire that swept around the world, or that it died and dried out, and then burned like a multitude of bonfires. However, this general interpretation is controversial—some geologists say the concentration of burning products is just a fluke of the way sediments were being laid down at the time, not really due to any increase in burning.

3.21 EMPTY SEAS

Another dramatic demonstration of the power of events during a mass extinction comes from comparison of the sea floor environment before and after the boundary event. In sites in China and Italy from the end of the Permian, the sediments and fossil record just below the boundary reveal an environment teeming with life. The rock which was once mud or sand on the sea bed is filled with traces of the burrows of all manner of different creatures, and thoroughly mixed up and turned over by them. Then suddenly, at the boundary, all of this stops. For tens of thousands of years there are no fossils, and no sign of anything burrowing or moving the sediment around. The fine layers of silt and mud that were laid down on the sea floor remain perfect, level, and undisturbed. The picture seems to be of a sea emptied of life. Only several meters above the boundary layer, a few forlorn burrows of small invertebrates appear, but still the picture is of a deathly quiet in the ocean. Similar changes are seen at the end-Cretaceous boundary in western Europe where the copious chalk that had defined that age suddenly halts, and is replaced only by bare clay. The plankton that had produced the chalk with their calcium carbonate shells had suddenly vanished, *en masse*.

3.22 CARBON-12 SHIFTS IN THE OCEANS AND WHAT THEY MIGHT MEAN

Further clues to what happened during mass extinctions come from traces of the chemistry of the oceans, which reflect wider processes on land and in the atmosphere too (Delaney, 1989). At the time of the end-Cretaceous, the end-Permian, and also the end-Triassic mass extinctions, there are sudden shifts in the isotopic chemistry of carbon worldwide, showing up within sediments that have now become rocks. Analyses of the sediments and fossils from these times show that the carbon-12 isotope (see Box 3.3) suddenly became much more abundant in the atmosphere and oceans, reflecting some sort of drastic change in global processes of the carbon cycle. There are several ways in which so much ^{12}C could end up pouring into the atmosphere. One way would be if most of the plants on Earth died and then rotted or burned, and the organic material in the soil underneath also rotted away to CO_2. The shift in ^{12}C across the Cretaceous–Tertiary boundary could be explained by about one-fourth of the organic carbon that is normally stored in vegetation and soils suddenly getting oxidized to CO_2 in the atmosphere (Arinobu *et al.*, 1999). However, the shift across the earlier Permian–Triassic boundary is too big to be explained in this way alone—there is simply not enough carbon in land ecosystems to ever release so much ^{12}C. Another way for ^{12}C to get out into the atmosphere is if shallow sea floor muds that contain a lot of organic carbon were broken down to CO_2 (e.g., by being stirred up into the water and oxidizing). The carbon might have left these oozes in the form of methane, a powerful greenhouse gas which carries a carbon atom within each molecule. Teasing apart what actually happened from among these possibilities is difficult and contentious. At present geologists think that the isotope changes that occurred in different mass extinctions might each have been caused by different processes, and they also tend to disagree among themselves as to what the main cause of each ^{12}C shift was.

Often it took a long time for the ^{12}C composition of the atmosphere, oceans, and living organisms to return to a more normal level. After the end-Cretaceous extinction, for example, there may have been the "Strangelove Ocean" (see below). Even so, on land the ^{12}C composition of plants seems to have returned to typical long-term levels within about 130,000 years of the impact (Beerling, 2000), suggesting that by this time most of the global carbon cycle had sorted itself out again.

At other times, it took longer for the carbon cycle to recover. Following the big burst of ^{12}C released with the end-Permian extinction, the carbon isotope composition of fossils and sediments did not quite return back to what it had been before the extinction, and in fact continued to oscillate quite widely for several million years afterwards into the Triassic (Benton and Twitchett, 2003). Exactly what was going on to cause these wobbles in isotope composition is not known, but it certainly shows that the global system remained somehow unstable.

3.23 THE STRANGELOVE OCEAN

Another thing revealed by carbon isotopes is that certain basic ecological processes in the ocean nearly ground to a halt following the end-Cretaceous extinction. Normally in the ocean, phytoplankton (tiny plants) floating in the sunlit surface waters take ^{12}C up selectively from CO_2 in the seawater and concentrate it into their living cells. This leaves the seawater around them rather depleted in ^{12}C. When the dead remains of these cells, and the fecal waste of the animals that eat the plankton, sink towards the sea floor and rot, it releases the ^{12}C-rich carbon back out as CO_2 that dissolves in the seawater. The release of so much ^{12}C deeper in the sea means that the water down there is enriched in ^{12}C, in contrast to the surface parts of the sea which have had a lot of ^{12}C taken out by the plankton. This all makes for a gradient in ^{12}C content: everything is rich in ^{12}C at the sea bed and low in ^{12}C in the surface waters where photosynthesis is going on.

The strange thing about the period just after the end-Cretaceous extinction is that for three million years (D'Hondt *et al.*, 1998), this gradient disappeared. To make it disappear, something quite fundamental must have changed about ocean ecology. When the disappearance of the depth gradient in ^{12}C was first discovered, it was suggested that there must have been very little photosynthesis, or very little decay, or almost none of both. Certainly, there must have been *some* biological activity, for it is the analysis of the remains of still-surviving populations of single-celled protozoa known as foraminifera that shows up the lack of a top-to-bottom gradient in ^{12}C in the first place. But the lack of any gradient in ^{12}C was taken to imply an ocean relatively devoid of life, without much photosynthetic activity and biological production to drive food chains. It was given the apt name of a "Strangelove Ocean", in honor of the demented Dr. Strangelove who in the movie (also titled *Dr. Strangelove*) ended life on Earth with a nuclear holocaust.

It is now generally accepted that the ocean was not as dead as first thought, since phytoplankton fossils turn up in abundance again soon after the Cretaceous–Tertiary boundary—so there must have been plenty of photosynthesis going on during the time of the Strangelove Ocean. The disappearance of the gradient in ^{12}C instead might have been due to the loss of many of the key forms of animal life that consume plant material near the surface and send it down as pellets of waste towards the ocean depths. A decrease in the particle size and compaction of organic waste from the surface would have allowed more of the stuff to rot on its way down and get recycled back into the ocean—and this meant less ^{12}C ending up near the sea bed (although there are some doubts as to why the extinction would actually have had quite this effect: Ellen Thomas, pers. commun., 2008). As the animal components of food chains re-evolved after about three million years, the gradient also reasserted itself.

Not all mass extinctions were associated with a burst of ^{12}C entering the general circulation. In fact one of them, the late-Devonian mass extinction, was associated with a worldwide decrease in ^{12}C—so this one went in the opposite direction. What was presumably happening at this time was that large amounts of organic carbon were being taken out and stored somewhere, probably in ocean sediments. How it might fit into the causes of this particular mass extinction is not clear, though.

3.24 SUDDEN TEMPERATURE SWINGS

Another aspect of the environmental upheaval during mass exinctions is the evidence for large sudden swings in temperature. For example, at the end-Permian boundary global temperatures soared by 6°C, according to evidence from oxygen isotopes in sea floor sediments and the clay composition of minerals washing off the land (Benton and Twitchett, 2003). At the end-Cretaceous boundary the signs of temperature change are more ambiguous (coming indirectly from the species composition of fossil floras, plus some rather controversial changes in oxygen isotopes in sea sediments), but some have taken them to indicate a sudden cooling followed by a big spike in temperature.

Box 3.3 Carbon isotopes as an indicator of ecological changes

Carbon mostly comes in two forms—or isotopes—which differ by the number of particles in the nucleus of the atom. By far the most abundant is carbon-12 (^{12}C), but a small percentage is carbon-13 (^{13}C). Atoms of the two isotopes behave slightly differently. ^{13}C is a bit more sluggish, less likely to get involved in a chemical reaction and slower to diffuse through air or water. This means that it is less likely to be taken up by a photosynthesizing plant, when the plant fixes CO_2 from the air into sugars. The result of this selectivity is that the carbon-containing molecules made by plants contain proportionately more ^{12}C (which gets into the plant more easily) than ^{13}C (which tends to get left behind), compared with the CO_2 in the atmosphere in general. The difference is not great, just a few parts in a thousand more ^{12}C than ^{13}C, but enough to measure reliably.

Microscopic plankton that are floating in the sea—the phytoplankton—also take CO_2 in from the ocean water through photosynthesis, and though they are not as selective about it as land plants, they too tend to get enriched in ^{12}C. This leaves more ^{13}C behind in the ocean waters. Not only the plants themselves are richer in ^{13}C, but also anything that is derived from plants. A herbivore that eats the plants also gets to be enriched in ^{12}C, relative to ^{13}C. And a carnivore that eats a herbivore also becomes rich in ^{12}C. Likewise, any dead parts of plants that do not get rotted down remain enriched in ^{12}C, and also any soil that is composed of them. If bits of land plants get washed down rivers and end up in sediments in the sea, they too retain this composition. The dead parts of phytoplankton that form a dark ooze in the sea floor sediment are also enriched in ^{12}C. Meanwhile, the ^{13}C that was not taken up by the plants accumulates in the atmosphere, and dissolved in the ocean waters where it forms calcium bicarbonate.

So, there are basically two main reservoirs of carbon on and near the surface of the Earth: the organic stuff fixed originally by plants, and the inorganic stuff that was left behind. The organic material is always enriched in ^{12}C, and the inorganic component has less ^{12}C and proportionately more ^{13}C. If the size of one of these two reservoirs changes, it will affect the composition of the other reservoir. For example, if suddenly a large amount of plant material on Earth dies and decays back into CO_2—the extra ^{12}C that was in it will show up as a shift in the inorganic reservoir in the CO_2 in the atmosphere and the bicarbonate in the oceans. Suddenly the atmosphere and oceans will become less rich in ^{13}C, diluted by the ^{12}C that came out of the plants. This in itself will mean that the CO_2 that plants around the world are now taking in through photosynthesis has more ^{12}C, and that will show up as a shift in the composition of the rest of the plant material around the world—it too will become richer in ^{12}C and poorer in ^{13}C. If the organic reservoir instead increases in size, because there are more plants, or more organic stuff is accumulating in sediments without rotting, this will mean that more ^{12}C is kept out of the atmosphere and oceans, and it will show up as a shift towards less ^{12}C (and so more ^{13}C) in these inorganic reservoirs. Hence, if there is a sudden big change in the global ecological system (e.g., a big die-off of vegetation, or a decrease in the amount of carbon stored in ooze in sediments on the world's sea floors) this will show up in the isotope composition of the atmosphere and seawater.

From our point of view here (i.e., trying to understand events in the distant past), how can we ever detect such a change within something as ethereal as water or air? It turns out that there are some clever indirect ways of doing it. When phytoplankton and other small organisms in the sea make calcium carbonate shells from bicarbonate in the seawater, the shells themselves tend to reflect not the ^{12}C enrichment of organic materials, but the composition of the rest of the seawater and atmosphere left behind. When the organisms die, these little skeletons rain down to the sea floor forming part of the sediment, in the form of calcium carbonate, which shows the amount of ^{13}C "left behind" after photosynthesis has extracted carbon into organic materials. All we need to do is analyze the calcium carbonate part of the sediment after it has turned into rock, to know if there was a change in the isotope composition of seawater. Likewise if we get hold of some well-preserved fossils of plant organic material, we can analyze them to detect changes in the overall amount of ^{12}C that was around in the atmosphere that the plant took its CO_2 in from. These sorts of methods have allowed geologists to track changes in carbon reservoirs over geological time (at least for the last 150 million years or so, though for various reasons they do not work well before then), and they have given some important clues to the devastating events of mass extinctions.

3.25 THE AFTERMATH OF MASS EXTINCTIONS: DISASTER TAXA

In the strange world immediately following the largest mass extinctions, certain survivors expanded out of all proportion, filling the ecological space that had been left empty by the loss of most forms of life. I have already mentioned the ferns which thrived after the end-Cretaceous extinction, but there are various other examples.

After the end-Permian extinction, a survivor among the mammal-like reptiles known as *Lystrosaurus* seems basically to have inherited the Earth (Benton and Twitchett, 2003). This was an ungainly creature with short legs, a stubby tail, and a bulky head. It was about the size of a pig, and it probably lived rather like one too—eating a generalized but mainly herbivorous diet. *Lystrosaurus* is known from the Permian but was obviously not particularly common, as only one specimen has ever been found. Suddenly, after the mass extinction it is everywhere, making up 95% of the land vertebrate fossils that have been found in the first million years or so. It looks like the removal of competition and predation had allowed it to take the opportunity to multiply to unprecedented levels of abundance. Possibly part of the reason that *Lystrosaurus* had been able to survive the mass extinction in the first place was that it was able to feed by grubbing up roots and suchlike when most of the plant material above ground was dead. Now that it had made it through, the world was wide open for it. It was only after millions of years that *Lystrosaurus* was joined by a wide diversity of newly evolving vertebrates, many from the broad group of lineages that would eventually include the dinosaurs. Also, for hundreds of thousands of years after the end-Permian extinction there were almost no large trees. Instead, most of the world's land vegetation seems to have been made up of weedy plants with mostly soft stems, such as the fern relative *Pleuromia*, the seed fern *Dicrodium*, and the grass-like pteridophyte *Isoetes*. In the absence of any competition for light from larger plants, these seem to have been able to blanket the landscape.

It is a similar picture in the seas following the end-Permian extinction. The primitive brachiopod *Lingula*, which still survives in the present-day world, had previously (then as now) been confined to muddy estuarine environments. Suddenly, after the big extinction event, *Lingula* was everywhere—along all sorts of sandy and clear-water marine environments, where it was not found before and has never been found since. At that time just after the great extinction, *Lingula* was almost the only burrowing shelly animal around and it had taken the opportunities presented by lack of competition to expand out into new habitats. In some of these places it made up around 85% of the fossilized shells (Rodland and Bottjer, 2001). As time went on and the diversity of marine life recovered, *Lingula* was squeezed back into its traditional niche.

Meanwhile, on tropical reefs the extinction of all the reef-building corals and most of the other life forms allowed another primitive form of life to thrive once again, as disaster taxa. Mound-forming "stromatolite" algae made up most of the bulk of reefs for a while, much as they had last done hundreds of millions of years earlier—in the days before the Cambrian Explosion when hardly any animal life had evolved. Adding to the picture, the only places we find lots of stromatolites nowadays are a couple of bays in northern Australia, where the seawater is too salty and hot for any animals to survive. So lack of animals, it seems, is what stromatolites need to do well. Lack of competition and lack of grazing invertebrates is probably what allowed the stromatolite algae to thrive again briefly after the end-Permian extinction, until diversity finally built up again and pushed them back into the marginal parts of reefs.

Something that disaster taxa teach us is the role that competition and predation must play in modern-day diverse ecosystems. Clearly, this is what keeps most species in check in our present-day world. Without such checks and balances, almost any of them would multiply up to dominate the world.

3.26 CAUSES OF MASS EXTINCTIONS

We know then that mass extinctions wreaked havoc on the world. But what could have caused such events? For a long time this question has fascinated geologists, and the debates are still as lively as ever.

3.27 DID METEORITE IMPACTS BRING ABOUT MASS EXTINCTIONS? THE END-CRETACEOUS IMPACT

Back before about 1980, wild hypotheses to explain the mass extinction at the end of the Cretaceous abounded. Among these was the view that a large lump of cosmic debris—a comet or an asteroid—had hit the Earth and destroyed most of its life. As there was no evidence either for or against it (which gave nothing to discuss, and no way of putting the idea to the test), it was largely ignored. But then suddenly everything changed.

This moment of change came when two scientists—a father and son team—took a fresh approach towards investigating the Cretaceous–Tertiary boundary. Walter Alvarez was a geologist, interested in finding out the processes that lead to clayey limestone rocks called marls forming. As a handy sample of marl, he just happened to choose the Cretaceous–Tertiary boundary layer (see Box 3.2), though he could equally well have picked some other marl layer of a very different and less significant age. Walter recruited his father Luis—already a Nobel Prize winner in Physics—to the challenge, and the two of them decided the best way to put a figure on how fast the boundary clay had gotten laid down, was to look at its content of iridium, which is a rare element rather like platinum. The reason they chose iridium is that meteorites are unusually rich in it, and such cosmic debris can be assumed to fall into the atmosphere at a constant rate over time, as the Earth drags in the countless rocky fragments that float in space. The iridium-rich dust from all these burned-up meteors drifts down to the Earth's surface and sinks into the deep ocean, where it ends up mixed with clays deposited on the deep-sea floor. So, the father and son (and several colleagues working with them) reasoned that if the boundary clay was deposited fast, it would not contain much iridium (it would have more clay, washed there from distant landmasses, and less iridium). If it was deposited slowly, with a very slow rate of clay washing in, it would be relatively rich in iridium because of a greater relative contribution from meteorites. When they took some of the boundary clay and analyzed it, the Alvarezes got a shock. The iridium content of the clay was way higher than they had expected: ten times higher in fact than any sample previously found on Earth—at least apart from meteorites themselves. They concluded that at the Cretaceous boundary, a lot of meteorite material must have been falling out of the sky, maybe one huge meteorite. Perhaps, they suggested, this could have devastated life around the globe by sending a powerful blast wave through the atmosphere, and immense tidal waves onto coasts around the world. Furthermore, the great dust cloud generated by the impact could have blacked out the Sun everywhere for months or years, shutting down photosynthesis in every ecosystem. Plants would have died, and the food chains that depended on them would have collapsed. This idea of a meteorite impact provided an elegant explanation for extinction on the scale that occurred at the end of the Cretaceous, encompassing land and sea, and both plants and animals (Alvarez *et al.*, 1980).

Science is a conservative profession, and it is not surprising that such a radical idea was at first met with skepticism. Up to this point, theories for the end-Cretaceous extinction had mostly concentrated on gradual effects such as a draw-down in sea level around the world, or a cooling of climate, or the spread of flowering plants through land ecosystems. Essentially, the way of thinking had been that events on the scale of a mass extinction must surely have occurred over millions of years. And now, the Alvarezes were saying that it had likely taken just a few months.

The initial objections to this theory included the reasonable question of why the huge crater from this impact had never before been found. There were plausible explanations for why a crater hadn't been located—although these were unlikely to satisfy the skeptics. For example, it was possible that the meteorite had fallen into the deep sea where the rocks are melted down and recycled every few tens of millions of years, or if it had formed on land that it had entirely eroded away without a trace. Nevertheless, it seemed worth looking for. Geologists searched the world with the aim of finding the traces of a huge crater, and before long they found what they were looking for. They were alerted to a possible impact site in Mexico by reports from oil company geologists of a glassy substance turning up in oil drilling cores. Drilling in the same places, they found it was the characteristic glass thrown out by a massive impact or explosion somewhere in the region that had melted rock.

Then the crater itself was found. It turned out that years earlier, surveys of the intensity of gravity across the Earth's surface had located a huge circular area of reduced gravity, due to its lower density rocks, in southeastern Mexico near a town called Chicxulub on the Yucatan Peninsula (interestingly, Chicxulub means "tail of the Devil" in the local Mayan language). Some 300 km across, the crater had the form and size that one might expect for a killer asteroid able to wipe out most of the life on Earth. The bowl-shaped crater that had been blasted out of the Earth's surface has long since filled in with limestone, and its edges have eroded down, which explains why it was not found earlier. Its identity as a real meteorite crater, and not something that coincidentally looked like a crater, was confirmed from coring into it. The bottom of the crater was full of smashed rock, known as "impact breccia". This had previously only been found associated with nuclear test explosions, and other younger meteorite impact craters, demonstrating that something extremely violent had happened there at Chicxulub. Furthermore, in a layer above the breccia and sprayed for hundreds of kilometers around was a mineral called "shocked quartz". This again was testimony to a very violent event; quartz sand grains had essentially been squashed by the shock waves that passed through them. Shocked quartz had also previously been found at the sites of nuclear explosions, and associated with other meteorite impact events.

But, the key question was whether the Chicxulub crater was of the right age for this to be the event that ended the Cretaceous. The eagerly awaited dates on rock samples cored from the crater showed that indeed, the impact that formed it *had* occurred right on the Cretaceous–Tertiary boundary, at 65.5 Myr ago. Everything now seemed to fit into place—a giant impact had occurred at exactly the time that the mass extinction occurred, along with all the signs of ecological collapse around the world. More observations that slotted into the general picture were made over the coming years; for example, rocky debris apparently left by giant tsunamis was found several hundred miles away in Texas, and it turned out to date to the Cretaceous–Tertiary boundary. Glassy spheres known as tektites were found scattered over a thousand mile radius (Sigurdsson *et al.*, 1991). These would have been formed when molten droplets of rock spattered out from the explosion rained back down to Earth, solidifying as they fell. A tiny fragment of what may be the meteorite itself was found, from the boundary layer in a core taken from beneath the floor of the Pacific Ocean.

From all of this, geologists have pieced together a probable series of events at the Cretaceous–Tertiary boundary, although no two researchers are likely to agree on every detail of what happened or how important it was. It is thought that the object that produced the Chicxulub crater was an asteroid, one of a myriad of lumps of rock that are orbiting within our solar system—apparently leftovers from the process of planet formation. Most of them stay safely well away from the Earth, in the space between Mars and Jupiter. But this particular lump of rock would have been one of a small proportion of asteroids that happen to cross the path of Earth in their orbit. On this particular day, 65.5 Myr ago, the Earth just happened to be in the wrong place at the wrong time. It is estimated that the asteroid must have been around 10 km across. Ripping into the Earth's atmosphere at 300,000 km/hr, it hit the surface from the southeast in a glancing blow. We know this because the spray of debris is elongated westwards from the impact site. The energy released, far greater than all of the nuclear weapons in the world's arsenals, would have been enough to vaporize the asteroid itself almost instantly. The explosion would have scoured a deep hole into the Earth's crust, a mile or more down, and sprayed everything out as a mixture of molten gases and debris. The asteroid hit in a shallow sea, underlain by layers of limestone and calcium sulphate, which would have momentarily been emptied of water by the impact.

For any animal standing within a few thousand kilometers of where the asteroid hit, the first sign of it would have been a blinding white light as the object entered the atmosphere, and then a few moments later hit the ground. The flash released by the impact itself was so intense that it would probably have roasted everything within several hundred kilometers all around. This would have been followed by the shock wave in the atmosphere, which flattened everything within about a thousand kilometers and then circled the world, several times over. Next would have been the glowing, lethal rain of molten droplets of rock, sprayed out from the initial fireball and now falling back to Earth, scalding everything for several thousand kilometers around. After that, a huge tsunami—generated when the asteroid had hit the sea—would have arrived at coasts all around the Atlantic Basin. This wave would have been at least 5 km in height, and capable of traveling hundreds of kilometers inland in many low-lying areas. It must have swept away anything living that was in its path.

All of this though, was probably just the beginning. Life could have recovered from the shockwaves and the fireball, given a few thousand years. But much more far-reaching effects had been set in motion by the impact. One of these would have involved the dust generated by the explosion, and a mist of sulfuric acid derived from sulfates in the rock and sediments that the asteroid had hit. Propelled up high into the stratosphere, it would have fed into upper-level wind systems and dispersed all around the world. The density of this cloud is estimated to have been enough to block out most of the sunlight for several months. Plants around the world—in the sea and on land—would have been unable to photosynthesize, and would have died. Ecosystems depend fundamentally on plants, and without any food source, herbivores would have starved, and so would the carnivores that fed off them.

In addition, the dust cloud would have caused a brief global freeze-up. The dust absorbed sunlight that would normally have reached ground level and warmed the Earth's surface. The dust particles themselves would instead have heated up in the sunlight, but the heat would have been lost quickly to space because there was a much thinner insulating blanket of atmosphere above them. It is rather like the way in which the top of a mountain stays cooler than the lowlands, even though it is receiving a lot of energy direct from the sun—the thinner blanket of greenhouse gases above allows it to cool off. The result would have been that the Earth stayed a lot cooler than it would normally have been. This would have resulted in a global "winter", with subfreezing temperatures everywhere, including the tropics. This winter must in itself have been the end of many species, especially tropical ones which were not at all adapted to cold. The "global winter" itself would have been short-lived—probably just a few years at the most—and though intense this would make it difficult to trace in the environmental record of the time. Nevertheless, there are at least some signs of cooling following the Cretaceous–Tertiary boundary, showing up in the distribution of the surviving types of single-celled foraminifera in the deep oceans (Galeotti *et al.*, 2004). The cooling in the ocean seems to have persisted for thousands of years, much longer than the dust cloud would have taken to settle out of the atmosphere. Galeotti and colleagues suggest that this ocean cooling was probably so long-lasting because cold water from the surface, chilled during the "global winter", had poured down into the depths (since cold water is denser) and filled up the ocean basins. Even after the atmosphere above recovered and a warm layer of water formed floating at the surface, the deeper water below, heavier and insulated from the world above, would have taken a long time to warm up. I am intuitively skeptical though that just a few months or years of cooling—all that the impact scenario seems capable of generating—was enough to make so much cold water. Other possible signs of cooling show up in the characteristics of the floras which survived on land (see below), and based on the frequency of different leaf margin shapes. However, I also have some doubts about the reliability of that particular strand of evidence, having worked on leaf margin–climate relationships myself. Nevertheless, even if we cannot find direct signs of it, the "global winter" remains a plausible scenario.

A few years after the impact, when the dust had finally settled out of the atmosphere, another climate shock may have awaited life on the Earth. When limestone—calcium carbonate—is heated it gives off carbon dioxide (CO_2). Because the impact occurred on a large limestone platform, it would have vaporized a huge amount of carbonate rock to yield CO_2—enough to raise the carbon dioxide level in the atmosphere many times over. This would have increased the greenhouse effect, and as soon as the atmosphere cleared enough to let sunlight through again, the heat-trapping would have begun in earnest. Temperatures globally would have soared, giving further stress to whatever species still survived. There are some clues suggesting that, indeed, just after the Cretaceous–Tertiary boundary the CO_2 level in the atmosphere was abnormally high. Beerling *et al.* (2002) looked at the fossil cuticles of plants across the boundary and used the frequency of stomatal pores (which is known to be sensitive to CO_2) to deduce that CO_2 shot up from around present-day levels—about 350–500 ppm—to a high peak of 2,300 ppm within 10,000 years of the boundary. Not everyone accepts this idea, however. For example, Peter Wilf and his colleagues (Wilf *et al.*, 2003) have argued that Beerling *et al.*'s CO_2 estimates of very high CO_2 after the boundary are unreliable, and that in fact conditions were generally cooler at that time.

What I have set out here is the most generally accepted scenario for what caused the end-Cretaceous mass extinction: a series of environmental shocks piled on top of one another, and all resulting from a single meteorite. The exact form that these shocks took, and how important they each were, is still uncertain. The impact may have been so bad because it landed in a particularly unfortunate spot—a huge mass of limestone with a set of sulfur-rich salty sediments underneath it. However, some have suggested that the single impact in Central America would not have been enough to cause so much disruption to life on Earth, and that perhaps several different impacts at about the same time (scattered around the Earth) caused it. If so, the impact craters have yet to be found and thoroughly dated.

Despite its popularity in the scientific world, the story that a meteorite caused the mass extinction is not universally accepted. In recent years another old and rival theory has been dusted off and brought out: the idea that the end-Cretaceous extinction was set off by massive volcanic activity (see below). One might wonder why anyone would bother to support a rival theory when the asteroid impact idea is so convincing and so complete, but strangely (and almost maddeningly) the volcanism theory also fits observations rather well. As we shall see, volcanism and the meteorite impact might even have been connected.

3.28 DID A METEORITE CAUSE THE END-PERMIAN EXTINCTION?

The scale of the end-Permian mass extinction certainly invites comparisons with the Cretaceous one. The devastation of life was in fact even greater at the end of the Permian. It is not surprising then that many geologists have suggested that this one too was caused by an asteroid impact. However, the evidence that can be used to make the case for an impact is much weaker.

There are several craters that might be candidates for an end-Permian impact event (e.g., the Australian Bedout Crater), but the precision of dating is much weaker than it is with the end-Cretaceous impact. Although one can know that the Bedout Crater, for example, was within a few million years of the mass extinction, it is not possible to tie it down exactly to that time and that is enough to leave it open to doubt (for as we are finding, there are many big craters of various ages scattered around the world). So far, no-one has found that any of the boundary layer sediments at the end of the Permian are rich in iridium, which tends to imply that there was no massive meteorite impact exactly at that time.

A few years ago, the idea of an end-Permian impact got a boost when Becker *et al.* (2001) reported finding some characteristic clues of an asteroid impact in the boundary layer. They reported they had found "buckyballs"—little spheres of carbon atoms that are found in some meteorites. Furthermore, they wrote that gases trapped within these hollow spheres had an isotope composition that was characteristic of meteorites. This brought about the interest of many, for it seemed to be the first good evidence for a major impact right at the time of the end-Permian mass extinction. In science, all important interesting claims should be tested, to see if they really are true. But in this case, testing of the same sediments and quizzing of the authors found that their paper had been misleading: the buckyballs had come from farther down in the rock sequence, below the boundary layer, and so nowhere particularly close in time to the extinction event. And in addition to this, it turned out that the gas atoms that had the characteristic isotopic signature of outer space were not actually trapped in the buckyballs at all, but were dispersed in the sediment. With this, the new and exciting evidence for an impact fell apart.

While it is hard to prove that the end-Permian extinction was not caused by a meteorite impact, the case for it is decidedly weak. So instead, attention has turned more to other alternative explanations for this particular mass extinction (see below).

3.29 OTHER POSSIBLE IMPACT EVENTS AT TIMES OF MASS EXTINCTION

There may have been other sudden changes in the environment associated with the other mass extinction events, although if they did occur we cannot be sure of their causes. I won't go into all them here, but just give one example. Across the Triassic–Jurassic boundary, there is a sudden spike in ^{12}C, although it is smaller than the ^{12}C spikes seen in the Biggest Two. It has been claimed that a strong peak in atmospheric CO_2 levels, to around three times the present-day level or more, is also detectable from the densities of stomatal pores in the leaves of plants across this boundary (McElwain *et al.*, 1999). One possibility is that the ^{12}C and CO_2 spike was connected to the devastation to land ecosystems caused by an impact event, although no crater has been found and other theories (below) could also explain a sudden jump in CO_2 and in carbon isotopes.

3.30 VOLCANIC ERUPTIONS AS A CAUSE OF MASS EXTINCTIONS

Intriguingly, the two biggest mass extinctions (the end-Permian and end-Cretaceous) both seem to have been times of immense volcanic activity.

Most of central India is covered by hundreds of meters thickness of lava rock, forming a large upland area known as the Deccan Plateau. Where it has eroded into valleys, the exposed edges of the lava plateau resemble a huge layer-cake. Each of the individual layers is a lava flow that solidified and was later covered by another lava flow. The name geologists give to these heaped layers is "traps", which comes from a Swedish word meaning a flight of stairs; and on a slope they do indeed look like an immense series of steps. These "Deccan Traps" were all formed in a relatively short span of time, geologically speaking—a few million years. Rather than any normal sort of volcano that we might see in the present-day world, they must have come out of some immense split in the Earth's crust. It just so happens that the time frame of this eruption includes the end of the Cretaceous, 65.5 Myr ago.

Thousands of kilometers to the north in the cold wilderness of conifer forests and tundra, much of east-central Siberia is covered by very similar-looking landscapes with layered lavas hundreds of meters in thickness—up to 3,000 meters thick in some places. When it was originally laid down, the mass of lava rock in Siberia apparently occupied an area several times greater than at present, about the size of Europe, but much of it has been worn away by weathering and erosion. The age of these lavas, the "Siberian Traps", neatly brackets the end-Permian boundary 251 Myr ago. The Siberian Traps seem to have been produced over an interval of at most 600,000 years, straddling the Permian–Triassic boundary (Benton and Twitchett, 2003). Again, lava must have been spewing out over a huge area at an almost unimaginable rate.

It all seems too much of a coincidence—that the two greatest periods of volcanic activity in recorded Earth history have both occurred at times of mass extinctions. If they really did not have anything to do with the extinctions, that is one heck of a false trail for nature to lay down for us. Given what we know of volcanoes in the present-day world, it is not hard to imagine how eruptions on this immense scale could have disrupted the whole ecology of the planet. When volcanoes spew out the sort of runny lavas that formed the Deccan and Siberan Traps, they also tend to produce CO_2 and sulphur dioxide (SO_2) (although I have heard varying opinions on this point—at least some volcanologists doubt that much SO_2 would have been produced by the trap lavas). The traps were extruded over thousands of kilometers, and often relentlessly one layer after another, and this would have meant a lot of CO_2 and SO_2 entering the atmosphere. CO_2 is a greenhouse gas, tending to trap heat and warm up the climate. SO_2 tends to have the opposite effect, forming acidic hazes that will reflect back the Sun's rays. Pouring into the atmosphere continuously for hundreds of thousands of years, these gases may have fought a battle between themselves for the control of the climate system. The SO_2 would have tended to push temperatures down, and the CO_2 tended to push them upwards. Intially during phases of the most intense eruption, the SO_2 haze from the eruptions might have induced global cooling and widespread frosts—but it could only stay at high concentrations in the atmosphere for a few weeks at a time. As the SO_2 washed out of the atmosphere, the accumulated CO_2 from many episodes of volcanism would remain behind, unleashing its greenhouse effect, and temperatures would have skyrocketed. This sort of instability between extremes of climate could have been enough to drive most of the life on Earth towards extinction. Other effects could have gone along with the volcanism. The sulfuric acid from all the SO_2 in the clouds could also have rained down, poisoning rivers, lakes, and oceans, and burning the leaves and roots of plants and the skin, eyes, and mouths of animals on land. The ash from the explosive phases of the eruptions could have filled the sky and obscured the Sun, cutting off photosynthesis for months at a time and causing ecosystems to collapse.

The idea that volcanism was to blame has found a lot of support among geologists who study the end-Permian mass extinction. With this particular mass extinction, there is little evidence of a meteorite impact, but plenty of good evidence for massive volcanic activity at exactly the right time. The evidence for oxygen-poor conditions in the oceans at the same time (see below) might also fit with warming caused by massive eruptions of CO_2 warming the climate.

In the case of the end-Cretaceous extinction, the existence of a big meteorite impact crater on the other side of the world rather complicates the issue. It is hard to ignore that this impact itself had immense destructive power, and yet the Deccan Traps eruptions also look like the sort of thing that could bring about worldwide extinctions. Could it be that one of these events was largely irrelevant, or that both simultaneously played an important role in the extinctions? Either way, it seems an incredible coincidence that two events as cataclysmic as this could have occurred at almost exactly the same time. Another possibility is that in fact the two were not a coincidence at all, and that what set off the Deccan Traps eruptions was itself a meteorite impact. It has been suggested that the shock of an asteroid hitting the Earth on one side would have produced a sort of split in the Earth's crust at the opposite point on the other side. This is rather like the way a ripe tomato dropped on the ground will also tend to split on the far side from where it hits the floor. Intriguingly, central India was just about at the opposite point on the Earth's surface from Central America, 65.5 million years ago. Another possibility is that there was not one, but two, big impacts on the Earth at the same time at the end of the Cretaceous. Perhaps, for example, as a result of some past collision in space, the asteroid had broken into two closely associated halves that were traveling together, and then hit the Earth right after one another. The Deccan Traps might have been produced if an asteroid that hit India punched a hole through the Earth's crust, allowing lava to spill out from the wound. Such ideas have an intuitive appeal to them, but geophysicists have difficulty in setting out a detailed scenario for making a wound through the Earth's crust, or a tomato-like split on the opposite side. It is certainly not the sort of event that they are used to dealing with in the present-day world. For now, the idea that both the Chicxulub Crater and the Deccan Traps are connected is hard to ignore, but it is also difficult to make any rigorous case for.

It is possible that another mass extinction was also down to volcanism. Berner and Beerling (2007) have made a strong case that the end-Triassic extinction was actually due to massive volcanic eruptions that occurred at precisely that time.

3.31 STAGNANT, BURPING OCEANS AS A CAUSE OF MASS EXTINCTIONS

Another set of explanations involves the oceans as a source of certain extinctions. For example, there are signs that both deep and shallow seas at the end of the Permian tended to be devoid of oxygen (Grice *et al.*, 2005). Even close in to shore, in seas shallow enough to be stirred down to the bottom by storm waves, the water was often lacking in oxygen (Wignal and Twitchett, 1996). In many different localities, old sea floor sediments from around that time are rich in the characteristic minerals that are produced by oxygen-free (anoxic, as geologists would say) conditions on the sea bed.

Things like, for example, iron sulfide which eventually forms shiny crystals of fool's gold when the sediments become rocks.

Why the seas became devoid of oxygen is an open question. It has been suggested that the precise arrangement of land and sea that was an underlying factor, hampering the circulation of oxygen-rich water down to the sea bed. How would this have occurred? Well, from what we know of the world around the end of the Permian, nearly all of the land was joined into one vast continent known as Pangea (meaning in Greek, "All Earth") (see Figure 3.7).

On the eastern edge of Pangea was an extensive shallow sea, blocked off from the vast ocean beyond by other smaller landmasses. Effectively this sea was imprisoned by a ring of land, that could have cut it off from the normal flow of ocean currents. This might have been part of the reason why it became stagnant, though it does not explain why the anoxic conditions apparently extended through other parts of the world's oceans besides (though rock sections from other parts of the world are few and far between). The lack of oxygen could have been caused by other factors too, perhaps acting in combination. For example, it could have happened if the Earth's poles warmed, cutting off the supply of sinking oxygenated water that normally

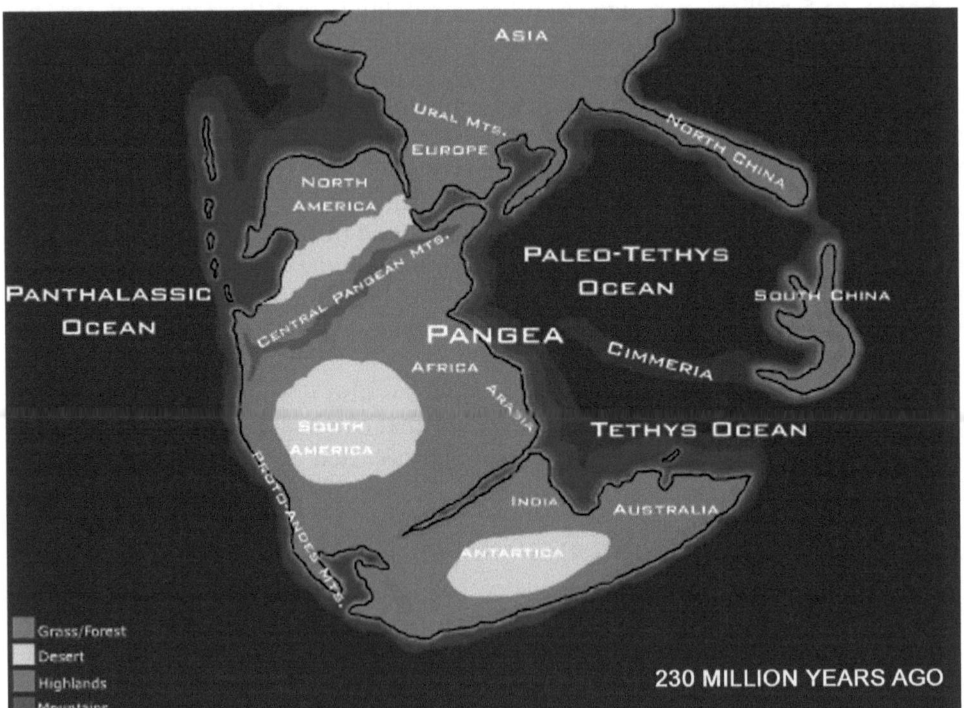

Figure 3.7. A map of the supercontinent of Pangea as it existed close to the time of the end-Permian extinction. Note how the lands seem to form a ring around the Paleo-Tethys Ocean: this might have been what set the scene for disaster, according to some hypotheses. *See also* Color section.

bathes the deep oceans. The anoxia was widespread over a long period of time, several million years across the late Permian and the early Triassic (Grice *et al.*, 2005). However, it seems to have been particularly intense around the time of the mass extinction, and this intensification might have been what finally pushed many species over the edge. Another source of the anoxia could have been the release of stored methane (derived from the ice-like substance methane hydrate) from the sea floor, as Ellen Thomas has advocated. Oxidizing in the waters, the methane could have taken up all the oxygen leaving an almost lifeless ocean behind.

One possible reason why the lack of oxygen in seawater suddenly gets worse around the end-Permian mass extinction is a sudden rise in temperature at that time. From the composition of oxygen isotopes in the Permian sea sediments, it looks like global temperatures shot up by several degrees Celsius right at the time of the boundary layer. Warmer water cannot contain as much dissolved oxygen, and this would have added to the suffocation. The warming could also have prevented the gentle "overturning" that brings cool oxygenated water sinking into the deep ocean from above. Instead, in a warmer world the surface waters just floated as a stable layer on the top, cutting the deeper ocean below off from this source of ventilation. However, oceanographers tend to doubt whether the oxygen isotope signal in the rocks is really a reliable indicator of warmth, and they also doubt that the warm water layer on top could ever have stayed stable enough to cut off the supply of oxygen to the deep.

It is possible then, that the late Permian world was primed for disaster by having oceans already rather short of oxygen. A sudden spike in temperatures (e.g.,caused by massive volcanic eruptions in Siberia pouring out greenhouse gases) could have been the final straw that led to the oceans suffocating. After they had died, the seas may then have taken the rest of the world with them (see below).

So these theories suggest that without the currents to stir things up and bring in newly oxygenated water, organic waste from land and ocean ecosystems would have accumulated on the sea bed, using up the meager supply of oxygen as bacteria attempted to break the material down. The oceans would have become almost entirely devoid of dissolved oxygen, even up close to the surface—leaving only a thin layer where oxygen-breathing organisms could survive. The Black Sea in the present-day world is like this, though on a smaller scale: it is landlocked, without deeper ocean currents to deliver oxygen to it. Its bottom waters remain still and airless without any animal life, and the black mud that gives the Black Sea its name covers the sea floor. Under conditions without oxygen, bacteria working away in the sediments produce a range of toxic by-products. One of these is hydrogen sulfide (H_2S), a very poisonous gas which smells like rotting eggs. It has been suggested that if a sea floor in the Permian became oxygen-free on a large enough scale, it could have bubbled out huge amounts of this rotten egg gas, poisoning seawater and the atmosphere above. If enough of the gas was being produced, it might have wiped out life in both the oceans and on the land.

Is there any evidence for the broader scenario, of poisonous gases leaving the oceans? One interesting clue comes from a change in the balance between two different isotopes of sulfur, which can be detected in sediments from areas that lay outside the main shallow sea. Just as with the isotopes of carbon, there are two main sulfur isotopes—^{32}S and ^{34}S—which behave slightly differently. For example, when bacteria break down something containing sulfur, molecules containing some ^{34}S are just a little more sluggish about getting involved in chemical reactions than those which contain ^{32}S. The result is that all living tissues, and even more so the gases like H_2S that come off from decomposition, are enriched in ^{32}S. Sediments from around the end of the Permian show that suddenly the sulfur within them became a lot lighter, richer in ^{32}S at the expense of ^{34}S (Hotinski *et al.*, 2001). It has been suggested that this is a sign of great quantities of H_2S bubbling out of the oceans. This could have been the agent that poisoned life around the world.

The idea adds up to a pretty convincing disaster scenario, but it depends on a lot of assumptions. For one thing, it is hard to explain where all of this sulfur that made the H_2S actually came from all at once (Hotinski *et al.*, 2001). Also, soil minerals eroding off the land surface after the end-Permian extinction do not show any signs of having been pickled in the acids that would have formed as the abundant H_2S broke down.

A variant on this idea is that some unknown mechanism within the Earth system caused the oxygen level of the atmosphere to plunge—and the anoxia seen in the sea could just have been something that went along with this global oxygen shortage. This plunge in oxygen would have killed animals on land just as in the sea, and might be the unifying factor that explains why extinctions occurred both on land and in the sea, and all around the world (Huey and Ward, 2005). The burst in abundance of the brachiopod *Lingula* following the mass extinction might have been due to its ability to tolerate low-oxygen conditions continuing into the Triassic. Another common seashell at that time was the bivalve mollusk *Claria* (Rodland and Bottjer, 2001) whose thin shell would have allowed it to take in whatever oxygen was available (there are similar-looking seashells that live in local areas of oxygen-poor waters off tropical shores nowadays, and their shells are so thin that they are transparent; Figure 3.8). It has been suggested that it was mostly the more metabolically active land vertebrates (such as the mammal-like reptiles) that perished at the end of the Permian, because their oxygen requirements were too high.

Oxygen-deficient oceans might have played a role in another, smaller mass extinction too. The extinction at the end of the Triassic seems to be associated with a widespread anoxia event, which started in the seas that then covered Europe and spread around the world (Little and Benton, 1995).

3.32 THE END-PALEOCENE EXTINCTION IN THE DEEP SEA

Another smaller, but widespread, extinction event (not one of the Big Five) has also been blamed on processes within the oceans. I shall talk about it here because it is part of the shaping of diversity in our modern world, and fits into the context of later themes. About 55 million years ago, at the end of the Paleocene period, there was a diverse array of foraminifera floating in the ocean depths. This group had come through end-Cretaceous extinction almost unscathed, but now it was their turn to

Figure 3.8. A modern-day "glass" bivalve from Malaysia. Its thin shell is an adaptation to allow the animal to breath in oxygen-poor waters (photo: Author, Malaysia).

get whacked, for suddenly around half these deepwater species vanished. The extinction happened within the space of a few thousand years at most, faithfully recorded by the excellent fossil record of the deep ocean sediments (Shackleton andThomas, 1996). It was not a mass extinction on the scale of some of the others, for it did not wipe out life on land and in the shallow seas, but it was a striking across-the-board extinction event in one particular environment. What could have caused this to occur, only in the deep oceans? The key seems to have been a sudden and extreme warming that shows up all around the world at this time (Huber, 2008). For perhaps 170,000 years, most of the world was tropical. Tropical rainforest extended up to mid-latitude locations which would nowadays have heavy frost and snow in winter, while parts of the tropics themselves might have been roasting hot—perhaps too hot for most forms of life (Huber, 2008).

Clues to the cause of the end-Paleocene deep-sea extinction also come from isotopes: in this case from carbon isotopes. Around the world, in rocks and fossils laid down at this time, there is a dramatic spike in the amount of ^{12}C relative to ^{13}C. The spike is so large that carbon coming out of soils, forests, or volcanoes could not have caused it—there is simply not enough carbon stored in those reservoirs. It must have been due to release of something exceptionally rich in ^{12}C, and most probably that was methane gas. Only methane can pack a concentrated dose of ^{12}C sufficient to cause such a change in the carbon isotope content of sediments. It must have been released from some form of storage, and in huge quantities. The only plausible reservoir of methane is methane hydrate, an ice-like substance that is embedded in the deep ocean sediments around the world. Methane hydrate is notoriously unstable: with a slight warming or release in pressure, it quickly decomposes to water, plus a huge volume of methane gas. It seems that something must have set off a wholesale destabilization of methane, causing it to bubble up into the atmosphere. There it would rapidly have oxidized to carbon dioxide, a greenhouse gas which trapped heat and sent the global temperature soaring. The more this warmed the atmosphere, the warmer the oceans got, and the more methane destabilized. It was what is known as a positive feedback loop: once given an initial push, the whole thing runs away, amplifying itself. Temperatures in the deep ocean, which had been around 10°C before, suddenly reached a milder 15°C. Oxygen does not dissolve as well in warmer waters, and the foraminifera in the deep sea suddenly found themselves without enough oxygen to live; hence, the mass extinction that occurred in the deep oceans at this time. Given what we know of the present-day deep oceans, there may be a huge number of species living on and in the deep-sea floor sediment, thousands of feet down. Presumably a diverse sea floor fauna of the time suffered a similar mass extinction to that which we see in the foraminifera, but because it does not fossilize we can see no trace of what was lost.

In contrast to extinction in the deep ocean, this was also a time of diversification in other environments. Coinciding with this big temperature spike, the first hoofed mammals and primates appeared on land, along with some other groups (Gingerich, 2006): it is essentially the time when much of the "modern" mammal fauna of the world began. Gingerich suggests that rapid evolution occurred actually within this brief warm phase, although the reasons remain unclear.

3.33 MASS EXTINCTIONS AND ICE AGES

Several times in the history of the Earth, global temperatures have dipped and ice has spread out across the planet's surface. We ourselves live in one of these times of ice: we are presently in a brief respite between glaciations that most of the time cover Canada and northern Europe with ice, the most recent one ending just 11,500 years ago (more about that one later in this chapter, though).

There was also that near-global freeze-up of the late Precambrian around 900 Myr to 700 Myr ago, ending before the first signs of animal life in the fossil record. In a sense this big freeze may have prevented and delayed diversification rather than actually destroying it. It seems that the warming and oxygenation of the oceans allowed by the meltback might have been the trigger for the first burst of diversification of animals.

Other ice ages have also occurred in the intervening time, and at least one of them has been blamed for a mass extinction event. This is a glaciation that occurred at the end of the Ordovician period, about 450 Myr ago. From what we know, it was a big one, icing up a large proportion of the Earth's land surface and sending sea ice well down into the mid-latitudes. As is typical of ice ages, it went through cycles of advance and retreat. It has been suggested that the initial shock of the glaciation starting, and the instability of repeated warming and cooling cycles, drove many life forms extinct. Almost all of life at that time was still confined to the sea: the colonization of the continents by animals and advanced plants had barely begun. Life in the shallow seas not only had to contend with dramatic changes in temperature, but rapid falls in sea level as the ice sheets growing on land stored up water that would otherwise have been in the oceans. In this case, the environmental instability and loss of habitats of repeated cooling and warming periods may have been what pushed so many species into extinction.

3.34 IS THERE A CYCLE OF MASS EXTINCTIONS?

Raup and Sepkoski (1984) have suggested that the timing and frequency of mass extinctions is not random: that there is a pacing to them. Using statistical analysis of the mass of data that shows when each and every species in the marine fossil record seems to have gone extinct, he found mass extinctions occurring on a cycle of 26 million years. Some of these events were far larger than others, but just about every 26 million years there was a peak in extinction rates around the world. Such a regular pacing seemed to require the clockwork-like movement of celestial objects—perhaps some sort of burned-out star stuck in a binary orbit with the Sun, that disturbs the path of comets and rains them down on the Earth each time it passes. Raup and Sepkoski dubbed this hypothetical object "Nemesis", after the ancient Greek goddess of retribution.

The idea of periodicity in mass extinctions, and an orbiting nemesis, is intuitively hard for geologists to accept. For one thing, why propose an object that we have never seen and have no knowledge of? Also if one looks on enough different time scales, can't we expect a periodicity to assert itself purely by chance? Raup and Sepkoski point out that even allowing for the fact they looked for periodicity on many different time scales, the chances of a cycle showing up so regularly over such a long period are enormous, so the periodicity is too much of a coincidence to be ignored. However, many other geologists dispute that the dating Raup and Sepkoski were using was correct enough to be able to claim a 26 Myr cycle. For example, in a recent reanalysis of the whole marine fossil record Alroy *et al.* (2008) change the dates of many mass extinction peaks, and that shifts many of them out of the regular cycle that Raup and Sepkoski claimed to have found.

There might be another mechanism within the diversity of life itself that makes mass extinctions more likely after something like 26 million years (Stanley, 1986), without needing to invoke mysterious objects floating in space. Let's say, for the sake of argument, that the life forms which survive a mass extinction tend to be fairly non-specialized—tolerant of a widely varying climate and other environmental conditions, and consequently occurring over large ranges. It makes sense, because lack of specialization and wide tolerance would be exactly the traits to allow certain species to survive the environmental upheaval when more narrowly adapted forms perished. Over millions of years, these generalist survivors would often tend to give rise by speciation to more specialized forms, suited to fit precisely into the intricate network of nature. Eventually, the niches of many species would end up so narrow, and their web of interconnections with certain other species so strong, that the whole thing would be primed to collapse if it took a hit. This hit could be the sort of environmental shock that occurs every few million years from varying causes, but it will only trigger a mass die-off if it hits long enough after the previous extinction, at a time when many species have now become precisely and narrowly adapted.

Stanley (1986) cites some examples which help to make the point that is the specialists that get hammered in mass extinctions. For example, with a sudden cooling of the Earth's temperature around 3 Myr ago (as the Earth entered its present phase of ice ages), many of the bivalves of North America's east coast went extinct. When the bivalve fauna of the east coast eventually recovered its diversity, the new species that had appeared were much more tolerant of a range of temperature conditions. Whereas before, each bivalve species had occupied a narrow range of latitude along the coast (constrained by a narrow range of water temperatures that it was adapted to), afterwards the species tended to be much more wide-ranging in terms of latitude and temperature—perhaps because they were descended from the few very generalized species that survived the mass extinction.

It is possible then that life tends to set itself up for each mass extinction by becoming over-specialized. The time taken for life to recover its numbers, diversify, and specialize again could be the pacemaker of mass extinctions that underlies the mysterious 26 Myr cycle.

3.35 DIVERSIFICATION AND RECOVERY

After each big mass extinction event, both land and sea were left depleted in life. Yet in time, the world's diversity always bounced back. Even though the end-Permian wiped most things out, by the late Cretaceous there seem to have been at least as many species in the world as there had been before. Despite the devastation of the end-Cretaceous, there are probably more species in the world now than there ever were in the Mesozoic—apparently continuing a broad upward trend in biological richness since the Cambrian.

Following a mass extinction, life seems to compensate, filling up the gaps in nature. It does not seem to be just a matter of chance that life comes back: it really looks like there is some driving force that pushes the survivors to diversify into more and more forms until the world's ecosystems are in some sense "full" again. We see this in rates of diversification of new genera and families which follow on from mass extinctions. After an initial lull of several million years after a mass extinction (when life still seems to be reeling from the shock), new forms start to appear (Erwin, 1998). Once it starts, the build-up of new life accelerates, with the appearance of new forms reaching its peak around 10 million years after the extinction event. Then, things gently slow down to a more normal pace, as diversity once again starts to approach what it had been before.

So, for example, after the end-Cretaceous extinction, the recovery of diversity took a few million years to really get started, but once it began it was rapid. Many fundamentally new groups—orders and families—of both animals and plants sprang up during the interval between around 60 Myr and 50 Myr ago, apparently filling up niches that had been vacated during the mass extinction events. In the first several million years after the extinction the mammals that appeared were very generalized: often omnivorous, and hard to place into any present-day orders. Then they begin to take on distinctive and recognizable forms and roles. Among the mammals, for example, we have the first appearance of the whales, the hoofed mammals, the carnivores, rodents, primates, and bats—plus several other orders of mammals that are now extinct, such as the creodonts and condylarths. Fundamentally, the modern fauna of large land animals was shaped after the end-Cretaceous mass extinction: 17 out of the 18 orders of mammals that we still have in this world have not been found before the Paleogene (the period that began with the end-Cretaceous extinction), and presumably only evolved into their distinctive form after the mass extinction (Fastovsky and Sheehan, 2007). Many new families of flowering plants also appear in this early phase (e.g., the oak family, the grasses, the fig family, and many others).

Sometimes, the evolution of new life forms does not wait for several million years to get started: following the end-Triassic mass extinction, the diversification began almost immediately (Erwin, 1998).

So, what drives the eventual rapid recovery from a mass extinction seems to be the abundant opportunity that comes from empty ecological niches. Ecologists tend to think that in many ecosystems there are only so many different "jobs" ("niches") that organisms can have, and if a species makes its living in a way that is too similar to how something else does it, it will tend to be pushed out. In the case of plants and the various bugs that chew on plants, ecologists are not sure that closely similar things always do get pushed out (Chapters 1, 2). But with animals larger than insects the general feeling is that yes, they probably do mostly need distinct niches. Under normal circumstances, a particular region or environment will be more or less full and it is difficult to shoe-horn in any more species: but after a mass extinction the world is full of vacancies. It may take time for things to get started, but eventually different populations of the surviving species begin to exploit the opportunity, evolving in different directions and filling up these empty niches. In time, nature becomes full again, and diversity stops building up. Instead it is replaced by more gentle rates of speciation and "background extinction", turning over species one by one.

3.36 "DEAD CLADES WALKING"

Merely surviving the boundary event of a sudden mass extinction was no guarantee of success in the aftermath. Very often, a genus seems to have staggered on for a while and then vanished sometime between a few hundred thousand or a few million years later. For example, after the Big Five mass extinctions, some 10% to 20% of the remaining marine genera were lost soon afterwards—a rate that is significantly greater than the rate of "background" extinction in times before the mass extinction events (Jablonski, 2002). Jablonski has dubbed these "dead clades walking" (a clade is an evolutionary "branch" of life)—genera that were already as good as dead at the time of the mass extinction. Several different causes have been suggested for the continued extinction of these surviving groups. One is that the physical environment might still have been too damaged—too extreme or too unstable—to support life forms easily. Food webs and other species interactions would have been unstable too, with the lack of diversity perhaps increasing the probability of extinctions. It could also simply have been that each genus which survived tended to have only one species remaining in it, which made it more likely that before long it would by chance go extinct, taking with it the name of its genus.

3.37 THE ROLE OF LUCK IN THE HISTORY OF LIFE

A lesson that comes from studying mass extinctions is how important luck is in determining the fate of evolutionary lineages. Most people tend to imagine that it was a matter of destiny that mammals took over from the dinosaurs and other extinct reptiles. After all, isn't it the case that the mammals simply proved to be superior? In fact, the dinosaurs were doing better than ever just before the end-Cretaceous extinction, and the mammals were showing no sign of going anywhere. If the mass extinction had not wiped the dinosaurs out, they would probably still be going strong, and we mammals would still be scurrying under their feet. The dinosaurs themselves had inherited the Earth by sheer luck, after the Permian mass extinction swept away the mammal-like reptiles that had dominated life on land. Some of the life forms that were lost in mass extinctions were probably competitively superior to their nearest equivalents in the present-day world. For example, those peculiar rudist bivalves were gradually pushing corals out of tropical reefs in the late Cretaceous, just before they got caught up in the mass extinction (Steuber *et al.*, 2002). If they had not been wiped out in the extinction event itself, rudists would probably still be dominating the reefs—and by now corals would be nowhere to be seen. As it was, the corals were saved in the nick of time. Some of them were lucky enough to survive the mass extinction, and came back to dominate reefs in the present-day world, and delight us with their bright and varied colors. It is hard to imagine that a bunch of rather distorted-looking clam shells would have quite the same appeal.

Although sheer luck plays its part in determining what survives and what does not, there may also be a sort of internal logic to the strange looking-glass world of a mass extinction. Certain characteristics that seem to have no great significance in normal circumstances—they are just details of adaptations to niches—suddenly take on a crucial importance in determining death or survival of a lineage. For example, in the end-Permian extinction, the ability to survive oxygen-poor conditions may have been the key, in a world in which oxygen levels in the sea (and maybe the atmosphere too) may have plunged. In the end-Triassic mass extinction, clamshells that lived by burrowing within the sea floor sediment survived a lot better than those that sat out in the open on the sea bed. Whatever it was that killed off so much life at the end of the Triassic—and we are still not sure what it was—it seemed to hit the species left out in the open much harder.

Another example of selection by mass extinction may have determined why ammonites perished in the mass extinction at the end of the Cretaceous, while their relatives the nautiloids survived. Both had coiled, chambered shells, eight tentacles, and sharp eyes, and lived by hunting in the ocean waters. But another thing about the nautiloids is that their young larvae float deep down in the ocean, where the adults also live. Ammonites, we know, lived near the sunlit surface waters and it would make sense for their larvae to do the same under normal conditions. Yet in the world of a mass extinction event, this would have exposed the ammonite larvae to the extreme conditions of temperature, oxygen starvation, or poisonous gases. Nautiloids, snug deep down in the ocean, may have fared much better than their cousins because they were away from the worst of it. Likewise, the foraminifera which floated near the surface of the ocean mostly went extinct at the end of the Cretaceous, while those species that lived down near the ocean bed mostly survived.

It may be a similar story for the animals that lived on land at the end of the Cretaceous. In this mass extinction, anything reasonably large—more than about the size of a four-year-old child—perished. The smaller things that survived might have been able to sit out the catastrophe in burrows (for only small animals are able to build burrows that don't cave in). Animals that could live off carrion, such as small crocodiles, were also able to make it through—whereas carnivores that required fresh meat mostly perished. Living in shallow freshwater also helped: crocodiles and terrapins survived the extinction event very well, perhaps because of the shelter that their ponds offered them. Also, the sorts of animals which could have eaten the grubs that live on rotting wood seem to have survived, in a world that consisted mostly of decay and not new growth. These grub-eating little animals probably include our own ancestors. If they had already been specialized into eating something more dignified, we would probably not be around today.

Erwin (1998) has pointed out that of all the species that have ever existed, probably fewer than 5% were lost in mass extinctions. The rest were lost one by one through ordinary background extinction. Yet, mass extinctions broke dominant patterns in the abundance and diversity of life which had prevailed for tens of millions of years beforehand (Bambach *et al.*, 2002). By opening up broad opportunities, mass extinctions have had a disproportionate effect on the evolution of life, and the form of life's diversity on the Earth today.

3.38 BEYOND THE MASS EXTINCTIONS: THE STORY OF TROPICAL RAINFOREST DIVERSITY

I will move on now to some aspects of the history of life's diversity beyond the consuming topic of mass extinctions, in a story that takes place mainly since the end of the Cretaceous. This is the history of the tropical rainforests—and the questions of how far back their diversity goes, and how it has changed over millions of years.

Until about 10 or 15 years ago, the story of tropical rainforest diversity was almost a complete mystery, a matter of speculation rather than any real evidence. For one thing, tropical rainforests do not often leave any detailed fossil record—unlike the cooler climate ecosystems which do, for example. Dead leaves, insects, and mammals that fall to the forest floor only rarely get preserved in the lowland tropics: they decay very fast in the heat and humidity. Not much pollen comes from the rainforest to end up in sediments, either. Most of the trees rely on insects to carry their pollen around, and they only need to produce small amounts (far less than the wind-pollinated species of the temperate zone). Pollen grains from insect-pollinated plants are also fragile, and they break and decay easily before they reach sediments that can preserve them.

However, just in the last few years, a lot of information on the plant diversity story has come from the study of some key fossil localities, where there is a slightly better record of tropical diversity in the form of preserved pollen. These are places close to the mouths of tropical rivers, where sediments were first laid down as offshore muds, containing pollen that had floated gently downstream from rainforests inland. The drill cores that have yielded this information were first taken by oil companies, and sat for many years unanalyzed from the point of view of diversity, until they were eventually released for study. The story from fossils remains largely biased towards plants, since hardly anything else fossilizes from the tropical rainforest environment. However, since the structure of the rainforest ecosystem is largely made up of plants, knowing the story of plant diversity might well tell us a lot about the diversity of the ecosystem in general.

Another very different source of information has come forward in the last 20 years or so. This involves comparing the DNA sequences of living lineages of organisms. The DNA story has added to what is known about plants, and provided some much-needed clues to the story in the case of animals. Overall, there are still many gaps and mysteries, but at least there seems enough now to weave into some sort of general picture.

Both the fossil and DNA evidence suggests that tropical rainforests as we think of them today—dominated by broadleaved angiosperm trees—first began in the Cretaceous period, before the big global extinction event 65 Myr ago (Morley, 2005). The first flowering plants seem to have lived in seasonally dry environments. The earliest fossil evidence of them actually making up a rainforest is from the late Cretaceous of West Africa. Back then, the rainforests also seem to have contained a lot of conifers, and other groups of gymnosperms, and it was not until later that the flowering plants managed to push them out almost completely from the tropical lowlands.

By the late Cretaceous, large seeds that are characteristic of the big trees that make up rainforest canopies begin to appear as fossils in Africa. A few million years after the end-Cretaceous mass extinction, similar-looking big seeds also appear in North America, showing that tropical rainforest was growing there too at that time. From DNA evidence of presently living plants, the late Cretaceous seems to have been roughly the time when some of the important broad groups of rainforest angiosperms started to diversify (e.g., the order Malpighiales which includes several rainforest tree families).

The big end-Cretaceous extinction might not have had very much effect on the plant diversity of rainforests. For instance, in Colorado, U.S.A. a diverse range of leaves from a species-rich tropical rainforest turns up only 1.4 million years after the boundary (Wilf *et al.*, 2006). These early days of the "Era of the Mammals" were already quite a lot warmer than the present-day world, allowing rainforests to become extensive. After a few million years, the Earth became much warmer still, reaching a peak at the end of the Paleocene 55 Myr ago. For a geologically brief time, perhaps 170,000 years, tropical rainforest reached as far north as Alaska, at 60°N at that time, and 57°S in Tasmania (Morley, 2005). It looks as if most of the world may have been covered in rainforest at this time, although it is possible that the equatorial tropics was actually too hot for it, and largely devoid of plant life (Huber, 2008). Because of the exceptional warmth of this time, groups of rainforest plants that had evolved in separate regions were able to spread across between far northern and far southern continents—adding to the diversity in each region. One example is the diverse tropical plant family Bombacaceae, which apparently managed to spread from North America to Europe across the high-latitude land links which existed at that time.

The signs from the fossil pollen record are that the diversity of the rainforests of both South America and Southeast Asia increased during the first 15 million years after the end-Cretaceous extinction. In the Americas it reaches a peak some time after the brief extreme warmth 55 Myr ago, during the globally warm phase 50 Myr to 40 Myr ago (Morley, 2005). In the South American forests it seems that the tree species richness reached during this interval was actually greater than it is at present.

From global warmth of 40 million years ago, there began a downwards trend in temperature that pushed the tropical rainforests back towards the equator. There was also a switch towards drier climates in both the temperate zone and the tropics, which gave the daisy family and other groups of weedy herbaceous plants their big break, as the forest shrank back into smaller remaining areas of moist climate. In the case of the tropical Americas, the diversity of the rainforest began a long halting decline, and the tropical forests of the region seem to be less diverse now than they were way back during the Eocene. It seems that the rainforests of Africa are also now a lot less diverse than they were back in the Paleogene (Morley, 2005). The fossil pollen record from West Africa shows that the region lost a lot of its diversity in several steps during the last 40 million years as the climate became cooler and drier. There was an especially big decrease in richness around 11.7 Myr ago in the late Miocene, and others around 7 Myr and 2.8 Myr ago.

At times, though, the diversity of rainforest plants received a sudden boost when continents collided (Morley, 2005). Different groups of rainforest species had evolved in isolation on separate continents, and when these continents met, these floras merged to give greater richness overall. In Southeast Asia, there was one increase in diversity when the northwards-moving landmass of India hit the southern flank of

Southeast Asia and added its own rainforest flora to the region. Another step-up in diversity occurred when there was a second collision, from the northwards-moving Australasian plate (the one that carries Australia and New Guinea). For example, with the arrival of Australasian groups of rainforest plants, the diversity on Borneo in Southeast Asia increased rapidly in the mid-Miocene. It also seems to have continued to increase since then, in fact, as more and more groups of plants have filtered across the narrow seas separating the continents. In the American tropics, there are signs of a slight boost to the diversity when the collision of North and South America blended in two different rainforest floras around 20 Myr to 25 Myr ago (Morley, 2005; Jaramillo *et al.*, 2006). Another thing that may have added to the species richness of the South and Central American rainforests is the rise of the Andes mountains, starting at around that same time. The varied, complex topography created by the mountains that literally rose up from underneath the rainforest probably fragmented their ranges, allowing them to diverge into separate species in isolation.

The overall picture from the plant fossil record is that the diversity of the world's tropical rainforests did not show any simple, uniform trend over the past few tens of millions of years. At least, it was not as monotonous as the decline that has happened in northern temperate forest diversity. At times it increased, especially when the world warmed and became moister. Other times it decreased, especially as the climate cooled and dried. In Africa, for instance, much of the rainforest diversity was lost as the climate dried out. Collisions of continents and merging of floras has also added to diversity. Broadly speaking, what seems to favor increasing diversity in a tropical rainforest region over millions of years is stability of warm and moist conditions, plus the occasional continental collision (Morley, 2005). However, because things have never remained completely stable for more than a few million years, we do not know whether diversity (given the chance) would simply have reached a plateau or carried on increasing indefinitely.

3.39 THE QUATERNARY ICE AGES

Two-and-a-half million years ago, the growing Greenland ice cap first spilled over the edge of the land, sending icebergs calving into the Atlantic. To geologists this point marks the beginning of the Quaternary—the time of ice that has shaped our modern world in so many ways.

The Quaternary has been a new phase of upheaval in Earth history. There have been broad, long cycles in global temperature: roughly every 46,000 and 100,000 years ice sheets several kilometers in thickness have grown to cover Canada and northern Europe, and then melted back.

Superimposed upon these long cycles has been a whole spectrum of different-sized climate jumps. Sometimes the climate has remained more or less the same for 10,000 years or more—but this sort of stability has been the exception, not the rule. Usually within a few thousand years there was a sudden, big switch in global climate. There is a growing consensus among Quaternary geologists that these various climate changes tend to be completed not over millennia or centuries (as used to be thought), but over just a few decades. This is the world that we still inhabit: we just happen to be living in a lull between major climate shocks, and so far we have been lucky that it has lasted for the past 11,500 years. Though we might delay it a while with greenhouse gases, most likely the ice will eventually grind its way south again over the land where most of the northern cities of Europe and North America now stand.

The changes in climate during the Quaternary were most dramatic in the mid-latitudes and high latitudes. Not everywhere was blanketed by the thick ice sheets, and the areas that remained uncovered had a much colder and generally drier climate than at present (Adams, 1997). For instance, areas of central Europe that are now naturally covered by broadleaved deciduous forest were at that time more like the Arctic in terms of temperatures, and mostly too dry for trees. In the place of forest was a rather strange open vegetation that in some ways resembled the dry margins of the Asian deserts, combined with species of tundra plants and even a few seashore plants thrown in as well. This vegetation type is called "steppe–tundra" to reflect its mixed composition. The steppe–tundra also extended across most of Siberia where there is now conifer forest, and into northern China and Alaska. To the south were other arid grasslands, and greatly expanded areas of desert in North Africa, the Middle East, and Central Asia. The broad glacial cycles were punctuated by frequent large climate shocks as temperatures plummeted, and then rose again thousands of years later. During the Quaternary, climates in the mid-latitudes have switched between temperate and Arctic, not just once or twice, but hundreds of times.

The cold phases did not only affect the higher latitudes. They were global events, cooling off the tropics by as much as 6°C. In most places in the mid-latitudes and tropics, these times were also significantly drier. At the same time during these cold phases, global sea level dropped as much as 130 m because of all the water stored in northern ice sheets—exposing large areas of shallow sea floor, and connecting lands to one another in many parts of the world.

Generally, during the low points of ice ages (known as glacial maxima), the tropical lowlands were about 5°C or 6°C cooler than they are today (Adams and Faure, 1997, 1998). That would be rather like ascending a tropical mountain by about a thousand meters. It is interesting to imagine how different the air would have felt in the lowland rainforests during ice ages. The steamy heat would have been moderated to some extent, and walking around in the lowland forest would have been a more pleasant experience, rather like it is on the lower slopes of tropical mountains nowadays. With the cooling of the glacials, we find various species of trees that normally grow up on mountain slopes coming down to grow in the lowlands, alongside the usual rainforest species. For example, pollen records show that certain tropical conifers such as *Podocarpus* became common during ice ages in the lowlands of Africa, Amazonia, and Southeast Asia. In the present-day world these *Podocarpus* trees are rare in the lowlands, generally being confined to the mid-altitudes of mountains. With the cooling, many of the more typical lowland rainforest species (which nowadays occur no more than a little way up into the foothills of the mountains) must have been squeezed down to the very lowest altitudes. If things had become any cooler, most of these lowland species would probably have been driven extinct. Aiding their survival, though, was the fact that sea level was much lower at that time—about 120 m lower due to all the water locked up in northern ice sheets. Extra bands of exposed land appeared along the tropical coastlines, and these were probably often forested. Those lowermost altitudes, down below present-day sea level, would have been about half a degree warmer than the other forests further inland, perhaps enough to make the difference between survival and extinction in many cases.

Another important aspect of the ice ages is that they tended to be quite dry in rainforest regions (Adams and Faure, 1997, 1998; Anhuf *et al.*, 2006). Pollen preserved in lake and offshore muds, plus the soils and minerals left over from ice age times, make it clear that in many areas the rainforest had shrunk back and that in its place there was grassland or dry scrub. In Africa, it seems that the majority of the present-day rainforest was gone (Adams and Faure, 1997, 1998; Anhuf *et al.*, 2006). In Southeast Asia and Amazonia, the general consensus is that the rainforest underwent a significant retreat (Adams and Faure 1997, 1998; Bush *et al.*, 2005; Kershaw *et al.*, 2005; Anhuf *et al.*, 2006) (Figure 3.9). Though opinions differ between one ecologist and another according to exactly how they interpret the evidence, my view and the view of many others in this field is that the world's tropical rainforest lost at least half the area that it normally had during the warmer, moister interglacials. This cycle of retreat and expansion of rainforest must have happened many times. There

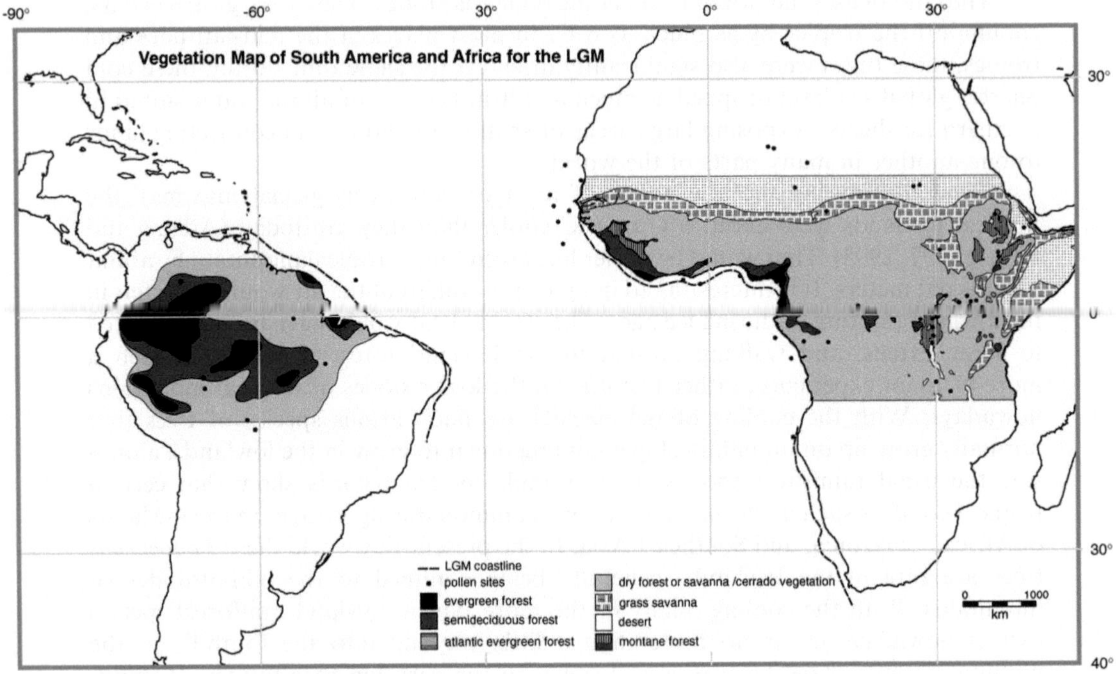

Figure 3.9. Reconstructed areas of rainforest in Amazonia and Africa during the Last Glacial Maximum 20,000 years ago (from Anhuf *et al.*, 2006, Fig. 3, p. 521).

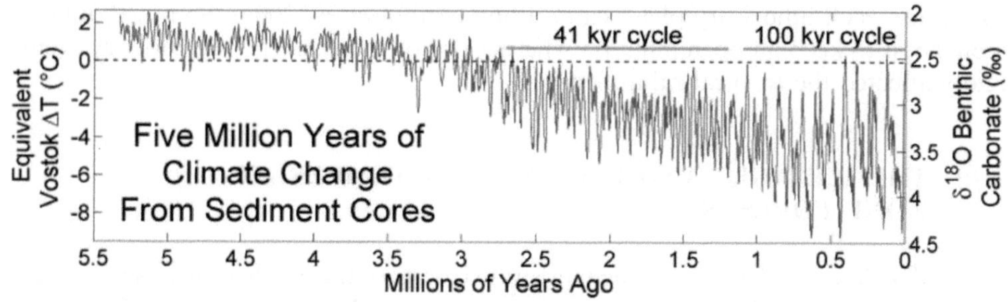

Figure 3.10. The unstable temperature history of the last two and a half million years, based on ocean sediment cores and ice cores. Upwards = warmer. Source: Wikipedia Commons.

were at last seven major ice ages within the last 700,000 years, and scores of slightly smaller fluctuations in climate going back over the last two and a half to three million years (Figure 3.10). We can imagine how a speeded-up movie of the Amazon Basin would have looked over the last two million years: the forest moving out and then retreating like waves on a beach.

As well as the big, broad glacial cycles, it seems that the tropical forests were also sometimes hit by shorter dry phases that started and ended suddenly. Some 12,500 years ago, a cold phase known as the Younger Dryas suddenly jolted the planet half way back towards an ice age (Adams, 1997), and isotopic tracers from the Amazon River indicate that there was at least 40% less water flowing downstream at that time. (Betancourt, 2000; Maslin and Burns, 2000). This reduction in river flow suggests that much of the Amazon rainforest would have experienced a severe drought for about 1,500 years—although we have nothing in the pollen evidence to indicate any change in the vegetation. This inconsistency in the evidence seems rather puzzling, although it might be because the vegetation that lived along the riverbanks and rained pollen into the water did not suffer so much from the drought.

A few thousand years later, about 6,000 years ago, frequent and lengthy droughts seem to have hit the eastern parts of the Amazon rainforest and allowed it to catch fire and burn frequently.

Such fluctuations in climate might not have been just isolated incidents. It seems rather unlikely that only these last two affected the region, and not others extending back in time. From what we know of the instability of global climates during the most recent ice age of the last 100,000 years, similar sudden cool phases of dry climate hit other parts of the world frequently, at least every 10,000 to 15,000 years. Since the beginning of the Quaternary, this all adds up to hundreds of dry phases knocking back the tropical forest, interspersed with moister phases. There are actually very few cores from Amazonia that go back as far the Last Glacial Maximum, just 20,000 years ago. For one thing, scouring away of the sediments offshore during the very low ice age sea level has eliminated any longer term record of the water coming down the Amazon River. In the few core records from inland that we do have, there are no signs of these earlier brief events affecting the region with severe drought. Perhaps the events were too brief to show up, or maybe there is a gap in the sediment record due to erosion each time a dry phase came along (Anhuf *et al.*, 2006). So far, though, it seems best to keep an open mind on whether or not these short-lived cool dry events really affected the tropical rainforests.

3.40 THE ICE AGES AND DIVERSITY IN TEMPERATE-ZONE FORESTS

Around 40 Myr to 25 Myr ago, the temperate forests of the northern hemisphere were far more diverse and extensive than they are today. From the fossil leaf floras of Europe, North America, and China there are long lists of tree genera that nowadays only grow together when they are specially brought in and planted in collections, such as at Kew Gardens in England (Tallis, 1991). A list of the types of trees found as fossils from that time in northern Europe illustrates the point (Table 3.2)—hardly any of these now survive in the wild forests of the region, even though many of them will still grow well if they are brought in and planted in the northwest European climate.

Table 3.2. Tree genera found in Tertiary fossil deposits of northern Europe (after Sauer, 1988).

A	***Genera still present in the region***	
	Gymnosperms	*Abies, Larix, Picea, Pinus, Taxus*
	Angiosperms	*Acer, Alnus, Betula, Carpinus, Cornus, Corylus, Fagus, Fraxinus, Ilex, Populus, Prunus, Quercus, Tilia, Ulmus*
B	***Genera still present in the region but surviving in these areas***	
	1 *Mediterranean region*	
	Gymnosperms	*Cedrus, Tetraclinis*
	Angiosperms	*Castanea, Celtis, Juglans, Liquidambar, Ostrya, Platanus, Pterocarya, Styrax, Zelkova*
	2 *Eastern Asia*	
	Gymnosperms	*Cathaya/Keteleeria, Cephalotaxus, Chamaecyparis, Cunninghamia, Ginkgo, Glyptostrobus, Pseudolarix, Sciadopitys, Thuja, Torreya, Tsuga*
	Angiosperms	*Actinidia, Ailanthus, Castanea, Cinnamomum, Cocculus, Corylopsis, Diospyros, Engelhardtia, Eucommia, Juglans, Koelreuteria, Lindera, Liquidambar, Liriodendron, Magnolia, Morus, Ostrya, Paulownia, Pterocarya, Sapindus, Styrax, Zelkova*
	3 *Eastern North America*	
	Gymnosperms	*Chamaecyparis, Taxodium, Tsuga*
	Angiosperms	*Asimina, Berchemia, Carya, Castanea, Celtis, Diospyros, Juglans, Lindera, Liquidambar, Liriodendron, Magnolia, Morus, Nyssa, Ostrya, Persea, Platanus, Robinia, Sabal, Sapindus, Sassafras, Styrax*
	4 *Western North America*	
	Gymnosperms	*Chamaecyparis, Sequoia, Torreya, Tsuga*
	Angiosperms	*Celtis, Juglans, Platanus, Sapindus, Styrax*

The tree genera that occurred all across the northern hemisphere during that time together form what is known as the Arcto-Tertiary flora, the "Arcto" part of the name coming about because it occurred up into Arctic latitudes in the warmer climates of those times, and "Tertiary" referring to the old name given to the geological phase that followed after the end-Cretaceous extinction.

Before the ice ages had even started, a slow decline in global temperatures began to eat away at the diversity of the temperate tree flora. Over millions of years, as drying and cooling set in across the northern hemisphere, the extent of these temperate forests began to shrink. One by one the tree genera dropped out, but often in a patchy way—surviving in some regions but not in others. Only a few of the old Arcto-Tertiary genera have gone completely extinct: most have by luck survived someplace or other. As is so often the case with the fossil record, we can mostly only talk in terms of the richness of genera back then, and merely guess how many individual species were lost. Many of these tree genera still have several species in the places where they survive nowadays, and probably they also had multiple species in regions where they occurred back then but have since been lost. In some cases we can actually distinguish several different (and now-vanished) species in the fossil record of a genus in the same region.

Though the long-term cooling trend had already done some work, the pace of extinction seems to have accelerated in the last 3 million years when the Earth began to enter the first Quaternary ice ages (see above, Figure 3.9). Oscillations in climate that had been going on for millions of years—and related to wobbles in the Earth's orbit—now became far more extreme. Many genera finally disappeared from the European flora at around this time, including the maidenhair tree *Ginkgo* (Figure 3.11, top) at around 2.5 Myr ago. The big glaciations of the Quaternary hammered back the European forests with drought and cold every few tens of thousands of years, interspersed with scores of other smaller climate changes. The coast redwood *Sequoia* (Figure 3.11, bottom) hung on in Europe during the early glacials, appearing back in the pollen record of northern Europe each time that the climate ameliorated for a while. But about 600,000 years ago, its luck ran out, and it too disappeared (Tallis, 1991). The trees that survived the Quaternary and make up the present-day European flora mostly seem to have survived in southern Europe, as scattered populations in moist mountain valleys (Tzedakis, 1993) although some may have persisted as populations farther north (Stewart and Lister, 2001).

As things stand at present, the original Arcto-Tertiary forest has shrunk back to four regions: Europe, eastern Asia, and the eastern and western sides of North America. All of these regions have lost tree genera that they once had, but they have suffered to differing extents. Europe came out especially badly—for despite being about the same size as the eastern U.S.A. or the east Asian temperate forest zone, it has only a fraction of the numbers of tree genera and species found in either area. Lowest of all in total diversity is the west coast region of North America, but it also occupies the smallest area—being hemmed in by mountains and aridity to the east and south. Partly compensating for its lack of species richness, it also happens to contain several unique Tertiary relicts such as the two giant redwoods, *Sequoia* (Figure 3.11, bottom) and *Sequoiadendron*.

Figure 3.11. (Top) The maidenhair tree, *Ginkgo biloba*, was previously far more widespread in the northern temperate zone, and grows well when planted in its original range. (Bottom) The coast redwood *Sequoia* survives in a small area of northern California and adjacent Oregon in the U.S.A. Millions of years ago, it too was common and widespread around the northern hemisphere. *See also* Color section.

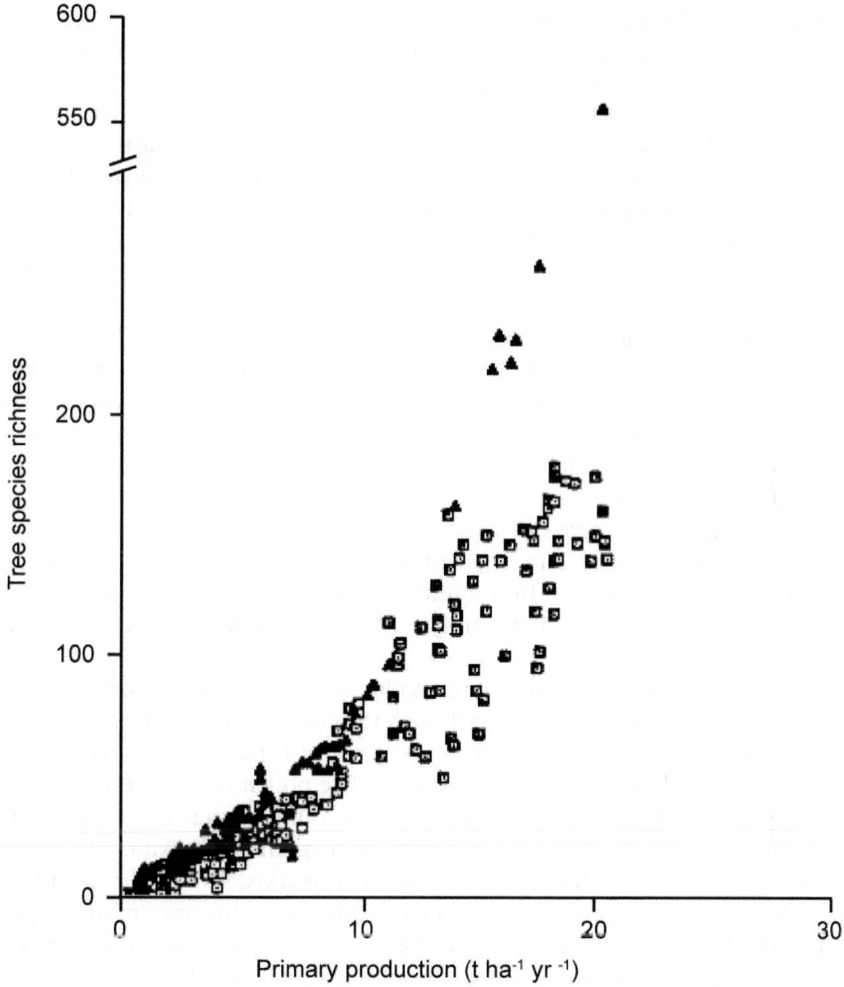

Figure 3.12. When compared in terms of similar climates, the tree species richness of temperate eastern Asia (▲) is much greater than North America (□), at least in the warmer more southerly climates (warmer climates have high NPP values, on the right of the scatterplot) (source: Author).

Eastern Asia (China, Japan, and Korea), then, comes out tops in terms of the present-day richness of its temperate tree flora (Figure 3.12). Many more genera are known from this region, including almost every genus that also survives in either Europe or North America, and many more besides. The individual genera in Asia also tend to have more species in them than their European or American counterparts: for example, the lindens or basswoods (*Tilia*) have around 20 species in east Asia but only 3 species in Europe and 1 in North America.

The key to the huge difference in diversity between east Asia on one hand, and Europe and North America on the other, seems to be the degree to which they were affected by aridity during ice ages. In both Europe and North America, ice ages were very arid as well as cold. Most areas of Europe that are now forested were reduced to arid grassland at the time (Figure 3.13)—and this was especially so in southern Europe, which would still have had climates warm enough for many types of trees during ice ages (Adams, 1997; Adams and Faure, 1998). Eastern North America was not quite as arid, but the tree cover was much more open than now with lots of bare areas because of the dryness. Florida, with its droughty limestone substrate, seems to have been more like a desert at times. This is not the ideal environment for trees to survive in. The cold and aridity made life impossible for most types of trees in northern Europe and the northeastern U.S.A., and the aridity in the south meant there were not many places for them to survive there either. The pollen record from the ice ages paints the picture: in many areas (especially in Europe) the more warmth-loving, more moisture-demanding forms that still survive were then so rare that they were almost invisible. They must have been confined to localized pockets of higher rainfall and moister soil, in sheltered valleys or gullies, hanging by their metaphorical fingernails on the edge of extinction. And indeed, the result of this happening again and again was that many types of trees did go extinct.

In Asia, by contrast, there are signs of much moister climates lasting throughout the ice ages (Liew *et al.*, 1998). In a number of places, pollen records reveal closed, moist forest—the sort of thing that simply isn't found in Europe or North America at that time.

It used to be thought that mountains killed off the trees in Europe, by blocking the way as tree populations shifted south with the cooling climate of each ice age. Yet ironically from what we now know, the mountains of Europe often saved the species which have managed to survive to the present. The pollen record from the cold, dry glaciations suggests that many species of trees only hung on in mid-altitude mountain valleys in the various ranges of southern Europe (e.g., in the Balkans of Greece, and the mountains of Spain including the Pyrenees; Tzedakis, 1994; Adams and Faure, 1997). After the ice age ended, they spread out down the mountain slopes and blanketed the lowlands of Europe. The reason they survived up in the mountains was that the climate there was moister: mountains tend to attract rain, because air blowing over them has to rise and cool—condensing out clouds and then raindrops. The cooler temperatures also meant less evaporation, even if growing conditions were not warm enough to be optimal for these species. In the very arid environment of ice age Europe, localized rainy slopes and gullies in the mountains may have been the only places that many species hung on, as little patches of trees in a mainly barren landscape. Now each time the European tree flora recovers from a glacial phase, spreading back out across northern Europe, it retains the legacy of the extinctions that have left it very depleted in species relative to the east Asian temperate flora (Adams and Woodward, 1989).

In Asia too, part of the key to the survival of so many Arcto-Tertiary trees seems to be the presence of mountains. The climates in some of the mountain ranges of southern China and Taiwan are incredibly rainy, reaching totals of several meters of rain a year. When this region cooled off during the ice ages, the climates also dried out—yet the drying was all relative. In many mountainous areas the climate is so wet

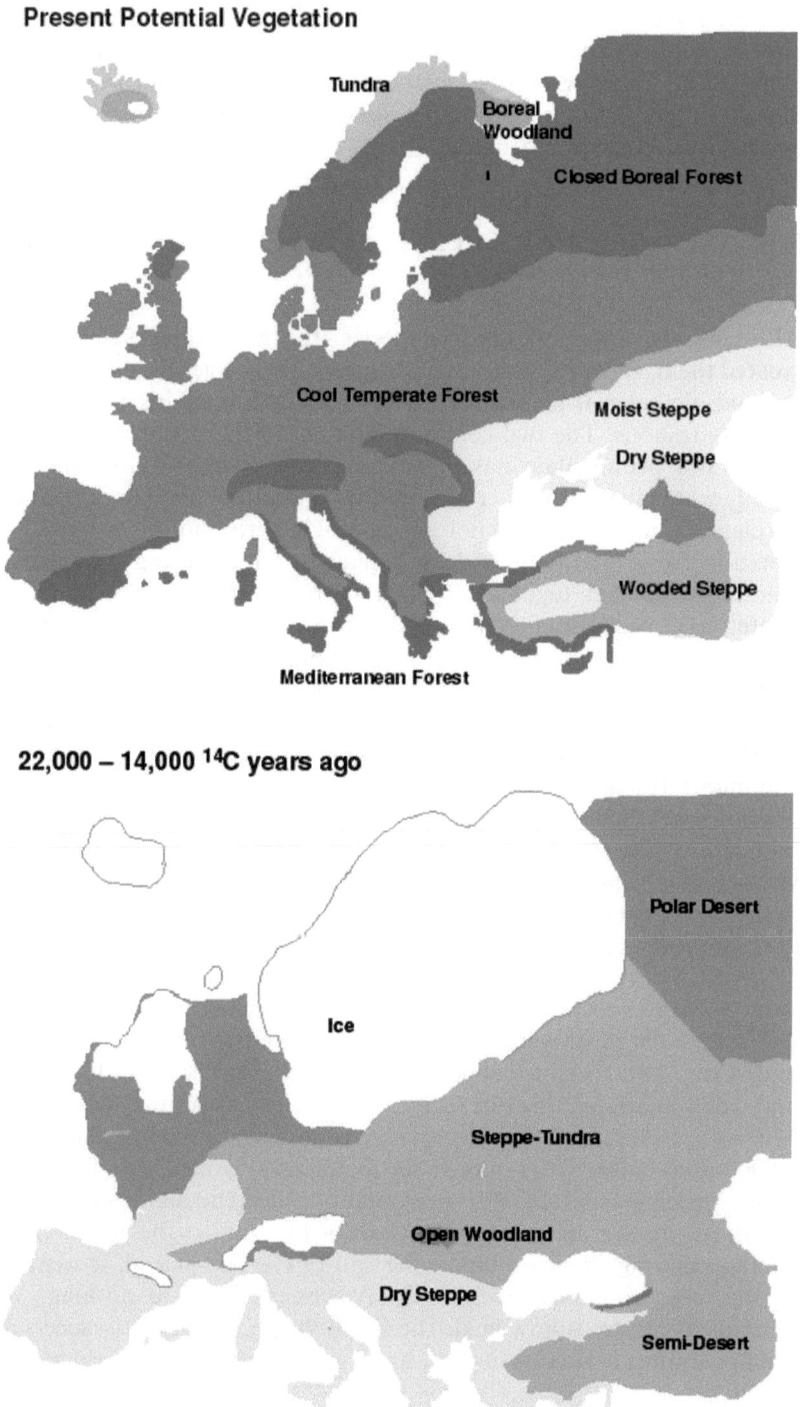

Figure 3.13. Reconstructed vegetation of Europe during (top) the present interglacial, before deforestation caused by farming, (bottom) during the last ice age. The ice vegetation was mostly devoid of forest cover due to a combination of cold and drought (source: Author). *See also* Color section.

that even with rainfall down to a half or a third of what it is now, there still would have been plenty to sustain forests. So, for instance, on Taiwan, in the pollen record from old lakes in the rainiest northern parts of the island, we find that moist forest hung on regardless through the ice ages—even though it had shrunk back from the drier slopes (Adams and Faure, 1997).

3.41 ICE AGES MAY CREATE AS WELL AS DESTROY TEMPERATE SPECIES

Although they often destroyed species, the glaciations sometimes seem to have created them. In various parts of Europe we see species that may have been produced by cold, dry glacial phases shrinking down and breaking up ranges that had once been continuous. The two common oaks of northern Europe—sessile oak (*Quercus robur*) and pedunculate oak (*Quercus petrea*)—are so close that they interbreed readily where they encounter one another (perhaps a good argument that they should be classified as subspecies not species). Their formation may well be a legacy of isolation from one another during glacials, while during interglacials they slowly tend to merge back together. In North America too, there are similar sets of closely related oak species that seem to be interbreeding and merging where they encounter one another (e.g., the *Quercus kelloggii* complex in the western mountains). In the east of North America, several members of the "black oak" group seem to be merging together by hybridization. For example, red oak (*Quercus rubra*) seems to be merging into northern pin oak (*Quercus ellipsoides*) (J. Romero, 1998, pers. commun.). It has also been suggested that the repeated rising and falling of sea level and arid–wet climate cycles in temperate eastern Asia have helped to add species to the already diverse tree flora. In a sense, there may be an optimum degree of environmental instability for promoting species richness: a certain amount of variation in the broad environment may help to generate species, but too much variability will destroy more species than it creates.

Not all of the tree species generated by ice ages survived to the present. An extinct spruce, *Picea critchfeldii*, is one example of a species that may have been both created and eventually destroyed by the climate variability of the late Quaternary (Jackson and Wang, 1999). Quite distinct from any living spruce in the look of its leaves, cones and wood anatomy, this tree turns up in ancient river muds from the Deep South of Mississippi, Alabama, and Georgia from the time of the last glacial, between around 30,000 and 15,000 years ago. It seems to have formed vast forests, thriving in the cooler, drier glacial climate—presumably a descendent of one of the other North American spruce species that nowadays live far to the north in the interglacial climate. During the warming at the end of the last glacial, it went from being one of the most abundant trees in the Americas, down to nothing. What this story perhaps illustrates is how fickle the fate of a species can be—success at any one time is no guarantee of success in the near future, especially in a world where climate varies wildly.

Animals too may have diverged as a result of isolation during ice ages. I do not know of any clear examples in the temperate zone of species produced in this way (though I am sure they do exist somewhere in the literature), but a number of populations or subspecies that might eventually give rise to fully fledged species if they were to be given enough time and go through enough glacial cycles. Both shrews and hedgehogs of Europe form two or more quite distinct populations, that occupy distinct areas but come into contact along a broad front. The hedgehogs of northern and western Europe are quite distinct in DNA terms from those of southern and Eastern Europe—they have apparently been separated for some 500,000 years, which is as far back as many species go. Likewise, there are geographically distinctive populations of grasshoppers (*Chorthippus parallelus*) in southern Europe which may have diverged during glaciations (Hewitt, 1996).

It is often difficult, though, to really be sure that any individual climate event of the Quaternary led to some particular species being generated. It had been widely thought that pairs of closely similar species of songbirds in North America ("sister species") had been generated during the most recent glaciation when ranges were shrunk down and fragmented, allowing the populations to diverge genetically. In fact, study of their mitochondrial DNA shows that the pairs of species are much older—diverging at various times over the last 5 million years (Klicka and Zink, 1997). These species pairs may well have diverged because of glacials or other earlier climate cycles, but it was certainly not during the most recent glacial!

3.42 WHAT ICE AGES DID TO TROPICAL RAINFOREST DIVERSITY

It seems a reasonable guess that the great ice ages of the Quaternary would have had profound effects on the diversity of tropical forests. For some years, ecologists were not sure whether the species richness of rainforests had increased or decreased overall during ice ages. There were clues that there had been cooler and drier conditions during the many glacials of the last 2.4 million years, alternating with relatively warm and moist conditions similar to the present. Exactly how cool and how dry things got during ice ages is still debated (Adams, 1997), but the general trend is beyond doubt. What did this do to tropical species richness? On one hand, the frequently changing conditions and long droughts could have been too much for many species, driving them extinct. On the other hand, the breaking up of ranges by cooling or drought could have allowed the separated populations to evolve in different ways, eventually becoming distinct species that each spread back out to enrich communities when conditions improved again.

This sort of process, involving species ranges shrinking, fragmenting, and then splintering into new species, is called a "species pump". In the species pump scenario, each time the forest shrinks it generates new life, that spreads back out as the forest expands again thousands of years later. Through the 1970s and 1980s, the dominant view was that fragmentation of ranges over and over again by ice age aridity had progressively enriched the tropical rainforests during the Quaternary (Haffer, 1969). Each time an ice age ended, the new species of plants, birds, and insects that had been generated either spread out, or stayed put within their former ice age range.

This idea of a species pump was first put forward for the Amazon Basin, to explain areas that seemed to have unusually high species richness of birds and butterflies. Haffer (1969) suggested that the reason for these "hotspots" (see Chapter 4) was that they had once been areas that provided a refuge for the rainforests during ice ages. The reason for their greater species richness was twofold. First, there were species that had shrunk back to these places during ice ages and never made it out again because their populations were so slow to spread. Second, there were new species generated in the hotspots each time the ranges of more widespread species were fragmented. Some of these had also not yet spread out from their places of origin, still revealing where they had been made. Prance (1987) suggested that areas of high plant species richness within Amazonia also represented former ice age refugial areas. This general idea caught on, and began to be applied to areas of higher than usual species richness in many other types of environments around the world. It became widely accepted that in many places repeated contraction of ranges, followed by expansion, has been acting as a species pump during the Quaternary.

The hypothesis of ice age species pumps and refuges within the Amazon Basin has fallen out of favor. The hypothesis was dented when a study by Nelson *et al.* (1990) showed that the areas of supposedly higher species richness were also areas where the collecting density by zoologists had been greater. More zoologists collecting more specimens are bound to yield more species from a particular area, leaving many other species elsewhere still undiscovered and unknown. However, many biologists still hung on to the idea that at least in part, these species richness hotspots were real, not simply an artifact of collecting.

The refugial theory has suffered more setbacks as geologists have learned more about the Quaternary history of tropical rainforests. At least for the last ice age, which reached its low point about 20,000 years ago, those who study the fossil pollen and other clues feel confident enough to draw some very general vegetation zones on a map—but different schools of thought come up with different pictures of where the forest was and how arid things generally were during ice ages (Anhuf *et al.*, 2006)! I tend to side with the viewpoint of Anhuf *et al.* (Figure 3.8) that the rainforest in the Amazon was indeed shrunken down and broken up by aridity during ice ages, occupying perhaps half its original extent. Some of the most likely remaining core areas of rainforest during these glacial phases do seem to me to correspond roughly to the areas that Haffer proposed on the basis of current diversity patterns. But some of Haffer's proposed refugia look likely to have been just grassland or scrub at that time. There is, however, a range of opinion about what the Amazon Basin was really like during ice ages. Other ecologists, Paul Colinvaux and Mark Bush, for example (Colinvaux *et al.*, 1989; Colinvaux & De Oliviera, 2001), suggest that there is no strong evidence of drying and that forest extent during the glacials remained similar to the present—in which case Haffer's supposed refugia were still just part of a big continuous mass of forest at that time.

So, overall, at present opinion in the world of ecology seems to have gone against the idea that the hotspots in diversity that Haffer noticed were ever really refuges for the tropical forest during ice ages. Some consider these hotspots as nothing more than a collecting artifact, while others think that they really do exist—perhaps for other reasons unrelated to ice ages.

Adding to the general picture that the Quaternary ice ages did not generate much new species diversity are the results of DNA comparison of genera that contain many similar and (apparently) recently diverged species. For example, a study of the South and Central American tree genus *Inga* suggests that its various species diverged several million years ago, not during recent ice ages as had been suspected (Dick *et al.* 2003). Tropical forest hardwoods and shrubs seem to have average species durations of somewhere between 27 and 43 million years, showing that most of their diversity is much older than the Quaternary (Kutschera and Niklas, 2004). Where we happen to find a very detailed fossil record of rainforest plant communities from millions of years ago, their species composition and overall diversity is very similar to the present. Mid-Miocene age coals, some 10 million years old, from northern Borneo show essentially the same combination of species as is now present in peat swamp forests in this area (Morley, 2005). The overall picture is of slow, slow change in the tropics, with species that were generated long ago and lasted a long time.

So it seems that ice ages did not generate much that is new, at least in the tropical groups that we have been able to study. But did they cause extinctions, comparable with the losses of Arcto-Tertiary trees in the higher latitudes? Again, lack of detail in the fossil record makes it difficult to know whether much has been lost—although the general picture is that very few of the pollen or leaf types we find just before the Quaternary are now extinct. The longest, most detailed pollen records from the tropical rainforest come from the small areas of rainforest that exist in Queensland, northern Australia. These northern Australian forests were particularly hard hit by aridity during the ice ages, only surviving in very localized areas. Yet despite the hard conditions that these Queensland forests had to endure, as far as we can tell from the pollen record, there has not been a lot of extinction since the last moist interglacial period 120,000 years ago. Only two of the many pollen types present in the Queensland rainforest during the last interglacial are now extinct—both of them types of conifers that still occur in New Guinea to the north. All the other numerous pollen types that have been found for rainforest back then are still present in the Queensland forests (Adams and Woodward, 1992).

The lesson from the fossil record we have available seems to be that the rainforest flora can stand quite a lot of disruption, without very many types of trees going extinct. Biologists used to think that the secret behind the diversity of the rainforests was that they had always remained undisturbed by environmental changes, even while the higher latitudes were being hammered by ice ages. What we can now see is that at least some areas of rainforests have been hit, again and again, by large environmental changes such as cooling, dryness, and sea level changes. Despite this, these forests still remain wonderfully diverse in both plant and animal life. It is of course possible, though, that if climates had remained stable, the tropical forests would be even more diverse that they are now. There are the signs (Jaramillo *et al.*, 2006) that indeed they were once more diverse in the past, and that the most species-rich areas of forest in the present-day world have in fact been relatively stable (Chapter 4).

Overall, what did the Quaternary ice ages do to biological diversity? It is surprising, really, how little evidence there is for anything having been lost during the Quaternary itself. The main losses of tropical and temperate forest diversity seem to have occurred well before the Quaternary, extending back over the previous 30 Myr or so, as the global climate began to cool and dry out. From the Quaternary, only a single species of tree (a spruce, *Picea critchfeldii*) from North America (Jackson and Wang, 1999) is known to have gone completely extinct during the last million years. Only a couple of regional-scale extinctions of tree genera can be found in the record from the rainforest of northern Australia (Adams and Woodward, 1992). Survival over time has been even more impressive during the last million years among the corals that formed reefs in shallow shelf seas, and in the seashells (bivalve mollusks) that lived on the sea bed (Adams and Woodward, 1992). Both have an excellent fossil record, living as they do in environments that are continually burying them, and their carbonate skeletons often preserve almost perfectly. Not a single species of coral or bivalve identified from the fossil record of the last million years is now extinct (Adams and Woodward, 1992). Russell Coope's work (Coope, 1987) has identified nearly two thousand species of insects—mostly beetles from the fossil record of the British Isles stretching back as far as the last interglacial more than 100,000 years ago. Every single one of these species identified in the fossil record is still present in the world.

But there are signs that the Quaternary itself started off with a step down in biological diversity. A burst of extinctions of marine bivalves, corals, and terrestrial mammals seems to have occurred as the first big climate fluctuations hit between around three and two-and-a-half million years ago (Adams and Woodward, 1992). Why were there so many extinctions back then, and yet so few later on? It may be that species which existed in the millions of years before the start of the wildly fluctuating Quaternary climates were often adapted to stable, predictable conditions. They had specialized into narrow niches to take full advantage of the opportunities available to them. They did not have the means of dispersal to migrate long distances if the climate suddenly changed, nor the mechanisms to allow them to survive at small population sizes and then expand back out when opportunities allowed them to. When the environment changed, these sorts of species were unable to either survive in place or migrate; so they went extinct. By the last million years or so, this winnowing out of species had finished. Only those species that happened to have the right characteristics to survive the unstable Quaternary world still remain with us today (Coope, 1987; Adams and Woodward, 1992).

However, one group was the exception: the large and medium-size mammals of several different continents. These survived most of the Quaternary very well, and yet just in the past 50,000 years most of them have gone extinct. In Chapter 5, I will look at the possible causes of this particular extinction event.

4

Hotspots and coldspots

4.1 GEOGRAPHICAL PATCHINESS IN SPECIES RICHNESS

Species richness varies across the Earth's surface, on a wide range of scales. It tends to follow grand latitudinal and climatic gradients, as we explored in Chapter 2. And it also shows local-scale patchiness (Chapter 1) on the scale of a piece of forest or meadow, or a hillside. Against this background, there is also a certain broader patchiness in species richness that is too large-scale to be local, fitting in more at a geographical scale, and complicating the latitudinal trends.

So, when we compare areas of similar latitude and climate to one another, some parts of the world seem anomalously richer in species than others. Some other places seem unusually poor in species, compared with the most closely comparable environments elsewhere. These are, respectively, the world's diversity "hotspots" and "coldspots".

4.2 HOTSPOTS

Areas that have anomalously high levels of species richness are called "hotspots", a term coined by the conservationist Norman Myers (1988; Myers *et al.* 2000). Myers first listed the world's hotspots for purposes of pinpointing places that should be given special priority for preserving biological diversity (Chapter 8), because there is potentially more to lose from these compared with less species-rich areas. Initially, Myers just concentrated on areas within the tropical rainforests. Another thing he factored in for designating his hotspots is how fast the destruction of habitat was occurring in each of these places.

The term "hotspot" for a center of high species richness has now become widely used in ecology and conservation. Following Myers' original paper, the definition has broadened now to include areas outside the tropics, and the habitat threat aspect of

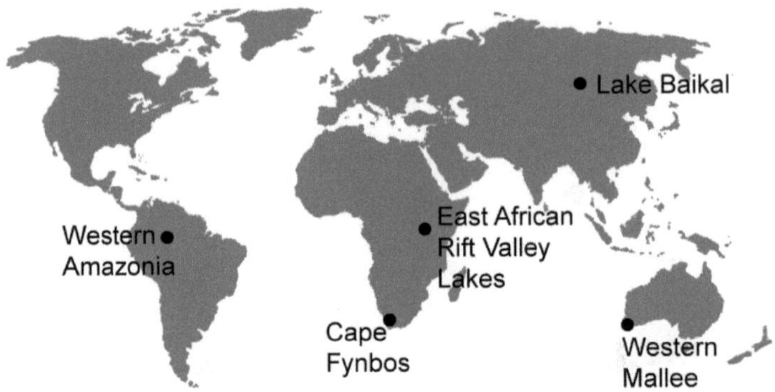

Figure 4.1. Some of the species richness hotspots discussed here.

the definition has rather fallen by the wayside too, with most of the emphasis on diversity. It is also a matter of opinion as to how diverse an area should be to deserve the label: some ecologists and conservationists are prepared to declare a lot more hotspots around the world than others! There is also some question as to how large a hotspot can be before it stops being a hotspot, and becomes a whole vegetation region or "biome". Some would label all of the world's tropical rainforest regions as hotspots, because these forests are almost always so rich in species. Others would say that we should only include anomalously rich areas within the forests themselves which stand out against the regional (but already incredibly diverse) tropical rainforest background. When the world's various hotspots are carefully compared with one another using the same criteria, only a few of those approach the levels of species richness of even fairly ordinary areas of tropical rainforest.

Nevertheless, the word "hotspot" continues to be utilized as a relative term. Ecologists tend to use it for any geographical-scale area that stands out against its surroundings (or other climatically similar areas) in terms of species richness. From the point of view of ecological processes, this is probably also the most useful definition of a hotspot. We can pinpoint hotspots to show where species diversity has for some reason risen up above its surroundings, and then try to understand why (Figure 4.1). As well as being very interesting in themselves, hotspots might also give us some broader clues to the controls on species richness over latitudinal gradients (Chapter 2), and over broader geological time (Chapter 3), as well as local-scale variations in diversity that occur everywhere (Chapter 1).

4.3 SOME EXAMPLES OF HOTSPOTS IN SPECIES RICHNESS

4.3.1 The big lakes of eastern Africa

On the eastern side of Africa, the Earth's crust is slowly splitting to give a new continent. The land has torn along the vast Rift Valley which stretches for several thousand kilometers. Scattered along the floor of the Rift Valley are three large lakes—Tanganyika, Malawi, and Victoria. And within these lakes, particular groups of fishes have diversified into hundreds of species. The most diverse group of all are the haplochromid cichlids, medium-sized fish between about 5 cm and 15 cm in length, belonging to a family that occurs in rivers and lakes throughout the world's tropics (Figure 4.2). Other smaller lakes scattered around the Rift Valley floor also have groups of endemic cichlids, often a couple of dozen species in each lake. For example, there is a ring of smaller lakes surrounding Lake Victoria (including, for instance, Lake Kivu) which each have their own cichlids as well as some they share with Lake Victoria.

Figure 4.2. Cichlids from the East African Rift Valley lakes: (top) *Metriaclima greshakei*, (bottom) *Labidochromis caerulus* (photos: Amy Lerch). *See also* Color section.

Within the Rift Valley lakes, there is a cichlid for almost every ecological role: indeed, many cichlids for almost every ecological role. There are herbivores, carnivores, omnivores, and feeders on dead detritus. Some scrape algae off rocks, and some eat mollusks in the lake shallows. Others are plankton eaters, parasite removers, or predators on other cichlids. There are even some species that specialize in sneaking up and pulling out a single scale from their victims, eating the bit of flesh still attached to it as they swim away. Some of the cichlids swim mostly in the open lake near the surface, some deeper down, others around the edges of the lake basin in the shallows. Very often closely related species seem to live in exactly the same way and are anatomically identical, differing only in coloration. Each species has its own characteristic color pattern, with patches or stripes of reds, greens, blues, or yellows.

It is surprisingly difficult to get hold of any straight answer from the scientific literature for how many species of cichlids there are in each of the big Rift Valley lakes (Martens, 1997). This reflects the confusion and incomplete knowledge that comes from their daunting diversity. But there is general agreement that the greatest diversity of cichlids is found in Lake Tanganyika, which has something between 700 and 1,000 species. Lake Malawi comes in second with at least 500, and Lake Victoria last with "only" around 400–500 cichlid species (though many of these in Victoria may now unfortunately be extinct: see Chapter 5).

There are also some other particularly diverse groups in each lake. For example, in Tanganyika there are clusters of locally unique (endemic) species of prawns and large freshwater crabs.

4.3.2 The western Cape of South Africa

At the southwestern tip of South Africa is an area of hills and low mountains, about 100 km by 150 km. European explorers named this the "Cape of Good Hope", encouraged by the sight of it. The dry seasonal climate there prevents trees from thriving, and instead the natural vegetation is a low scrub, known as fynbos. Seen from a distance during the dry summer when not much is in flower, the fynbos might look unpromising as a reservoir of biological diversity. But in fact it is spectacularly rich in species of plants. There are at least 600 species of heathers (*Erica*) (Figure 4.3), with more being discovered all the time. Heathers are widespread in the cooler parts of the world, but usually just a few species in any one place, certainly nothing like this. Also in the fynbos are some 300 species of proteas (Proteaceae) (Figure 4.4), and several hundred species of bulbs of the lily-like family Restionaceae. In all 9,000 species of plants are packed into this small area, and about 6,000 of them are endemic to (only found in) this southern tip of Africa (Figure 4.4).

The Cape region has a typical Mediterranean climate, with a cool rainy winter and a hot dry summer. It is only one of several such areas scattered around the world, all of them including areas of evergreen scrub vegetation. Yet in the Cape, species richness among the plants has risen far higher than in the other comparable areas in Chile, California, eastern Australia, and in the Mediterranean Basin itself. One exception is the Mediterranean climate area in southwestern Australia, which rivals the Cape in terms of species richness.

The exceptional diversity of plants in the Cape is not paralleled by animal species. Birds, for example, are no more species-rich there than in other climatically similar areas of the world.

4.3.3 Lake Baikal in Siberia

In the midst of southern Siberia, a location about as remote as anywhere on Earth from the bustle of our modern world, is a huge freshwater lake about the size of the island of Britain. Lake Baikal occupies a huge tear in the Earth's crust, and extends deeper than any other lake in the world (to about 1,620 m below the water's surface).

Baikal has a peculiar fauna that includes, for example, the world's only freshwater seal, but perhaps its most remarkable characteristic is its shrimps. Gammarids are a group of little brownish shrimps, common on the beds of cool streams, lakes, and rocky sea shores of Eurasia and North America, where they mostly feed on detritus. When I was a boy, on picnics by the river in rural Herefordshire, we used to catch them in little nets and put them in jam jars full of water to look at for a while. In such a river in England, there will usually be just a handful of species in a particular place—but in Lake Baikal the gammarids have diversified into at least 260 different kinds (constituting one-third of the world's total of gammarid species). The gammarids of Baikal live in various ways, many filter-feeding in the water or scavenging off the lake bed, including one blind species that lives in the

Figure 4.3. One of the several hundred species of heathers, *Erica*, endemic to the Cape (photo: Maille Neil). *See also* Color section.

Figure 4.4. A protea endemic to the Cape (photo: Maille Neil). *See also* Color section.

permanent darkness of the depths. Other groups in Baikal have also produced smaller bursts of endemic species (e.g., the cottoid fishes have around 30 species that are found only here, in just this one lake).

4.3.4 The mallee scrub of southwestern Australia

In Australia, the Mediterranean climate scrub goes by the name "mallee", an old Aboriginal word. There are actually two quite distinct zones: the eastern mallee in southeastern Australia and the western mallee thousands of kilometers away in the southwest of the country. I have spent a couple of weeks in the eastern mallee, sleeping under the stars in the dry cool nights, and it is a delightful place. Variously colored shrubs grow nicely spaced out, separated by areas of open sandy red soil—the whole impression is of one vast garden. I have never visited the western mallee, but I am told by those who have that it looks "boring" by comparison. But if it does, that appearance is deceptive, for it contains many times more species of shrubs and wild flowers than the eastern mallee. In fact this is an area that seems to rival the Cape in terms of plant species richness. For example, the proteas are here in abundance too, reflecting an ancient link between the southern continents that goes back to the time of the dinosaurs. There are 58 species in the *Banksia* genus alone, and many other genera in that family besides. Another very diverse group is the genus *Acacia* (in the pea family Leguminosae), with 250 species in the western mallee. The Australian "gum" genus *Eucalyptus* is also diverse in the western mallee (Figure 4.5).

4.3.5 The western edge of Amazonia

At the west of the vast Amazon rainforest, where the foothills of the Andes begin to rise out of the flat river plain, is probably the richest hotspot anywhere on Earth. Just the normal background level of diversity found across Amazonia would be enough to make it a hotspot elsewhere by comparison with most grassland, scrub, or temperate forest ecosystems. For example, 80 or 90 species of trees are commonly found in just a hectare of Amazon rainforest. But in the western part of the Amazon Basin and the low foothills closer to the Andes, species richness rises to 150, 250, or even 300 species of trees per hectare. Furthermore, many species seem to be highly localized, adding further to the broad-scale (gamma) diversity. For example, different hilltops often seem to possess their own endemic species of plants. There is another hotspot of tree species richness farther east, in a small area of central Amazonia near to Manaus.

There have been claims of other hotspots within the Amazon region. Haffer (1969) noted that certain scattered parts of the region seemed to be especially diverse in terms of birds and mammals. According to Prance (1987), plant diversity also seemed to follow the same patterns. However, as I mentioned earlier the intensity of collecting efforts has been greater in these same regions, so this might be the reason why more species have been found there. If similar effort had been put into collecting in other areas, they might turn out to have the same levels of species richness as the other hotspots. While Haffer's scattered animal diversity hotspots are to some extent in doubt, as are some of Prance's, the plant species hotspots of western and central

Figure 4.5. A western Australian mallee species of *Eucalyptus* (photo: Maille Neil). *See also* Color section.

Amazonia are still thought to be real. They have been established more rigorously than the previous sets of hotspots in Amazonia, on the basis of a widespread, standardized sampling program. In all, more than 400 hectare-sized plots scattered across Amazonia have been meticulously counted for tree species richness (Silman, 2005).

Box 4.1 Bursts in diversification beyond hotspots

Explosive diversification is not confined to hotspots. In certain times and places, a particular group of plants or animals has diversified into hundreds or even thousands of species—not in a localized area but across a region or even globally. Sometimes the trigger seems to have been the evolution of a new and highly successful adaptation for survival, that allowed species in that group to expand into many different niches. At other times it seems to have been a combination of environmental changes, and then new traits evolving to exploit this new environment.

Across the world there are many thousands of species of wild flowers and weeds that live for only a year or two (annuals and biennials as botanists call them), or at most just a few years (these are short-lived perennials). They fill any waste or open ground, and often delight us with their pretty colors. They are also the ancestors of many of the world's crops and vegetables, such as wheat, barley, rice, cabbages, carrots, and lettuce. These short-lived plants succeed by exploiting brief opportunities, living fast and dying young after putting their energies into making seeds.

Particularly successful among them are the daisy family Compositae (also known as Asteraceae) with 23,000 species (Figure 4.6), and the mustard family

Figure 4.6. Some examples of species from the daisy family, Compositae (source: Wikimedia Commons). *See also* Color section.

Figure 4.7. A member of the cabbage family Cruciferae: *Arabidopsis thaliana* (source: Wikipedia Commons). *See also* Color section.

Cruciferae (also called Brassicaceae) with 3,700 species (Figure 4.7)—and there are several other diverse families besides these which have similar roles in nature. It was not always like this—in evolutionary terms these flowers that color meadows and roadsides are relatively recent arrivals. For instance, both the daisy and the mustard families have not even been found in the fossil record before 42 million years ago, and they have only reached their present levels of richness within the last 10 million years or so.

The key to the explosive diversification of these little flowers seems to have been the drying and cooling of global climate that pushed back the tropical and temperate forests which had previously blanketed the world. Short-lived wild flowers need open, sunlit environments, without the shade of a forest canopy. They are well suited to living in arid climates, though paradoxically they themselves cannot survive drought. Instead they make the most of sparse and unpredictable rainfall, getting in during each year's brief wet season or following a chance desert storm. With long-lasting dormant seeds and rapid growth, these weedy species are already waiting in place to sprout quickly and complete their life cycle within just a few weeks on the moistened soil. After apparently starting off in arid environments, many of these weedy plants have now even diversified into moister climates—adapting to live in gaps in the forests, where they exploit the burst of light and nutrients that comes after a big tree comes crashing down, or colonizing the sandbanks left after a river has flooded. The weedy species often disperse on the wind, carried by feathery plumes—in effect "searching" for opportunities.

Another burst of diversification also began in a very different group of plants at about the same time as the wild flowers. Instead of being small herbaceous weeds, these are long-lived tropical trees: and many of them are the biggest in the forests where they grow. They are the dipterocarps, the family Dipterocarpaceae. The

Figure 4.8. Dipterocarp seeds, Malaysia (source: Author). *See also* Color section.

name comes from Greek words meaning "two-winged cup", and refers to their seeds which each have two little helicopter wings to send them spinning hundreds of meters on the breeze down from the mother tree (Figure 4.8). Presently, the dipterocarps dominate the forests of Southeast Asia, both in terms of diversity and sheer abundance. In any tropical forest in Malaysia, Indonesia, or Thailand, they tower all around, often reaching far above the surrounding trees—billowing up above the canopy like green thunderclouds. Whenever I get identifications on specimens of leaves or shoots from my fieldwork in the Malaysian rainforests, it seems that half of them are various species of dipterocarps. In all, there are about 475 species in this family across the Southeast Asian region. Before 30 million years ago they seem to have been much less common and diverse in Southeast Asia. Then suddenly about 25 million years ago their pollen appears in a great burst, in offshore sediments throughout the region. What had originally triggered the arrival of the dipterocarps was the collision of India with the main landmass of Asia. They would have been carried as passengers on India's long drift northward after it had split off from Africa about 60 million years ago. The Asian dipterocarps still have relatives in Africa, reflecting the ancient southern connection. But the African dipterocarps are all savanna or dry woodland trees, quite unlike the Asian ones which thrive particularly in the moist forests. We have no good fossil record of what went on as India moved northwards, but it is likely that the dipterocarps made the transition to rainforest trees during this time. Some time after India hit the southern side of Asia, they spilled out into the forests across the region, diversifying as they went.

In the case of the dipterocarps, success seems to have been achieved simply on the basis of their own merits in living as forest trees. There are no signs that the environment in Southeast Asia changed to suit them (more the opposite in fact, because it began to dry out). The only thing that changed in their favor was India's collision with Asia, providing them with their big break. The dipterocarps were somehow primed to exploit the rainforest environment and seasonal monsoon forest, although we do not know the secret of their success.

Other analogous success stories have happened elsewhere. An example is the gum tree genus *Eucalyptus* which fills the forests, woodlands, and scrublands of Australia (Figure 4.5). Occurring only on Australia and the islands to the north, there are about 700 species, most of them confined to the Australian mainland. It is a mystery what could have made *Eucalyptus* so successful there, although it may have something to do with their adaptation to coping with the continent's nutrient-poor old soils and the frequent fires that burn through the dry vegetation.

4.4 WHAT CAUSES HOTSPOTS?

Ecologists are fascinated by hotspots, just as they are by most other aspects of species richness patterns. But as is typical, we are not really sure why most of these hotspots occur. Several widely discussed ideas have been put forward to explain why they exist. One hypothesis is that hotspots are more stable, and that the stability has allowed diversity to build up, without many species being taken away by extinction. Another (apparently contradictory) idea is that hotspots have been relatively *unstable* over time, with the repeated fragmentation of populations allowing them to evolve in different directions and eventually form separate species. Other ideas emphasize some peculiarity of certain types of organism present in the hotspot, suggesting that it is this which has led to them branching off lots of species (this idea seems plausible in some hotspots, such as those that are mostly only rich in species because one particular group has proliferated into many species). Lastly, there might be some other characteristic of the local ecology that marks the hotspot area off as different from similar but less diverse places elsewhere. For example, in the case of plants it might be something about the soils there which allows more species to coexist without a few species pushing all others out.

But there is no reason why only one of these factors has to be all-important. Hotspots could be the product of quite different processes dominating in different places. In fact, all of the factors I have mentioned above could be important in the same place: where there is precisely the right balance between long-term stability on one hand and instability on the other, plus some characteristic of the groups of organisms involved that make them liable to produce a lot of species, and an environment that can comfortably support many species living side by side.

4.4.1 The stable environments hypothesis

Probably the most popular idea to explain hotspots is that they are ancient. If the environment in a particular place goes back very far in time and has stayed fairly stable throughout, we might expect it to have accumulated a lot of species. It takes time for lineages of organisms to adapt to a particular environment, and it also takes time for them to give rise to large numbers of specialized branches that end up as different species, dividing up resources very finely. Even if narrow specialization is not required for them to coexist, species might also just tend to accumulate over time due to occasional chance splitting of populations. It is surely a lot easier to destroy species than to make them: it might only take one large environmental upheaval to eliminate a large proportion of the species in a particular region, wiping the slate clean.

Because the stability idea invokes processes that presumably occurred over millions of years, the only way to test it is to find ways to look back in time. One method of looking back is to use the environmental record preserved in the chemistry and texture of rocks to look back at the history of the climate and other factors. We can then ask: Did this place really remain stable, or did it go through dramatic changes of climate or topography?

Fossils that are present in the same rocks can also reveal the presence of the ancestors of the groups that have now diversified; showing how long the same lineage has been there, and how long it took to proliferate into so many different types. If the stability hypothesis is correct, we should expect to see diversity in the fossil record building up gradually over many millions of years in this steady environment.

Another method for looking far back in time is to use DNA sequence evidence from within each related cluster of living organisms that inhabits the hotspot. The DNA sequences of the various species are compared to estimate how long it is since they split off from one another. If the different species in a group began to diverge a long time ago, they will have DNA sequences that have accumulated a lot of changes, one species compared with another. On the other hand, if the species all formed from one ancestral group quite recently, they will not have very many changes in their DNA.

4.4.2 The story of the Cape hotspot

The remarkable species richness of the fynbos vegetation of South Africa has attracted a lot of attention from ecologists, and many different theories to explain it. In the past few years, the fossil record of the region has begun to yield up clues to the time scale and processes which have led to so many species.

The fossil record of the fynbos is not ideal, but it is much better than we have for some other areas such as the tropical rainforests. The somewhat drier and cooler environment seems to favor deposition and preservation of fossils more than the wet tropics, where everything breaks down rapidly. Fossil-containing sediments that were once under the sea contain a record of pollen blown and washed off the land surface around South Africa since 30 million years ago. The sediments were compacted into rocks, and folded up to levels where geologists could reach them at the surface, or where they were drilled into by mining companies.

What the fossil record shows is that the very species-rich flora of the Cape (e.g., the heathers—genus *Erica*—and the proteas—family Proteaceae) built up its diversity gradually, starting around 15 million years ago as the climate dried and cooled from a time when the region was forested. Molecular evidence, comparing DNA sequences of living proteas in the flora, indicates that the process of making new species was fastest in the first few million years. It then slowed down, but continued to the present as a slow gentle rise in richness (Sanderson *et al.*, 2004). So whatever the cause, the diversification did not occur all in one sudden burst, nor was it recent. This gradual increase in diversity does not give specific clues about what the mechanism behind the diversification was, but it may at least narrow down the range of possibilities. More about that later, though.

What about the general look of the vegetation cover: how long has this scrubby, evergreen fynbos existed, and has it always stayed in place there? Though the Cape's environment seems to have gotten drier over time, the general combinations of plant types suggest that something resembling fynbos has been in place there for several million years (Deacon and Lancaster, 1988). Fossil pollen evidence shows that even though the world went through its climate convulsions of the last two and a half million years, the fynbos vegetation persisted despite the drier and cooler ice age climate (Deacon and Lancaster, 1988). Part of the reason that the fynbos stayed in place may be that there were no other types of plants to push in and replace it. Fynbos plants are adapted to the particular rocky, well-drained, and very nutrient-poor soils of the Cape. Even if the climate changes, they are still likely to be the best-adapted plants around to cope with this environment, and this is probably why they have hung on. Most other places around the world did not stay as stable as this and ended up with a very different vegetation cover during ice ages. Thus, it could be that continuity is part of the secret of the extraordinary diversity of the Cape hotspot. The Cape species did not all appear suddenly in a burst—instead they accumulated gradually over time. Both the fossil record and DNA sequence evidence shows that many of the lineages of plants there have diverged progressively over the last several million years. Presumably then, if the environment had been less stable, diversity would not have reached the heights that it has.

However, time alone cannot be the whole story. The geological record shows that the several other Mediterranean regions of the world also have a long history, and the Cape region is only about average among them in terms of the length of time that it has existed. The Mediterranean Basin of Europe, for example, has had a similar climate for nearly twice has long. And yet despite their long history, those other regions mostly have far fewer plant species than the Cape. So while stability might have been a necessary condition for the build-up of diversity in this place (seeing how gradually diversity built up over time), it cannot provide the whole explanation for what makes the Cape so exceptional.

4.4.3 The story of Baikal

Another hotspot that has probably offered a stable environment is Lake Baikal in Siberia. The broad-scale geological evidence from Baikal suggests that the lake has existed for at least 17 million years, and may be much older (Freyer, 1991). While the climate of southern Siberia was much colder during glacial periods, Baikal was never covered by the mile-thick ice sheets that blanketed much of Europe and North America. And although the water temperature would have been colder during glacials and the lake would have been covered by floating ice for the greater part of the year, hot springs deep within the lake basin could have helped to keep it warm and provide local refuges for species that require more warmth. Hence despite the big climate changes, parts of the essential environment of the lake could have remained intact because of its sheer size and depth. It is possible that populations of the shrimps and other animals of the lake began to diverge from one another when they became isolated as separate populations around each of these springs. The lake also consists of three large, deep basins which might perhaps have been separated in the past during times of drier climate (including the glacials, which tended to be drier in this region). This too could have allowed populations to go their separate evolutionary ways, eventually forming distinct species (Freyer, 1991). So, overall stability, with a modicum of instability, might be what it takes to hold on to species richness—and help it to build up too.

4.4.4 The story of the African Rift Valley lakes

What about the big lakes of Africa? How long have they been there, and how has their environment changed? The geological record suggests that all three lakes formed in the last 9 million years or so, with a rifting of the Earth's crust in that region (Martins, 1977). For the first part of their history they were probably not much more than shallow swamps, and it is only during the last several million years that they have become the great lakes that they are now. The actual fossil record within the East African lakes is negligible, but the deep layers of sediments that they have accumulated tell stories of their environmental history.

The idea which has long prevailed to explain the very high fish species richness of these big lakes is that they offered a stable environment for a very long period of time—and that this stability prevented extinctions of emerging, incipient species that might have occurred elsewhere. The stability would not only have allowed species to originate but would also have provided long-term predictability in the environment, which allowed species to gradually become focused into narrow niches.

We now know that this picture is not entirely true. For instance, Lake Victoria—the northernmost of the three rift valley lakes—seems to have had a dramatically unstable history. Victoria is much shallower than the other two lakes and thus more susceptible to drying up when the climate changes. Geological evidence shows that it probably dried out during the last ice age only about 20,000 years ago, and most likely during each of the several ice ages that preceded it. If the lake dried out or became very small and saline, any freshwater fishes such as cichlids would have died, and all of the species that are now confined to it must have arrived and then evolved in place since the lake began to fill after the end of the last ice age. DNA evidence confirms that, indeed, most of the cichlid species that are found in Lake Victoria evolved only within the last 12,000 years or so from a couple of founder species that colonized from nearby Lake Kivu: a remarkable feat compared with the slow diversification seen in some other hotspots (Schliewen *et al.*, 1994). More recent evidence, though, suggests that at least some of the cichlids in Lake Victoria are older species derived from groups that lived there before the last ice age, and actually did survive the shrinking of the lake (Verkeyen *et al.*, 2004). Probably they hung on in a few very small freshwater lakes that survived within the dried-out basin. Those earlier survivors do not go very far back in time either, though. They mostly seem to have split off from one another as species within the last 100,000 years (Verkeyen *et al.*, 2004)—most likely because other earlier ice age droughts really did wipe out Victoria's cichlid fauna completely.

Lake Malawi has also had an unstable history, though not so extreme as Lake Victoria. During the aridity of ice ages, it probably fragmented into several smaller basins (Martens, 1997). Even in just the last few thousand years, it was severely affected by long-term cycles in rainfall, with the southern basin of the lake drying up completely. DNA evidence from the present-day fishes of the lake shows a mixed picture. Many of the cichlids of Lake Malawi diverged there millions of years ago but with an overlay of much more recent diversification. We see signs that the processes of species formation are still active, whenever environmental circumstances give them half a chance. For example, many of the subspecies of fish that inhabit the southern end of Lake Malawi seem to have formed several centuries ago when the level dropped, breaking it up into two separate basins for a while.

Lake Tanganyika is so deep (1,435 m) and extensive (like Baikal it is about the same area as Britain) that it has been fairly resilient against climate change, including the ice age cycles which chilled and dried the African continent. Nevertheless, it too was to some extent affected by these changes. The glacial periods cooled off Lake Tanganyika by several degrees, and aridity dropped the lake's surface level by more than 600 m. While the lake remained large and very deep, the fall in its water level would have meant that it shrank back around its edges and broke up into several large basins. Nevertheless, the prevailing picture for Tanganyika is stability. The DNA evidence in this case suggests a long, slow, and steady diversification over millions of years among the fishes there—but with an overlay of young species from the last 100,000 years or so, which might have formed as the lake broke up during the last ice age. It seems that Tanganyika itself is actually the long-term source for most of the cichlids of the other two big lakes.

So it really is a mixed picture for the role of stability in explaining the diversity of the three rift valley lakes. One of the lakes was apt to dry up almost completely, while the other two retained large lake basins in place but were severely affected by climate change—and one more so than the other. Tanganyika, the most stable, has the greatest diversity of fishes. Malawi is second in diversity and somewhat less stable. The most unstable, Victoria, has the least diversity. So stability does seem to help, since Tanganyika is the most diverse of the three.

And yet despite the fact that it almost vanished in the recent past, Victoria is not really far behind the other two lakes. It still has hundreds of species of its own—which shows that instability is not as much of an impediment as we might have expected. In fact in the two biggest lakes, Malawi and Tanganyika, it seems that a moderate degree of instability in the environment could actually have helped to generate bursts of new species. This has certainly been the case in Malawi, where fish populations have diverged as far as subspecies in only the last few centuries as a result of it being broken up by drought. In Lake Tanganyika too, the most severe ice age droughts would have fragmented it into three main basins, and in each of these new species may have formed. There are signs nowadays that the distributions of many cichlid species are still clustered around each of the three deep basins within Lake Tanganyika (Martens, 1997). So while stability helps, a certain amount of instability may well have helped to form new species, by splitting up populations from time to time.

On the subject of big lakes, ironically the best fossil record we have anywhere of species diversification in a lake is the long-vanished Newark Basin of New Jersey, more than 100 million years ago. Right where I sit writing here in my office in Newark, there was a subtropical lake that periodically dried up and then expanded along with wet and dry cycles in the climate. What we see in its sediments is the generation of six new species of fish (of the extinct genus *Semionotus*) within several thousand years after the lake had dried up into several separate basins that then merged as it refilled with water (Martens, 1997).

4.4.5 The story of the southwest Australian mallee

The history of the diverse southwest Australian flora has been traced back over several tens of millions of years from the fossil record of pollen and leaves. The region shows a slow progressive drying of climate, from vegetation dominated by cool temperate rainforest species in the early days, to forms adapted to warmer seasonal arid conditions. It is not entirely clear what aspect of its environmental history has made the western mallee so much more diverse than its eastern Australian counterpart, or most other Mediterranean vegetation zones around the world. It could be, for example, that the western mallee has been unusually stable during the climate cycles of the Quaternary. Or it might have had just the right amount of instability to repeatedly break up plant populations and allow them to form new species—without quite wiping them out. Hopefully, DNA and careful examination of the geological record will eventually come up with some answers.

4.4.6 The story of the Amazon hotspots

The western and central Amazonian hotspots have only a very incomplete and recent fossil record, which is typical of tropical rainforest regions. Judging by the limited amount that is known of the last ice age, the Amazon Basin as a whole cooled and dried out somewhat, shrinking down the area of rainforest, and thinning out many of the areas that remained. However, it is reasonable to believe that the western and central parts of the Basin would have fared better. The western hotspot area presently has exceptionally high rainfall, several thousand millimeters a year, and even with a large decrease in precipitation during the ice ages, there would still have been plenty to sustain rainforest plants. Other shorter lived drought events are known to occur periodically across the Amazon region due to times of warmer temperatures in the seas surrounding South America, and once again the very wet climate of the western and central Amazon may mean that they are well placed to survive any dip in rainfall with more than enough left over to sustain diverse species. Only towards the eastern

Amazon do we find evidence of large droughts disturbing the ecology of the forest. Some DNA evidence from the western Amazon also supports the idea that the forest and populations of animals and plants within it have remained stable through repeated glacials and interglacials. One study compared within-species DNA diversity found in various North American rodents with species in the western Amazon, looking for the effects of bottlenecks in population. In North America, all the rodents showed the signs of population bottlenecks having squeezed out the past genetic variability in the past thousands of years. By contrast, the rodent species from the western Amazon seemed to have remained steady in numbers, retaining high levels of variation in its DNA. This supports the idea that the western Amazon has long offered a stable environment for the plants and animals within it.

More support for the stability of the western and central Amazonian forests comes from the fossil record. The record in sediments off the east coast of Brazil, near the mouth of the Amazon, suggests that there was a major dry phase around 12,000 years ago at the same time as the Younger Dryas cooling that hit other parts of the world (Maslin and Burns, 2000). However, there is no sign of the Younger Dryas hitting the western and central Amazon, where good pollen records exist. It looks as if these two hotspots offered a stable environment for rainforest to persist despite the Younger Dryas, and maybe through the many other sharp climate changes that have hit the world in the last couple of million years. For example, we do know that a site close to the central Amazon hotspot also remained as forest during the Last Glacial Maximum, a time of dry climates around most of the world (Anhuf *et al.*, 2006; Adams and Faure, 1997, 1998). Two other sites near the western Amazon hotspot show no sign of loss of rainforest around the glacial maximum, although gaps in the sediment sequences could perhaps disguise a dry phase around that time (Anhuf *et al.*, 2006; Adams and Faure, 1997, 1998). As I mentioned above, the stability of western and central Amazonia can perhaps be understood in terms of its very high present-day rainfall: more than 4,000 mm annually over most of these areas. Even if the rainfall was reduced by a large amount—say half—during the last glacial, there would still be enough to keep rainforest in place there.

By contrast, there are tentative signs of dry conditions across many other areas of the Amazon Basin during the last glacial. In lakes across the Amazon Basin, deposition of pollen-bearing sediments stops during the Last Glacial Maximum, a sign of more arid conditions, before starting again as the cold phase ended. Ecologists and geologists who study the last ice age argue over what all this means: I tend to favor the view that the rainforest shrank back a lot in the eastern Amazon during each ice age, with shorter more intense drought phases scattered through. If the forest there did shrink back, similar arid phases repeating many times over the last two and a half million years could have been responsible for limiting the species richness of these parts of the Amazon Basin.

The persistent moistness of the west and central Amazonian climate over millions of years may then be the key to why these areas have been able to build up and retain such high levels of plant diversity. Lengthy dry phases in the eastern Amazon and in other rainforest regions may have caused periodic waves of extinction that knocked back their species richness.

4.5 SOME CONCLUSIONS: HOW IMPORTANT IS LONG-TERM STABILITY FOR HOTSPOTS?

Overall, what can we conclude about the environmental history of hotspots: Do they turn out to have been particularly stable environments? In a sense the answer is both "yes" and "no". Some of these hotspot environments have been in place for a very long time—several million years or more (e.g., the western Amazon, the Cape of South Africa, Lake Baikal). Others, however, have been remarkably unstable and have shrunken down or vanished at times within the last few tens of thousands of years (e.g., Lake Victoria, Lake Malawi)—and yet despite this, they are remarkably rich in species.

It seems likely that in some of these diversity hotspots, a moderate degree of instability has actually helped promote species richness. For example, in Lakes Tanganyika and Malawi, drought phases that broke the lakes up into smaller basins may have allowed species to diverge from one another. In Baikal, the cold of ice ages may have forced shrimp and fish populations to cluster around hot springs, isolating the populations and allowing them to form into new species. This is an example of a "species pump" in action. But too much instability would certainly have destroyed species. If Baikal had frozen right down to its bed, we can be sure it would have very few if any endemic species now—just as in the big freshwater lakes of North America which formed only after the last ice age. And yet, on the other hand, there is Lake Victoria—which paradoxically dried out and lost all or most of its fishes, then regained a diverse, endemic fauna in just a few thousand years.

4.6 PECULIARITIES OF LOCAL ECOLOGY: ARE THESE WHAT IT TAKES TO SET OFF A HOTSPOT?

Although the right general background of history might well be necessary for diversification, it seems that this alone is not sufficient. After all, for some hotspots there are similar environments elsewhere which have much lower diversity, even though they seem to have had a broadly similar history. Take, for example, the various Mediterranean regions of the world: Why did the Cape diversify so spectacularly whereas California or Chile did not? Or why did the western mallee in Australia become so much more species-rich than the eastern mallee? For plants in Mediterranean environments, a long and stable history might be more or less a precondition for species richness to explode in a hotspot, but a certain something extra might also be necessary to actually make it happen.

Likewise, even within a particular hotspot, some groups diversify and others do not: Why has it been the gammarid shrimps in Lake Baikal, and why the cichlids in the east African lakes? The very uneven way in which groups diversify within their hotspots suggests that the key is often something peculiar about the biology of the group which undergoes the diversification. If their characteristics had been just subtly different, they might never had speciated so wildly. If by chance there had been no cichlids in those rift valley lakes, would some other family of fishes have diversified in their place? My guess is that some groups would have been able to speciate like this, but others would not, so that it would depend upon the luck of the draw as to whether the big lakes started off with fishes or shrimps capable of diversifying explosively.

So we should consider the possibility that most of the world's hotspots are best explained not primarily in terms of stability, but in subtle details the local ecology or landscape, plus particular features which are unique to the clusters of species that have diversified there.

4.6.1 What is peculiar about the Cape?

In all Mediterranean scrub regions, fires caused by lightning frequently burn through the vegetation. This burning kills the top growth, so that the plants must grow back from near ground level or from seedlings sprouting in the exposed ground among the dead branches.

In the Cape diversity hotspot, what may be unusual is not the frequency of disturbance, but the slow growth of the shrubs on the shallow infertile soils. The concentration of important nutrients such as nitrogen and phosphorus in the soils of the Cape is generally much lower than in the less diverse regions that also have a Mediterranean climate: Chile, California, southern Europe, and southeastern Australia (Figure 4.9; Huston, 1993).

The important principle here may be that it is the balance between growth rate and disturbance that determines how species-rich a plant community can become. This is just a scaled-up version of the way that local-scale (Chapter 1) patchiness in species richness may be caused by the balance between these same two factors. In the Cape, then, the very slow rate of recovery of the vegetation from each disturbance may hold back the more aggressive competitors, and allow many more species to coexist as part of the vegetation.

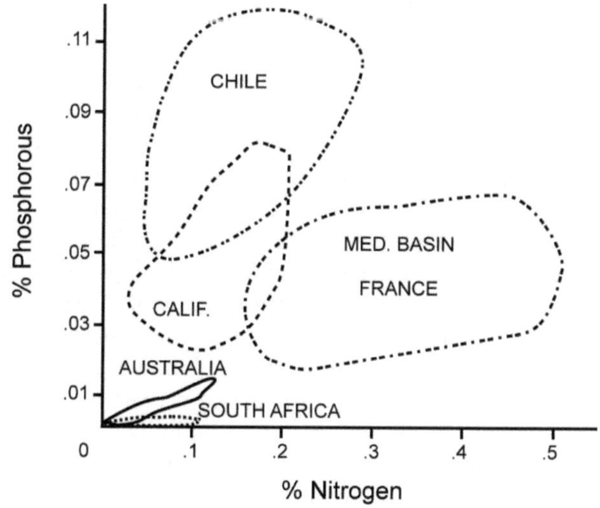

Figure 4.9. Nutrient content of soils in different Mediterranean climate regions around the world. Note the paucity of nutrients in the very diverse South African Cape (fynbos vegetation) and the west Australian mallee (after Huston, 1993).

Thus in a nutrient-poor and slow-growing vegetation, when new species occasionally form there is still room for them in the plant community. The disturbance allows lots of species that have very similar, overlapping niches to carry on coexisting in the Cape vegetation, without any one of them outcompeting the others.

The same pattern that contrasts the Cape with other Mediterranean regions may also occur on a smaller scale within the Cape itself (Chapter 1). Locally, the highest species richness is found on soils which are relatively low in nutrients and also shallow and droughty: both aspects which would tend to mean that the vegetation recovers slowly from disturbance, reducing the amount of competition between species.

The situation may be similar in the western Australian mallee. The soils in many areas of the western mallee are also very nutrient-poor (Figure 4.9) and this may be what allows so many species to coexist without some of them outcompeting all the others.

For a hotspot to accumulate a lot of species, there must also be a mechanism which generates these new species in the first place. A factor that may have helped so many species to form in the Cape hotspot is their relationship with certain species of ants which disperse the seeds. Many of the plant genera that are most diverse in the Cape have fatty lumps (known as oil bodies) on their seeds. The ants pick up the seeds below the plants and carry them a short distance to their nest. Inside the nest the ants prize off the oil body, which they eat, and then throw the seed out near an entrance to the nest mixed in with soil and other debris. There it may germinate and grow, with a bit of extra fertilizer from the waste of the ants' nest. Since seeds only get carried a short distance from where they fall, it has been suggested that the plants tend to form locally distinct populations that diverge and form new species. Certainly in the fynbos, many species have very localized distributions, being found in an area only a few hundred meters across, and nowhere else—as one might expect if they had formed in this way.

4.6.2 What is special about the cichlids in African lakes?

The spectacular diversification of cichlid fishes in the large African lakes has been the source of much discussion among evolutionary biologists. One characteristic that may have allowed a lot of species formation in the cichlids is that females are very picky about the detailed color patterns of males, and the color pattern they prefer is wired genetically within their brain. If there is anything slightly "wrong" with the pattern a male has (compared with her genetically ingrained ideal), a female will refuse to mate with him. In theory, speciation could easily occur if a very small population of a cichlid is temporarily isolated. Chance fluctuations in gene frequencies in the small population—genetic drift—may gradually cause the males to shift to a slightly different color pattern, or gradually cause shifts in the females' genetically fixed preferences (either following or leading the change in the male color pattern). As long as the changes are very small and incremental, most of the females will not balk at mating, but over many generations the cumulative changes in both sexes turn out to be quite large. In the end, any female who has the "old" genes for color preference will be unable to mate with the males that now predominate in the population, and her genes will not be passed on. All the females will now have the "new" set of color preferences. If the isolated population comes back into contact with the main population of the species it once was part of, the females of each population will refuse to mate with males from the "other" population. Effectively, a new species has formed, and the two populations now slowly diverge further in coloration and in their ecological lifestyle, never to interbreed again (Freyer, 1991).

This is an elegant explanation, but a question that has perplexed biologists is how the different populations of a species ever manage to get physically separated from one another, in a large but rather homogeneous lake. What sort of barrier can possibly prevent fishes from one population swimming across and meeting fishes of the other population before the divergence in colors and preferences has gone too far, and thus swamping out the process that would lead to a new species forming? Because of this problem of how to get the populations physically separated, some researchers have dared to think of a previously heretical idea in evolutionary biology: that new cichlid species actually form in the midst of their ancestral species, continuing to coexist while they each change to a point that they can no longer interbreed. Known as sympatric speciation (see Box 4.1) this process might be necessary to explain species formation in environments that are not dramatically broken up. Advocates of this idea have refined their case with computer models, which help make the general argument more rigorous.

However, we might not really need to invoke sympatric speciation, if we view the lake from a cichlid's point of view. What looks like an open and mixed-in environment in the lake may actually consist of many discrete breeding populations, located in little bays around the edges. Cichlids in the Rift Valley lakes have a habit of staying close to where they were born, and this lack of mobility will tend to allow populations to accumulate differences that may ultimately lead to separate species being formed. Even species that swim out into the open waters of the lake often return to the same ancestral area in the shallows to breed—and mating only with their home population will tend to keep that local population discrete, and let it diverge from others over time (Freyer, 1991).

4.6.3 What could be peculiar about the western and central Amazonian forest hotspots?

In the super-diverse western and central Amazon rainforests, with their very high levels of plant species richness, it could be that the secret of diversity is a combination of the precise ecological conditions that allows many more species to fit into the ecosystem.

One factor is likely to be the climate of these areas. Compared with other parts of the Amazon Basin, and other parts of the tropics in general, these rainforest hotspots have an unusually high rainfall—more than 4,000 mm a year on average, and in some places more than 9,000 mm. They also have no real dry season, with plenty of rain through all parts of the year, and not very much year-to-year variability in rainfall either. So these rainforest hotspots have a very predictable, steady moist environment. Why could this be so important for species richness? There are various possible reasons. One is that it may simply be "easier" for many slightly different shapes and growth strategies of plants to survive in physiological terms. As a result, evolutionary lineages of plants can easily "fall into" such an area and proliferate, with few constraints on their function when they produce many slightly different forms. This then is just a continuation of the latitudinal gradients in species richness that follow warm, moist environments of a global scale (Chapter 2). Here the "latitudinal" gradient just bends round east–west to follow differences in the climate. This east–west gradient in species richness in Amazonia is predicted by the JeDi model (Chapter 2) which also predicts latitudinal gradients in plant species richness. The JeDi model does not get it exactly right, and predicts the very greatest diversity should be slightly farther west on the other side of the Andes mountains. Nevertheless, it is interesting to consider that this model might be grasping at some fundamental truth, even if it misses it slightly.

Another possible reason why a predictably moist climate may favor diversity is that it allows species to adjust themselves to narrow and specialized ways of living. This again is an argument that is applied to tropical environments in general (Chapter 2), and the Amazon hotspots just represent an extreme scenario. Where there is no dry season, plants will grow, flower, and fruit all year round. This may allow specialized pollinating insects to evolve only to visit the flowers of certain species of trees (because there are always some in flower), or birds or mammals to specialize to take and disperse the fruits of a certain limited range of plants (because there are always some trees around in fruit). Plant-eating insects or fungi that attack trees can also narrow in to a very limited range of host plant species, because they provide such a dependable resource—always growing, always shedding dead leaves. The plants may themselves respond by adaptations to each of these types of organisms they interact with. They can evolve to encourage their specialized seed and pollen dispersers, and to fight off the relentless attack by their enemies. They may also evolve in response to the other plants around them, which are always active and growing, always a factor in the environment. Specialization like this seems likely to lead to splitting of species, as different populations evolve off in their own specialized directions in response to the more intense threats and benefits around them.

Against the background of a very stable moist environment that allowed many specialized species to build up, another factor that may allow huge numbers of plant species to coexist is the soil of the western Amazon. Silman (2005) states that the western Amazonian soils are relatively fertile compared with those elsewhere in the Amazon Basin. However, apparently contradicting this, Huston (1993) emphasized the idea that the key to the hotspot of the western Amazon is that it has relatively infertile soils. He suggests that in the warmth and the perpetual dampness of this climate, the chemical processes that break down minerals in rocks and soils (known as weathering) have leached away most of the nutrients, leaving only inert end-products such as clays and oxides. Huston cites chemical analyses of western Amazon soils revealing that they have very low concentrations of nutrients, despite the lushness of the vegetation and despite the potential of the rocks underneath to yield nutrients. Essentially, as soon as nutrients within the ecosystem are released from the breakdown of plant or animal tissues, they are taken up again by the greedy fungi (mycorrhizae) that serve as nutrient-gatherers for the plants. So in western Amazonia, according to Huston, the abundance of greenery is a deception; it does not indicate fertility. Plants are chronically starved of the nutrients they need, and because of this most of them are growing slowly (except where they scramble to exploit the short-lived burst of nutrients released when a large tree crashes down and its leaves, fine roots, and branches start to rot). If most species of plants are growing only slowly, the strongest competitors in the forest do not get much chance to build up their numbers and push out the weaker ones, before some random disturbance such as a landslide or a treefall jumbles everything up. In the ecological chaos that follows such an event, there may be a free-for-all on the open ground which allows seedlings of just about any species to establish irrespective of how well they would fare in a long and grinding contest against the best-competing tree species around there.

Thus according to Michael Huston and some other ecologists, one of the keys to the richness of the western Amazonian forests is that their soils are fairly infertile. Soil infertility equals slower growth of trees and longer generation times (as they take longer to reach maturity). This means that the competitive battles between populations of stronger and weaker species in the forest are not fought out to their conclusion, before the game ends and the clock is restarted by a disturbance event. In a sense, the slower growing the trees are, the more frequent the disturbance events *seem* relative to their own lifespans, even if these disturbances are spaced just as many years apart in western Amazonia as in other parts of the region.

Paul Colinvaux has emphasized a complimentary idea for the high diversity of the western Amazon. I remember he came to give a very interesting seminar at Oxford when I was an undergraduate there in the 1980s, and it was one of the things that most inspired my early interest in biological diversity. His suggestion then was that in the western Amazon, disturbance is more frequent. The very high rainfall and rapid breakdown of rocks means that deep soils form rapidly in this environment, and on a slope they can easily become unstable and slip downhill. This tends to give a much higher frequency of landslips, and may also mean that large trees topple over more often. The rivers too are often swollen torrents that burst their banks and shift channels, sweeping away areas of forest and leaving bare mud that can be colonized by seedlings in a new free-for-all. So according to this idea, what tends to prevent competitive contests between species from reaching their endpoint in the western Amazon forests is disturbance being more frequent. On reflection, I am rather skeptical now that disturbance rates are really that much greater in the western Amazon, especially as most of the forest is not on river floodplain—but it would at least be interesting to see some careful comparisons of disturbance rates in different areas of the rainforests.

Even if slower growth and more frequent disturbance are what allow so many species to coexist in the western Amazon forest, another mechanism might be necessary to explain what generates all these species in the first place. Botanists have noticed that many plant species in the western Amazon hotspot have very localized distributions (e.g., confined to just a single ridge or hilltop). It may be that the topography of the area, with many fairly small hills and ridges, breaks up the ranges of species frequently—allowing them to evolve in different directions and form separate species. Stability of climate, lack of nutrients, and frequent disturbance might be necessary to allow so many species to coexist in the western Amazon: but what actually generates the species that take advantage of this moist climate in the first place could be the splitting up of populations by varied topography.

4.7 DO HOTSPOTS HAVE MORE ROOM FOR SPECIES, OR HAVE THEY JUST BEEN GIVEN AND RETAINED MORE SPECIES?

Two different threads run through the various explanations for species hotspots that I have discussed so far. One of these suggests that a key to many hotspots is that they have more room for species, compared with other less diverse areas. This may be either because there are more different ways of gaining a living—more niches—or because niches can overlap more in hotspots.

The other thread suggests that it is not so much a question of greater capacity in hotspots, and that in fact other similar but less diverse environments could also fit in just as many species. The key factor may instead be how many species have originated in that area, through the process of speciation. Perhaps if speciation had not worked overtime, there would be just ordinary levels of diversity? It may be just a matter of luck then, that in certain places some lineages have exploded into a mass of different species, because of some peculiarity about their biology (e.g., the breeding system of cichlid fishes).

If it is not greater speciation that is the key to hotspots, then maybe the rate of extinction is less, allowing a higher number of species on balance despite the inevitable turnover. Very low rates of extinction might be the result of an unusually stable environment, for example. Or it could be some peculiarity of the types of organisms there too which allows their populations to recover well from occasional disasters.

In fact both aspects (higher capacity *and* more speciation/lower extinction) might have to operate together to make a hotspot. So a hotspot may require greater species production or better retention of species to fill up a greater available capacity. If either of these is missing (i.e., if a place has low rates of speciation/high rates of extinction), or a low overall capacity for species—then that particular area will not become a diversity hotspot.

4.8 COLDSPOTS

Though they do not get as much attention, another very interesting set of places are those which have ended up with far fewer species, compared with what is normal in broadly similar environments elsewhere. They do not seem to have been given a general name, so here I will label them "coldspots"—being as they are the opposite of hotspots.

Some of the most obvious examples of coldspots are islands and peninsulas, and there is a large and venerable branch of ecological theory that seeks to explain what controls the species richness of these places. Other anomalously species-poor areas also occur, often in areas of particularly unusual climate (e.g., very damp cool summers on the edges of oceans).

4.9 EXPLANATIONS FOR WHY DIVERSITY COLDSPOTS OCCUR

4.9.1 Island coldspots

It has long been known that islands tend to be species-poor, compared with equal-sized areas of mainland. One very obvious explanation for this is that if an island forms anew out of the sea without any land-living species on it (as many islands do, built up, for example, by undersea volcanoes), then it will be difficult for life on the mainland to reach it.

Species tend to reach islands by chancy processes. Land birds blown out to sea may reach islands by luck before they succumb to exhaustion. Small mammals and reptiles can be carried across on rafts of debris that float down rivers and out into the ocean, and be washed up on the beach of an island. Presumably often it is a fertilized female that turns up, able to establish a ready-made founder population with her offspring. For every land-living animal that manages to make it to a remote island hundreds of kilometers from the coasts, there must be many millions that are not so lucky and end up drowning. Plants too may arrive against huge odds when their seeds are blown out to sea, mixed in with rafts of debris, or carried in the stomachs or plumage of birds.

There are some species, however, that seem quite expert at making the leap across large stretches of ocean water. The seeds of some shoreline plants such as the coconut (*Cocos nucifera*) and beach morning glories (*Ipomea violacea* and related species) resist seawater and float, and can easily establish on the salty soils of beaches. Certain seabirds do not particularly need land, and can feed out at sea and rest floating on the waves for weeks at a time: one example is the wandering albatross (*Diomedea exultans*). This makes it especially easy for seabirds to reach remote islands thousands of miles from any shore. However, such species are very much the exception, and there are not enough of them to contribute much diversity to islands.

It is no surprise then that islands tend to be poorer in species: they are simply hard to reach. And the smaller and more remote they are, the less likelihood that life from the continents will wash up on their shores. This fits in with the generally held observation that more isolated and smaller islands tend to have fewer species than islands that are larger or closer to the mainland.

In the 1960s two ecologists, Robert MacArthur and Edward Wilson (1967), thought of another factor that might tend to make islands less diverse than the mainland: extinction. In the confined area of an island, the total population of an organism is going to be fewer than on the much larger total landmass of the mainland. MacArthur and Wilson knew that, simply in terms of mathematics, chance fluctuations in numbers are far more likely to drive a population extinct if it is always held down to low numbers by lack of food or territory. As we will consider further in

Chapter 8, a small population of just a handful of individuals is likely to go down to zero (i.e., to go extinct) within a few generations, simply due to chance fluctuations in individual health, accidents, and male–female ratios. This is known as "stochastic" extinction, from a Latin root-word meaning "chance". Stochastic extinction comes from within the population itself, essentially independent of the environment.

Operating quite aside from this is another related factor, the probability of a small population being wiped out by disasters. If the population size is smaller, there is more danger that chance events will stamp it out to extinction. For example, imagine that there is a big storm or a drought which kills 99% of the individuals of a species on both the mainland and the island. If the population on the mainland was 10 million before, there will still be thousands of individuals left to re-establish the population afterwards. Even if it dies out in local areas, individuals can easily spread back in from surrounding areas of mainland and re-establish these local populations. By contrast, if the island population has only 50 individuals, there is a good chance that there will be none left and the population will have gone extinct—and it is not likely that there will be more individuals arriving anytime soon from elsewhere to re-establish the population.

Frequent extinction then is a second factor that may suppress the species richness of islands. Even if a species manages to reach an island and establish itself there, its long-term survival there is not as assured as it would be on the mainland. Generally, smaller islands should tend to lose species more often than large ones, because populations are smaller and thus more susceptible to extinction events.

MacArthur and Wilson visualized the actual numbers of species that occur on each island as a balance—an equilibrium—between the rate at which species arrive (or recolonize, if they went extinct already) from the mainland, and the rate at which they go extinct. In their scenario they mainly considered stochastic extinction, but "disaster" extinction could act in the same way too. So the actual number of species we find on an island depends not just on what reaches there, but on the *relative* rates of the two opposing processes: arrival and extinction (Figure 4.10). If 100 species of animals establish populations on an island in a century but 98 of those populations go extinct, there will only be 2 species present despite all the colonization that went on.

If there is even slightly more colonization than extinction on an island, what should happen over time is that the number of species there grows and grows. Eventually though, its species richness will tend to level off at a fairly stable level. This is not necessarily because it is in any sense "full" of species, though. According to MacArthur and Wilson's simple idea, the island stops accumulating species only when the rates of chance extinctions match the rates of chance colonizations, not because any true "capacity" in species richness has been reached.

Though MacArthur and Wilson did not emphasize it in their early scenarios to explain the species paucity of islands, it is possible to see that other factors may kick in beyond the simple balance between arrival and extinction. For example, it makes sense that there *could* actually be some upper limit on how far species richness on an island can rise, because there are only a certain number of "vacancies" (niches) available within the animal and plant communities on the island. Just like in a community on land, this may limit how far the island can ever go on loading up

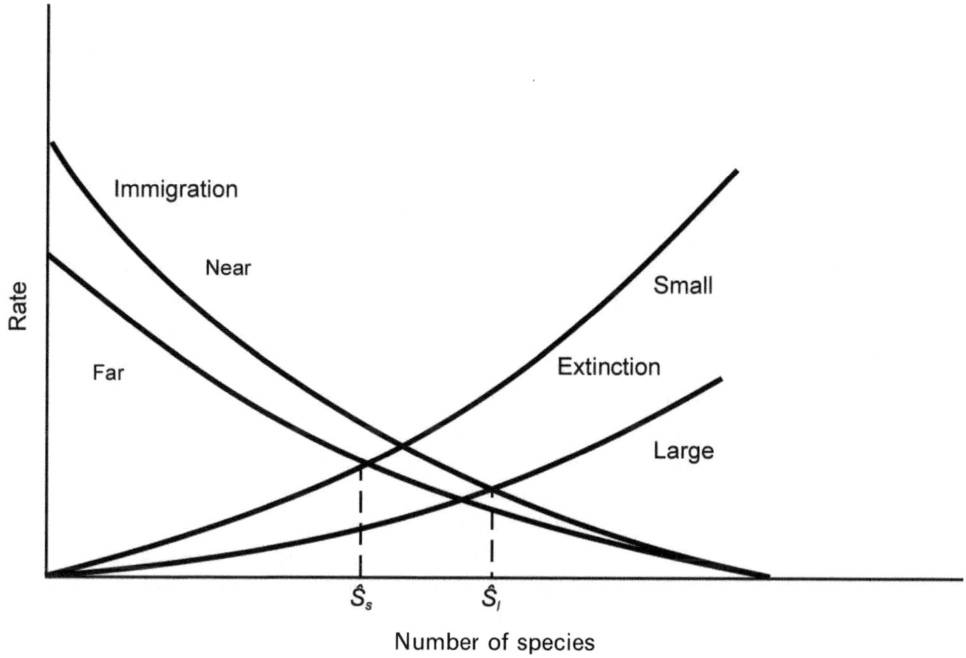

Figure 4.10. The balance of factors determining the species richness of islands according to MacArthur and Wilson's original theory. Small islands suffer more extinctions than large islands, dropping down their level of species richness (after MacArthur and Wilson).

with species. As more and more species establish on a newly formed island, the latecomers may have more difficulty fitting in, or if they do form a stable population they may have to push other species out in the process. But bear in mind that the very idea that such a ceiling on richness exists in ecosystems anywhere is still to some extent controversial (see Chapter 1).

In any case, in MacArthur and Wilson's scenario, whatever stable level of diversity the island tends to reach, it is still one that would be higher without the extinctions of populations that constantly occur there. The smaller the island, the more often things go extinct, and so the lower the plateau level of diversity.

So MacArthur and Wilson's idea basically states that island diversity is a balance between arrival and stochastic (i.e., chance) extinction. It is the most popular explanation of island coldspots, explaining not only why islands tend to be poorer in species than similar-sized patches of the mainland, but also why some islands have more species than others. It has really grabbed the attention and imagination of ecologists, and just about every introductory college course in ecology mentions it. Perhaps part of its appeal is the understandable craving that we ecologists have for neat, simple rules in nature: a sort of "physics envy". Because of all the enthusiasm about it, a lot of effort has gone into testing whether MacArthur and Wilson's scenario actually occurs—with encouraging but somewhat mixed results.

How is it possible to test an idea that deals with something on the scale of whole islands? For a start, in a general way we can compare different islands and see how their pattern of diversity matches what we should expect from the theory. One thing to expect is that if islands of differing size are compared, the larger islands should have more species (because extinctions are less common, given the larger populations of each species there). And indeed, when the size of islands in the Caribbean is plotted against the number of bird species found on each island, there is a close relationship. Although this is consistent with what one would expect from MacArthur and Wilson's theory, a complicating factor here is that a larger island is likely to have a greater range of habitats, each with its own set of species, so a larger island could be more species-rich for this reason alone, unrelated to the processes described above that balance arrivals and extinctions. Nevertheless, at least the data do not clearly refute MacArthur and Wilson's theory. It is always good to have an idea survive a test, even if it is not a particularly rigorous test.

While this example agreed well with the expectations of the theory, when more Caribbean islands were added into the data analysis the relationship became far more scattered than the classic graph that is usually shown (Gilbert, 1980). And another study of the Caribbean islands found no relationship at all between *Anolis* lizard species richness and island size. Likewise, when the diversity of bats on Caribbean islands was compared, or the diversity of ants on different islands in the Seychelles, no relation to area was found (Gilbert, 1980).

In a much more recent and extensive study, Holger Kreft and colleagues (2008) compared the plant species richness of 488 islands around the world. They found that about two-thirds of variation in species richness among islands was predictable by island size—apparently demonstrating just how important island area is. I am a little wary of the simplicity of this conclusion—since small ocean islands also tend to be more distant from anywhere else, although I guess that the isolation aspect is also part of MacArthur and Wilson's theory (see just below).

Another test is to compare islands of similar size but different distances from the mainland for their species richness. MacArthur and Wilson's theory predicts that since dispersal of species to an island is an important limit on its diversity, more isolated islands should have fewer species because they are harder to reach. Comparing small islands in the eastern Pacific, located at differing distances from the large landmass of New Guinea, the eminent ecologist Jared Diamond found that the most isolated islands have fewer species. However, complicating things in this instance is the fact that the more easterly and isolated islands tend to be drier, and in many parts of the world drier climates are less rich in species (Chapter 2). Once again it is difficult to be sure what influence this additional factor might be having in producing the pattern we see.

The different climate of islands, compared with the mainland, may also be at play in certain cases in the high latitudes of the northern and southern hemisphere. These are islands that are simply devoid of certain important life forms, even though they seem to have a moderate climate and are not too isolated. One example is the Faroe Islands, north of Britain and west of Norway. The Faroes completely lack the species of trees that are so abundant on the continent, even though they were connected to the mainland after the end of the last glacial before the sea level rose. In this case the problem seems to be in the detail of the climate of the Faroes. This is a very moderate climate, but a little *too* moderate. Although it has mild winters it also has very cool summers dominated by the ocean breezes: there is not enough of a seasonal temperature cycle to warm trees up to grow in the summer. Even when trees are planted there, they simply die or end up badly stunted unless they are coddled and protected. In this case, lack of species richness in trees is not so much a legacy of history as the product of an unsuitable environment.

In warmer climates, something else that can differ from one island to another is how often they get hit by hurricanes. Some islands have been lashed by hurricanes particularly frequently in the past couple of hundred years, either by sheer bad luck or because their location puts them within certain "hurricane belts" (Whittaker, 2000). Volcanic activity is another thing that differs from one island to another. Both of these can wipe out populations of species directly, and also affect the types of habitats available—which influences species richness. So variation in hurricane frequency or volcanic activity could perhaps cause islands to have differing levels of species richness, independently of the factors that MacArthur and Wilson advocated. It is not hard to find factors that complicate the comparisons between islands.

A prediction of the island diversity idea of MacArthur and Wilson is that each island should keep a relatively constant level of species richness over time, but that the detailed species composition of the flora or fauna should be in constant flux as species come and go. In some of his early work in ecology, Jared Diamond compared his own observations with records of the bird fauna of the California Channel Islands, off the coast near Los Angeles, for signs of this constancy with turnover. He found that over nearly 60 years, from 1917 to 1967–1968, there had indeed been a considerable amount of turnover—as many as 62% of the species on some islands—yet overall species richness on the islands had not changed much.

This then seemed to be in agreement with the predictions of the theory. However, Diamond's study also found that turnover did not depend on the size of an island or its distance from the mainland, which contradicts what the theory expects. Another complicating factor is that the original 1917 data were not a simple two-year tally (as Diamond took in the 1960s) but a compilation of all records of species over the *previous* 50 years or so (Gilbert, 1980). This seems to make the basis for comparison of numbers look a bit suspect. And despite the turnover that occurred, there is obviously a substantial core of bird species that do not turn over, because they belong to quite separate subspecies found only on the California Channel Islands. This again is a complication beyond the idealized system that MacArthur and Wilson had in mind (where any species on the island is fairly susceptible to extinction), and yet these "permanent residents" are certainly a feature on many islands. One must also consider the likelihood that massive brush fires, and the extensive use of DDT, on the islands from the 1950s onwards also contributed to the turnover of bird species. Adding further to the mixed picture for MacArthur and Wilson's theory, the "steady state" number of bird species on each island was only loosely correlated with area and isolation, with the number of plant species actually turning out to be a much better predictor (Gilbert, 1980).

Looking at birds on Caribbean islands, the great ornithologist David Lack (1972) made the point that much of the variation in species richness seemed to be due to differences in the number of habitats on each island. The birds didn't seem to have very much difficulty getting there, and once they had arrived they didn't seem to be going extinct, as they were nearly always present as long as the right habitat was there on the island. So in this case, it does not look much like bird diversity on the Caribbean islands is at the mercy of the chance processes of colonization and extinction. Instead it is a predictable, repeating pattern (Lack, 1972; Whittaker, 2000).

Jared Diamond (1974) reached somewhat similar conclusions from observing the bird communities of different ages on Pacific islands around New Guinea. He studied how bird communities reassembled themselves after big volcanic explosions swept each island clean of life (as tends to occur quite often in that geologically restless area). At the early stage there is a lot of chance in what arrives, and a fair amount of extinction as MacArthur and Wilson would suggest. Even so, the birds which tend to arrive early on are drawn from a particular set of species that are constantly on the move and good at finding islands—species which Diamond calls "supertramps". Later on a fairly predictable set of species arrives, and these tend to stay on with less turnover. So this is not quite what the island biogeography theory predicts—which would be random arrival and continual arbitrary turnover.

These and various other studies I haven't mentioned here suggest that things are not nearly as simple when it comes to island species richness as MacArthur and Wilson had initially supposed when they came up with their theory back in the 1960s. Ecology textbooks do not tend to mention the complications and exceptions, just focusing on the neat examples which do work. This is understandable, because their aim is to illustrate a concept to students who are only just beginning—but it has been a shock for me to realize that the principle that I took in as an undergraduate as a sort of universal law in ecology, actually has many glaring exceptions.

4.9.2 Island biogeography on land and in lakes

There are times when we unavoidably have to skip across scales, from geographical to local: and this is one of them. For it may be that the geography of islands is all around us in miniature, in every little patch of habitat. It seems reasonable to suppose that isolated bits of any habitat could behave like islands (e.g., in terms of reaching a dynamic balance between species immigration and loss). Lakes and ponds surrounded by land, remnants of forest in a mainly denuded landscape, or isolated mountain peaks surrounded by warmer lowlands, all may in a sense resemble islands from the point of view of species that are specialized to live in each of them. The rather island-like nature of these places is often considered in nature reserve design, in attempts to try to hold on to populations of as many species as possible (see Chapter 8). Various studies have also tried to determine whether such "inland islands" show the patterns of variation in species richness that would be expected of them from

MacArthur and Wilson's scenario. Overall, the picture is mixed, but in general only mildly supportive of the idea that random immigration and random extinction dominate variation in species richness among such islands.

A study by Barbour and Brown (1974) in the U.K. found that the number of fish species in ponds and lakes showed no relation to the size of the water body (in effect, the size of the "island"). And in fact, another study (Scheffer *et al.*, 2006) on 215 small lakes in the Dutch countryside found that the smallest and most isolated ones tended to have *higher* species richness in terms of birds, plants, amphibians, and invertebrates. This is precisely the opposite of what would be expected from the theory! Scheffer *et al.* suggested that this pattern emerged because the most isolated little ponds are less likely to be found and colonized by fish, which eat back the submerged aquatic vegetation and have all sorts of cascading effects on the ecosystem. It seems to emphasize the point, that to really understand islands we need to consider the complex nature of species interactions—it is not simply a numbers game of identical species coming and going.

Comparing patches of temperate forest of different sizes in North America, Galli *et al.* (1976) found a significant relationship of bird species richness to the size of forest patch. However, a similar study on patches of forest in Britain found little correlation between the number of breeding bird species and patch size, or the distance to neighboring woodland patches (Heliwell, 1976).

MacArthur and Wilson's idea has done better overall in explaining the species richness of smaller creatures on island-like patches of plants. From example, Brown and Kodric-Brown (1977) looked at the variety of arthropods (insects, spiders, etc.) on thistle patches, and found more species on larger and less isolated patches. Ward and Lakhani (1977) found that the number of species of plant-eating insects on patches of juniper (*Juniperus*) was related to the size and density of the "islands" of juniper in the landscape. Kruess and Tscharntke (2000) found that the species richness of insects attacking another wild plant, vetch (*Vicia*), was also greater in larger less isolated patches.

4.9.3 Some experimental tests of MacArthur and Wilson's hypothesis

Studies such as those I have mentioned, which compare natural variation in nature with test hypotheses, are often called "natural experiments". However, they are not really experiments—just sets of observations. Always when comparing such large-scale natural systems, there are many different things that vary in parallel. This usually makes it rather difficult to draw firm conclusions from them.

In science, true experiments are the ideal way of testing an idea, because we can keep everything that might affect the results more or less constant, except for the factors we actually want to study. Unfortunately for the present purpose, processes that control species richness tend to occur over large distances and long time periods, so it is not often that we get a chance to do such experiments. However, an early test of island diversity theory *did* actually consist of a real experiment, dismantling island communities and then watching them reassemble. This study, which has become a classic in ecology, was conducted in 1969–1970 by the same two ecologists (MacArthur and Wilson) who came up with the theory that island species richness is a dynamic balance between the rate of colonization and the rate of extinction.

MacArthur and Wilson conducted their experiment on a set of small and relatively manageable islands, just off the coast of Florida. In the shallow waters off southwestern Florida, islands of salt-tolerant mangrove trees often establish on the mud flats that are exposed at low tide. These islands are all sorts of different sizes and shapes, and each one has its own community of insects that live among the branches of the trees. The basic strategy of the two ecologists was to log the insect species richness of some of these small islands to see how it varied according to size and distance from the mainland. Then they "wiped the slate clean" by enclosing each island in a tent and fumigating it with insecticide that killed all the insects on each island. In the months afterwards, they watched the insect fauna re-establish itself on each island by colonization, and listed the species that they found.

What the theory would predict, then, is that species richness would initially vary according to island size and its degree of isolation from other islands and from the mainland. What MacArthur and Wilson actually found was that in the first place, before spraying, smaller and more distant mangrove islands had fewer species of insects. So this, indeed, is what their theory predicts.

The theory also predicts that, after spraying, each island should gradually return to about its former level of diversity, dependent on size and isolation. And, very importantly, the actual list of species in the fauna of each island should differ before and after spraying and recolonization. This is because, according to MacArthur and Wilson's theory, it is not the identity of the species present which is important: they each come and go randomly and independently of one another. The list of species is then just the product of chance colonization and extinctions. Over time, the list of species on an island should change—a process of turnover that can be accelerated if all of the previous species on an island are wiped out, and the island is then allowed to re-accumulate species by chance.

So what actually happened? Following spraying, when the insect faunas of the islands were recovered, each island was found to have roughly the same number of species as before. And the list of species was different even though the total number of species was about the same. This showed then that the species richness of an island can be a predictable function of its size and isolation, and that the richness is actually produced by a balance between continuous extinction and continuous colonization of some of the many different species that could potentially arrive there.

The experimental approach was taken a step further on the same set of mangrove islands by Daniel Simberloff, a graduate student of Wilson. He tested a prediction that follows on from MacArthur and Wilson's theory: that if you could somehow reduce the size of an island, its species richness would decline as a result of extinctions becoming more frequent among the reduced populations. Simberloff recorded the insect fauna on various mangrove islands, and then made some of them smaller by cutting parts off with a chain saw. After giving the chopped-up islands time to recover, he counted up the insect faunas and found that whereas the untouched islands had about the same species richness as before, the reduced islands now had fewer species—just as would be predicted by the theory (Simberloff and Wilson, 1970).

Other experiments have also tried creating or depopulating islands of different sorts, and watching them recover their species richness. For example, Patrick (1967) set up glass microscope slides in the current of a stream and watched them accumulate species of diatoms. As the theory would predict, larger slides accumulated more species. But on larger slides, there was actually an increased rate of species turnover—the opposite to what the theory predicts!

Davis (1975) sprayed insecticide to kill off all the insects in patches of nettles in England, and found that the numbers of insect species that returned on recolonization did not depend on the size or isolation of the nettle patch, thus contradicting what the hypothesis would predict.

So, some of the various other experimental studies seem to support MacArthur and Wilson's idea, while others seem to contradict it. It is fair to honor MacArthur and Wilson's idea with the term "theory", because it has survived some important tests which could have disproven it. But combining its results with the picture from observing the natural history of groups of islands, it seems that we should not simply accept that the simple processes of chance arrival and random extinction (which this theory involves) are the single overwhelming control on species richness patterns on islands.

Instead, island species richness probably depends just as much on many other things such as habitat diversity, and interactions between species. Each island is likely to have a core of species—some that colonize easily and do not often die out (e.g., coconuts), in contrast to certain others that come and go frequently (e.g., supertramp bird species). There may also be others that do not colonize easily, but also do not often die out once they have got there. And in addition, in the longer term there is evolution of endemic species on islands, which dominates the species composition and richness of many distant islands (see below). In his critical review, Gilbert (1980) concluded with a quote from MacArthur, which was in turn taken from Picasso: "A theory is a lie which makes you see the truth." This seems too harsh a way of describing MacArthur and Wilson's theory of island species richness: the forces it invokes are surely part of what determines the diversity of islands, even if they are not the whole story. MacArthur and Wilson were both quite capable of realizing that things were not so simple with islands, and they loosened up their theory over time to allow for other factors coming into play (as MacArthur recognized in his book, published in 1972). But admittedly it is true that, with MacArthur and Wilson's theory as a target to aim criticism at, we have ended up with a clearer picture of the various factors that control the species richness of islands.

4.10 THE PENINSULA EFFECT

For a long time, ecologists have also noted a "peninsula effect" (MacArthur, 1972). Just as islands tend to be poor in species, so do peninsulas, though their diversity is not as low as islands. One apparent example of a "peninsula effect" is the lower tree species richness on the large peninsula of Kamchatka, which hangs down from the top of eastern Siberia, far north of Japan (see Figure 2.20).

What could cause peninsulas to be less species-rich? It may be a dose of the same two processes that MacArthur and Wilson invoked for islands. Since a peninsula is relatively cut off from the rest of the mainland, it is harder for populations to reach it. That is one thing that will tend to damp down its diversity. Also, because their populations tend to be relatively small and isolated, despite some connection to the mainland, species on a peninsula may be more susceptible to dying out there. So this is another factor suppressing diversity, just as on an island.

There may be another factor too, in some high-latitude places. This is that peninsulas are very much influenced by the sea and tend to have cool, damp summers. This is something that might be expected to decrease the species richness of trees, butterflies, or other groups needing summer warmth. Generally, the evidence for a real "peninsula effect" as MacArthur would explain it is rather weak. Lower latitude peninsulas might well tend to be less species-rich because their climate is just different from the adjacent mainland. For example, Wiggins (1999) suggested this for the bird diversity of the Baja Peninsula of California, which is drier and has different vegetation.

4.11 BURSTS OF SPECIATION ON ISLANDS

Although islands in general are poor in species, there are some cases where nature has bucked the trend and produced a whole set of unique species. This sort of thing only tends to occur on the most remote islands, where the animals and plants that do manage to reach there and establish populations are rarely disturbed by new arrivals. Endemic species will only tend to form on islands when the flow of individuals (and the genes they contain) from other lands is too small to prevent the island population from becoming different. In time, the island population can eventually become a new species. And, presumably, the time taken for island species to evolve must often be lengthy, which requires that the island offers a stable environment lasting millions of years.

In terms of producing endemic island species, it also helps if most of the niches on the island are left unfilled, because only a very incomplete range of species has managed to make it across. This means that the species which *have* managed to reach there can eventually produce offshoot populations which evolve to fill these niches—and these each become new species. Speciation to fill a range of different niches, either on an island or a hotspot on land, is called "adaptive radiation". This is an apt term, for the evolutionary tree looks just like a burst of rays spreading out from the central point of one ancestor, ending up in a range of very different forms.

There are many examples of adaptive radiation on islands, but perhaps the best known is the finches on the Galapagos Islands in the eastern Pacific—made famous by Darwin who was the first to recognize the evolutionary forces which had produced them. From a single seed-eating finch population that arrived from the South American mainland millions of years ago, they have produced a great diversity of species, with different shapes and sizes and behaviors, adapted to a wide range of niches.

Several other examples of this process come from the Hawaiian chain of islands. For example, the little fruit flies of the genus *Drosophila* have produced something over 1,000 species there, all descended from one common ancestor which arrived millions of years ago. On tiny St. Helena in the middle of the Atlantic there are also the daisy trees that evolved from some little weed that arrived millions of years ago (Chapter 5).

Hence, within the world of island coldspots, we do sometimes find hotspots of a sort, where the circumstances are right.

4.12 COLDSPOTS MADE THROUGH GLACIAL EXTINCTIONS

Some areas of the world seem to have low species richness that has been depleted by the effects of glacial history. Europe is much poorer in tree species than eastern North America, and especially temperate eastern Asia (Chapter 3). In the late 1800s, the botanist Asa Gray pointed this out and suggested a possible explanation for this difference in species richness. He pointed to fossil evidence that all three regions had a very diverse flora several million years ago before the ice ages started, and suggested that Europe had suffered more extinctions because its mountain ranges run east–west forming a belt (running from the Pyrenees, the Alps, and the Balkans to the Caucasus) that divides Europe into southern and northern halves. During ice ages, Gray suggested, European trees could not make it to potential refuges in the south because the mountains blocked their way. The result was that they went extinct. In eastern North America and Asia, by contrast, mountain ranges have a north–south trend, with broad coastal plains that could allow species to escape southwards during glaciations, and then spread back north as the climate warmed. Hence in these other two regions, many more species survived. I remember being utterly fascinated by this story as a bookish teenager reading *Bellamy's Europe*, which is a wonderful popular account of the natural history of Europe. In fact I was so enthralled that I ended up doing my master's degree on the topic a few years later.

Piecing things together, it now looks like the time-honored explanation (compelling though it had seemed) is not quite correct. For one thing, western Europe too had a broad flat coastal plain that was exposed during low sea levels during ice ages, and that could have allowed species to migrate south by slipping down along its western edge. So there is no reason why mountains had to be a problem for the trees moving southwards as the glacial began, at least on the western side of Europe. And there is another problem for the traditional explanation: it actually looks like the mountains themselves were a cosy, hospitable refuge that saved many tree species, rather than a barrier that caused extinctions! There was severe drying as well as cooling in southern Europe during the glacials (Tzedakis, 1994; Adams and Faure 1997, 1998) (Chapter 3), which could well have been the reason why fewer tree species survived there. Pollen records show that many of the types of trees that Europe still does have were absent or nearly absent from the lowlands of southern Europe, but

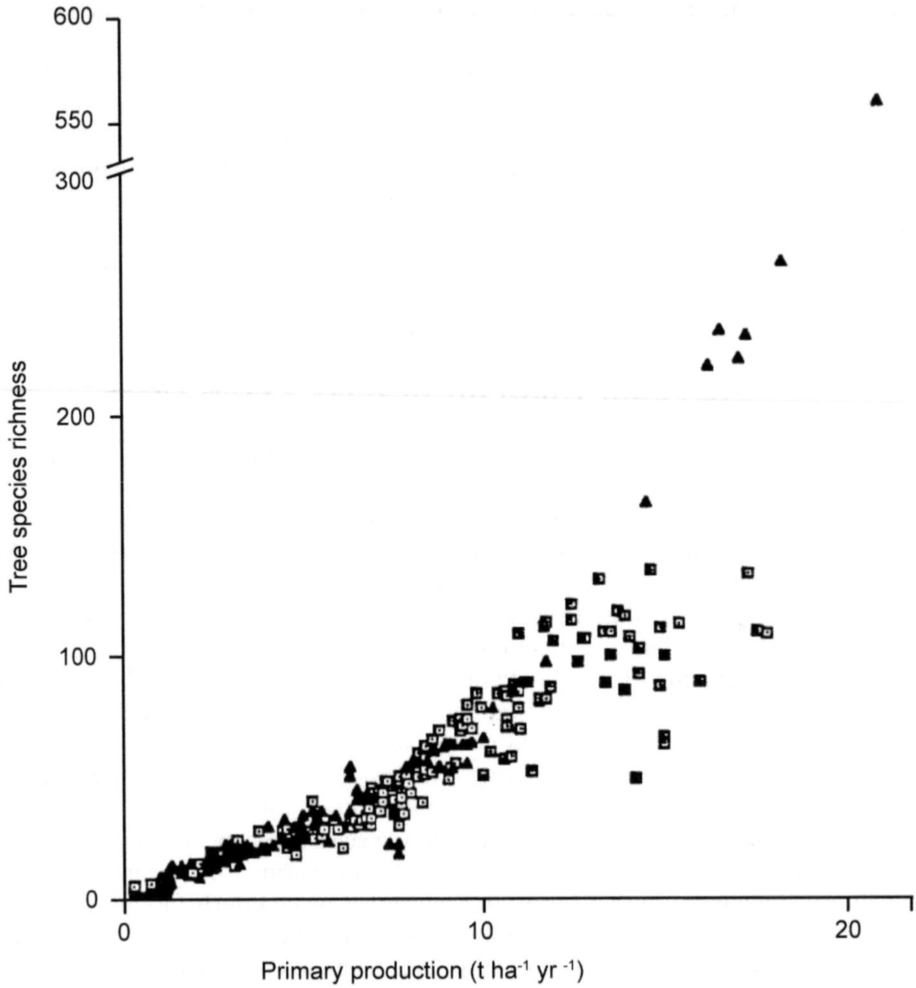

surviving in the rainier mountain valleys of the Balkans and Apennines (Tzedakis, 1994). Uplands tend to be wetter because air is forced to rise as it blows across them: this cools the air causing moisture to condense out as clouds and rain—so that is the reason the mountains stayed moister while the lowlands dried out.

In the case of eastern Asia, the reason it has so many tree species now may be that during the glacials it had mountain areas that were both wet enough and warm enough to allow many more tree species to survive. For example, the only place anywhere of the three regions that we find pollen evidence of really moist forest surviving during the last Glacial Maximum is in northern Taiwan, which has ridiculously high rainfall nowadays and probably still had plenty back then. Pollen evidence from the last glacial, for example, directly suggests more extensive survival of moist forest in that region (Adams and Faure 1997, 1998).

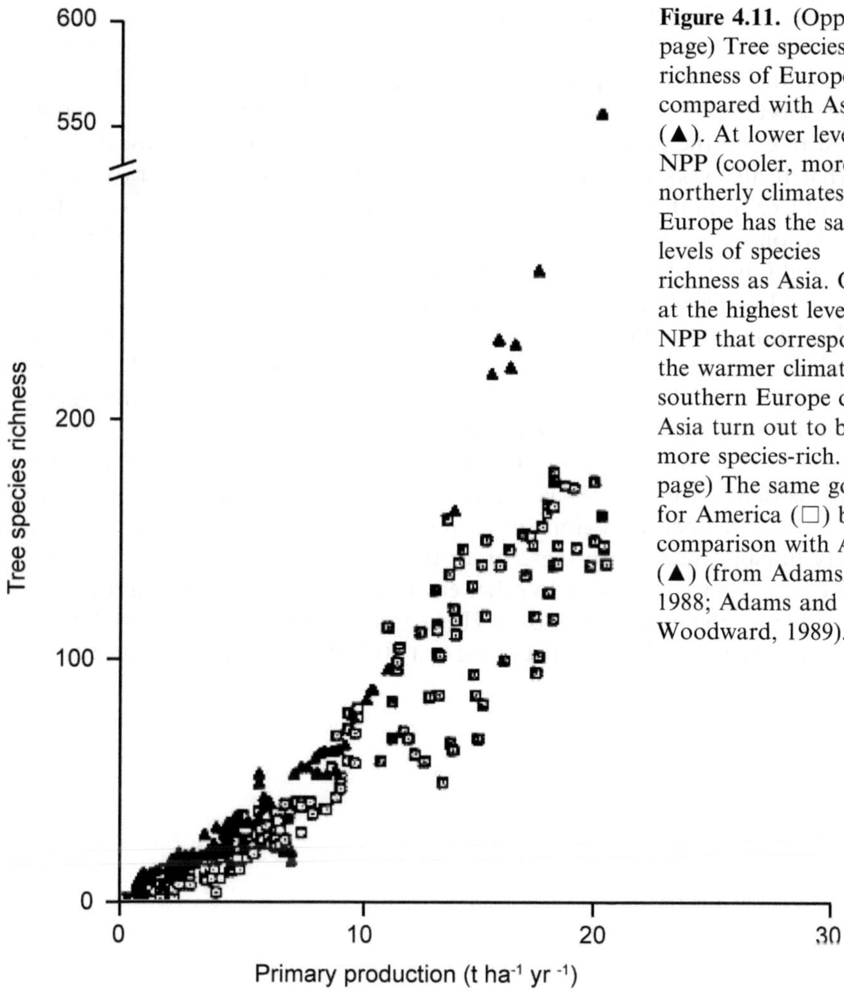

Figure 4.11. (Opposite page) Tree species richness of Europe (□) compared with Asia (▲). At lower levels of NPP (cooler, more northerly climates), Europe has the same levels of species richness as Asia. Only at the highest levels of NPP that correspond to the warmer climates of southern Europe does Asia turn out to be any more species-rich. (This page) The same goes for America (□) by comparison with Asia (▲) (from Adams, 1988; Adams and Woodward, 1989).

We can compare the species richness of three temperate regions by the same criteria, using the climatic factor that best explains variation *within* each region: an index linked to actual evapotranspiration that predicts net primary productivity or NPP (something from the world of latitudinal diversity trends, Chapter 2). A surprising result that comes out of this is that the cooler northern and central parts of Europe are no poorer in species than areas in Asia and North America which have a similar climate (Figure 4.11; Adams and Woodward, 1989). So when we compare the cooler parts of different regions on the same terms as one another, there is no sign of any glacial extinction effect at all.

A difference between the regions does show up, but only when the warmest most southerly parts of each region are compared (these are the areas towards the right-hand side of the axis, with high predicted NPP). Here eastern Asia stands out clearly, with levels of species richness that soar high above those of America and Europe. And in another unexpected twist, America is actually no richer than Europe in tree species richness if the two are compared under similar climatic conditions. The only reason America has areas that are richer is that it has climates that are warmer overall as well as moist (e.g., in the Deep South of the U.S.A.; Adams and Woodward, 1989).

What it all looks like then is that in Europe and North America nearly every species of tree from the colder more northerly parts of each region was able to survive the glaciations by parking itself in a mountain valley (or even north of the mountains; Adams and Faure, 1997; Stewart and Lister, 2001) during the ice ages. It was the trees of warmer climates which lost out heavily in these two regions, perhaps because there were few moist mountain valleys that also stayed warm enough for these more demanding species to survive in. In Asia, by contrast, moist upland areas occurred far enough south that the crucial combination of enough warmth and enough moisture on their lower slopes and valleys enabled most of the original Tertiary tree flora to survive. Considering all of this, Europe looks less like a coldspot, since it is actually as rich in tree species as comparable areas of North America. On the other hand, warmer temperate regions of eastern Asia now seem to stand out as being a diversity hotspot, relative to these other two regions!

In Chapter 5 we will move to other events that took place during the last ice age, continuing into the more recent past. Huge areas of the Earth's surface have in effect been made diversity coldspots just in the last 50,000 years. And the cause of it all may be humans.

5

The march of Cain: Humans as a destroyer of species

5.1 THE HUMAN SPECIES

About two and a half million years ago, the Earth entered the first of a series of ice ages. That point marks the beginning of the span of time that geologists call the Quaternary. In East Africa, soon after the Quaternary began, there are the very first signs of our own human genus *Homo*—in the form of a layer of stone tools, the oldest ever found. It seems true to say that we humans are "children of the ice ages", for the timing of our origin, evolution, and spread around the world have all taken place against the shifting environments of the Quaternary.

During the Quaternary, there has been a succession of human-like forms. Some of these have been confined to Africa, but others have managed to make it to other continents. In the present-day world only our own human subspecies (*Homo sapiens sapiens*) survives: this is the type known to biological anthropologists as "modern" humans. The repeat of the *sapiens* epithet emphasizes that we are the true human subspecies, the typical form that the biologist Linneaus first described in the 1700s (Chapter 6). The name *Homo sapiens sapiens* applies not only to ourselves in the "modern" world of fridges and televisions but also our ancestors back into the Stone Age tens of thousands of years ago. Civilization is only the latest of many inventions that were brought about by modern humans—we moderns were the first to develop art, and farming, for example. Compared with other, earlier human species we are small-boned, more delicately built, and physically weaker. But there is ample evidence from the archeological record that from the start we made up for our lack of brute strength in some indefinable quality of creative agility, and perhaps in the ability to focus obsessively on a problem until it was solved. These have apparently been the secrets of our success as a subspecies: brains, not brawn (Stringer and Gamble, 1993).

Though we may pride ourselves on our sophistication, we wimpy, nerdy moderns seem to have been rather a curse on the rest of the planet. Increasingly, it is clear that our ancestors caused waves of extinction rolling around the world: and that this is a process which continues right up to the present. Ironically, some of the first to suffer extinction at our ancestors' hands may have been our closest evolutionary relatives.

5.2 HUMANS AND THE EXTINCTION OF OTHER HUMANS

The story of how we came to be the only surviving form of human reminds me of the ancient tale of Cain and Abel, where a man kills his own brother in order to get what he wants (Figure 5.1). Perhaps there is also a hint of the old fairy stories of goblins living in the woods, and ogres being slain by heroes. I sometimes wonder whether these other forms of humans, mistrusted and eventually killed by our ancestors, have left traces in folk memories around the world.

For most of the time since the beginning of the Quaternary, there were several different human-like species in the world. As well as our own genus *Homo* with large brains and clever tool use, there was the group of species known as *Australopithecus*—smaller-brained, apelike but upright-walking, and probably the genus from which we *Homo* are derived. Nowadays many who study the evolution of humans also mark off another genus, *Paranthropus*, from within the more ape-like *Australopithecus*. A mass of different human and human-like species came and went during the Quaternary, in a story which is still confused in its details. But from all of this, what leaps out is that it is unusual for us now to find ourselves as the only human-like species around. At times, there may even have been several of these humanoid (or "hominid" as biological anthropologists call them) species in the same geographical region. For example, in southern and eastern Africa, various species of

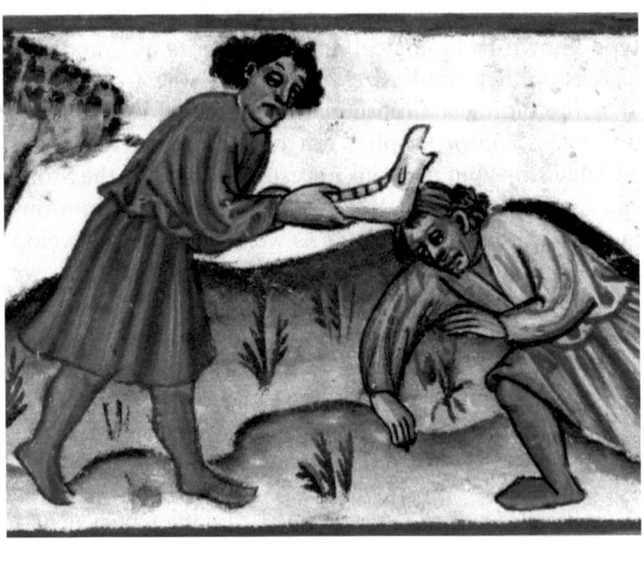

Figure 5.1. Cain kills his brother Abel, depicted in a 15th-century manuscript (by Butko, in Wikipedia Commons). *See also* Color section.

Australopithecus or *Paranthropus* seem to have coexisted with two or more larger brained species of *Homo* for nearly 2 million years, before the last *Australopithecus* finally went extinct about half a million years ago.

The timeline and shifting geographical mosaic of human species is complex, but in addition to coexistence there seems to be at least some tendency for species of *Homo* to replace one another in sequence (Rob Foley, pers. commun.). One species appears, and at around the same time or not long after, another drops out.

In Africa, which has the longest and most diverse fossil record of human species, the earlier and smaller brained species such as *Homo habilis* seem to have been replaced by later forms with progressively larger brains and more sophisticated tool use—such as *Homo erectus* and then later on by early "archaic" subspecies of *Homo sapiens*. Eventually around 130,000 years ago our own modern subspecies *Homo sapiens sapiens* spread throughout Africa and out around the world, taking with it more innovations such as art and body adornment, and an ever-changing stone and bone toolset. All of this makes sense in terms of classical ecological theory. It is generally held that if two species occupy too similar a niche then one will eventually drive the other extinct (though as the other chapters make clear, in some other groups such as plants the idea is open to doubt).

Before about 250,000 years ago, the human fossil record is very fragmentary, and it is not detailed enough to tell us much about when or why the earlier forms of these hominins died out. For instance, did each new species of humans spread and wipe out earlier forms by competition and warfare? Or were earlier species instead eliminated by drought or other climate shocks, leaving vacant territories that new forms of hominins later spread into.

On the other hand maybe there was no sudden replacements and instead a slow, gradual transition from one form to the next by evolution? After all, a species can in theory go "extinct" if its whole population evolves into something else. There are signs of overall trends of increasing human brain size and certain other aspects of anatomy during the Quaternary, but it is hard to tell whether such changes actually took place in the same continuous set of populations. It might instead have happened through a series of sudden stepwise replacements by "new" species, that spread out of localized areas where they had evolved.

During the last quarter of a million years or so, the fossil record of the various human species improves a great deal: there are many more fossils, and more precise dating of the fossils that we do have. From this better record, at least one clear story emerges of the events surrounding the replacement of a human species by its close relative. In Europe and Central Asia there was once another, parallel strain of humanity—the Neanderthals (*Homo neanderthalensis*). The people are something of paradox: they had very large brains (equal to and sometimes larger than those of modern humans), and apparently all the anatomical equipment for complex speech. Yet their culture seems to have been lacking in the sort of sophistication and adaptability that modern humans have shown for tens of thousands of years. The archeological evidence shows that Neanderthals used fairly simple tools that barely changed over hundreds of thousands of years, and crude, dangerous hunting techniques (possibly jumping onto the back of the animals as they tried to kill them) that often resulted in injury to themselves. They also seem to have lacked any form of art, at least until the very final days of their existence after they had encountered modern humans (Stringer and Gamble, 1993).

Modern humans spread out from Africa sometime in the last 80,000 years, and they started moving westwards into Europe from the Middle East within the last 45,000 years. All across Europe we can clearly see signs in the archeological record that they stepped into a gap left by the Neanderthals. In each part of Europe, within a few thousand years of modern humans turning up, the Neanderthals vanish but the moderns stay in place. It did not occur as a sudden, simultaneous extinction, but instead as a westward-moving wave that took about 15,000 years to cross the continent. In each locality, what we see from the record is that modern humans first arrive, and within a few thousand years the Neanderthals are gone.

Watching this westwards wave of movement of the moderns, and the progressive disappearance of the Neanderthals, it certainly looks like the moderns caused the end of the Neanderthals. But could that perhaps have been coincidence? Maybe the Neanderthals were unlucky enough to be wiped out by some sort of progressive climate change, and the moderns just stepped into the void they left behind?

If it was climate change that killed off the Neanderthals, it seems pretty remarkable that it waited to occur until exactly when moderns were spreading westwards through the region. An important thing to bear in mind is that the Neanderthals had been around for a very long time before modern humans turned up. Neanderthals seem to have evolved during a warm phase around 250,000 years ago, but they hung on stubbornly through various cold phases (the glacials) that reduced most of Europe to a frigid wasteland, as well as drastic sudden warming phases (interstadials) that at times made the regional climate similar to the present. By contrast, there was nothing especially extreme about the climate at any time during the period between 45,000 and 30,000 years ago when the Neanderthals went extinct. Europe was colder than it is now, and drier, but it was nowhere near as cold and dry as it had been a few thousand years before, nor as cold as it would be a few thousand years later after modern humans had established themselves in the region. It is hard to make any case that the demise of the Neanderthals was a result of climate change—they actually went extinct at a time of relatively moderate climate (Adams and Faure, 1997), unlike the extremes that had marked much of their tenure in Europe.

Given the lack of any convincing reasons for thinking that climate played a role, and the striking coincidence of the Neanderthal extinction with the time of arrival of modern humans, there cannot be much doubt that modern humans were the main cause. In some areas of Europe—such as south-central France—the Neanderthals seem to have coexisted in the same general area as the moderns for several thousand years, and they even seem to have traded some art with them. But the end result was just the same; complete extinction for the Neanderthals. Tellingly, the last place where Neanderthals hung on was Gibraltar, at the southern tip of Spain, up until 30,000 years ago. This also happened to be the last area that modern humans colonized in southern Europe, after they spread in from France and down southwards through the Iberian Peninsula.

But did the Neanderthals *really* go extinct? Perhaps they were just absorbed by intermarriage with the population that spread in? It could be that the Neanderthals still survive, scattered through the blood of modern Europeans. Most archeologists do not agree with this viewpoint. The anatomical difference is too striking, and the replacement too sudden, to accept that there is much Neanderthal ancestry in people surviving today. Predominantly, they feel, the Neanderthals died out without leaving descendants. This view is backed up by the general picture from DNA evidence, taken from many thousands of modern Europeans and compared with those of the Neanderthals. Modern humans in Europe and everywhere else in the world possess a close cluster of mitochondrial DNA sequences, that have not changed much since our recent common origin in Africa (Krings *et al.*, 1997). It has also proven possible to obtain DNA from well-preserved Neanderthal bones and teeth and compare it with present-day humans. DNA sequencing from the remains of several different Neanderthal individuals has shown that they possessed a quite distinct set of sequences, related to but also quite different from anything found in modern humans throughout the world (Krings *et al.*, 1997). The conclusion must be that the Neanderthal population of Europe was not absorbed, and was instead completely replaced wherever the moderns spread in.

A similar story of extinction through replacement seems likely to have occurred with an earlier subspecies of our own species, which is known as "archaic" *Homo sapiens*. The "archaic" people existed before about 130,000 years ago throughout Africa. Then, by 80,000 years ago (and perhaps earlier), our own "modern" *sapiens sapiens* subspecies was present across all of Africa south of the Sahara—and the archaic people had vanished. Again, DNA evidence gives us some clues to what happened. Since all the world's modern humans possess a very limited set of DNA sequences that center on Africa, they must all have come from some small founding population (at most 2,000 people, maybe less) within Africa. The probable "founding date" for our subspecies from that one population is some time in the last 150,000 years or so. There is no trace among modern humans of older populations with more distinct DNA sequences. Since the founding population was so small and recent, we can guess that what most likely happened was that our ancestors spread out as a wave of population into areas already inhabited by "archaic" *sapiens*. Rather than interbreeding with them, our ancestors completely replaced the archaic people.

Roughly the same occurred in eastern Asia, where long-established populations of an earlier human species *Homo erectus* were widespread from China south to Indonesia. They were, like us, of African origin but their ancestors had left the homeland much further back in time—more than a million years ago. Modern humans seem to have arrived in eastern Asia by around 60,000 years ago, and at this point the regional *Homo erectus* populations vanished. The last few erectus populations on islands in Indonesia seem to have died out 30,000 years ago, after the archeological record shows moderns moving into the same general area. The DNA of modern humans in Asia always shows the same familiar set of recent African sequences, with no sign of the very different DNA that we would expect if the older blood of *Homo erectus* had been absorbed into modern populations. Perhaps the most recent victim of our ancestors was *Homo floriensis*, a strange dwarf offshoot of the *erectus* lineage, only 3 feet tall (Brown *et al.*, 2004). "Flores man", as this species is called, inhabited the island of Flores in Indonesia up until perhaps 18,000 years ago according to radiocarbon dates—around the time when modern humans first turn up in the archeological record there.

Generally, then, we can see that our recent ancestors did not interbreed with the other human species around them. It is no surprise that they kept to their own kind. Modern humans tend to choose their mates partly by visual appearance, and partly by cultural affinity. Other human species would probably have been far too different in looks, culture, and thinking to have much appeal. There may have been some exceptions at the time, though. A few clues from the archeological record of Europe suggest that hybrids between Neanderthals and modern humans were sometimes produced, even though the lineages ultimately died out. Several skeletons that have been found from the time when moderns were sweeping across Europe have a mixture of features that might suggest a hybrid origin. So far, however, no DNA evidence has been obtained from the bones, so we cannot be sure whether these individuals really were hybrids.

There is, however, stronger evidence from deep within our own DNA of a far more distant hybridization event, which perhaps occurred before our own genus *Homo* originated. The DNA of most of our 23 chromosomes shows that we diverged from the ancestral line of the chimpanzees 5 or 6 million years ago. Yet there is just one chromosome—the X chromosome, which in a double dose determines female gender—that shows much closer affinities to that found in chimps. Our X chromosome seems to have separated only 2 million to 3 million years ago from the chimp lineage, while in contrast the Y chromosome of every human male has been diverging from chimps for a similar time to the rest of our DNA (Patterson *et al.*, 2006). What it looks like, then, is that at some stage a male of our main ancestral species mated with a female ancestor closely related to modern chimpanzees, and produced a fertile offspring that is the ancestor of us all. Most of the DNA of that child has been lost after it blended back into its father's species—but the X chromosome alone survived and prospered there. Where was this chimpanzee-related ancestor? Possibly on the east side of the great African Rift Valley that runs down through Kenya and Tanzania, where our own evolutionary line seems to have originated. If so, it is an extinct species, perhaps vanquished by our own, for there is no chimpanzee now east of the Rift Valley. For our brother the chimpanzee, the final extinction has waited until the present-day world, where humans are now steadily reducing chimps' populations through persecution and habitat destruction.

5.3 THE SECRETS OF OUR SUCCESS OVER OTHER HUMAN SPECIES

Homo erectus, *Homo neanderthalensis*, *Homo floriensis*, "archaic" *Homo sapiens* ... What factor could explain why modern humans have managed to eliminate all of their closest relatives and competitors? Humans make their place in the world mostly through inventiveness, and it is likely that the success of our own subspecies had something to do with us being more innovative than all other types of humans. The archeological record makes it clear that even from early on, there was something different about the behavior of our modern human ancestors, compared with all the other types of humans that previously existed (Stringer and Gamble, 1993). Other human species and subspecies seem to have relied on nearly unchanging sets of stone tools, with fairly much the same shapes and materials persisting over hundreds of thousands of years, just about wherever in the world they went. Modern humans have by contrast invented and changed tools constantly according to needs and fashion. The "moderns" also used additional materials that their ancestors had not (e.g., bone) to make finely crafted implements such as needles and fishhooks. Another striking and unique feature of our own subspecies seems to have been art in its various forms. Although there have been a few disputed claims of art by the earlier Neanderthals and *Homo erectus*, it generally seems that no other form of human produced art before modern humans came on the scene. The earliest known modern human dwelling site, 120,000 years old in a cave near the coast in South Africa, had sticks of red ocher for body painting stashed away on ledges next to where people were buried in the cave. By 80,000 years ago humans elsewhere in southern Africa were making necklaces of beads made from small seashells. By 30,000 years ago, modern humans in Europe were producing stunning cave paintings of ice age animals, and carved human figurines. It seems likely that the imagination and innovation that shows up in the art and finely honed tools of the moderns has had something to do with their success over other human species. Part of this victory must have been brought about by indirect competition, depleting the food resources that these other types of humans relied on. But undoubtedly from all that we know of human societies around the world, much of it was finally decided by cunning in warfare.

So the circumstantial evidence suggests that modern humans started out by destroying all the most closely related species around them. This would not have been an unusual thing to happen in nature, and in fact it is probably quite normal when similar and related species come into competition for the first time. But in the case of humans it was only the beginning of a phase of destruction that spread all around the world, even to the farthest islands. The inventiveness that had stood humans in such good stead against their closest relatives also allowed them to launch an onslaught against nature. Humans have become such a force in depleting species richness that we, and our effects, have to be the central topic of several chapters of this book.

5.4 SURVIVAL OF SPECIES DIVERSITY DURING THE QUATERNARY

The last two and a half million years has been a time of huge climatic instability, the time of ice ages which geologists call the Quaternary. Even though much of my academic training is in the field of the Quaternary (it is what my Ph.D. was about), and though I have spent years living and breathing the information on it, there are still times when I stop in wonderment at the scale of the changes that have occurred in the Earth's most recent past.

Only 16,000 years ago (just a few thousand years before the first town—Jericho—was founded in the Middle East), Canada was covered in ice sheets as thick as a mountain range, that extended down as far as New England. Another vast ice sheet covered northwestern Europe, and to the south of it instead of the present forest vegetation was a vast arid plain that stretched all the way to the Pacific shore of Russia. In Africa, the Sahara had expanded hundreds of kilometers farther southwards. The great rainforest blocks of the world in the Amazon Basin, central Africa, and Southeast Asia seem to have been reduced to a fraction of their present size: in their place was dry scrub and grassland. Due to the sheer volume of water that was held in ice sheets on land, sea level was some 120 meters lower. Consequently, in many places low plains extended hundreds of kilometers out beyond the present coastline, covered in scrub, grassland, or whatever else the regional vegetation was at that time. Shallow-water marine environments were hit by both the colder water temperatures and the fall in sea level. The sea level drop had taken coastlines far out to the edge of the continental shelf in many areas, eliminating most of the shallow marine habitat that sits between the land and the deeper ocean (Adams and Faure, 1997).

This was not just a one-off event—far from it. Essentially the same thing had previously occurred six times within the last 700,000 years. Each time, at the end of a 100,000 year cycle driven by oscillations in the Earth's orbit, there was a dramatic warming into conditions similar to the present-day world, starting what is known as an interglacial. The warm phase would then last maybe 10,000 or 15,000 years before temperatures began to decline into next glacial. Slightly smaller, but similar, fluctuations have also occurred scores—even hundreds—of times since the Quaternary began (Figure 5.2).

This overall picture has emerged from meticulous study of the clues that we have to the past world. These clues include lake beds and undersea muds that have lain

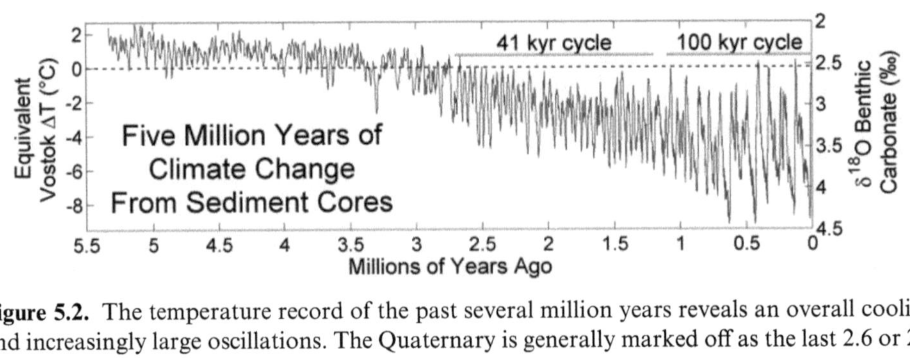

Figure 5.2. The temperature record of the past several million years reveals an overall cooling and increasingly large oscillations. The Quaternary is generally marked off as the last 2.6 or 2.4 million years of this phase.

undisturbed, sometimes in annual layers that have recorded the events and climate of each individual year. There are similar layers in the ice sheets of Greenland and Antarctica, and in some glaciers high in tropical mountains—because the ice that forms in summer has a different texture from the ice that forms in winter, giving layers that can be counted down from the surface like the rings of tree. In the ice, each layer contains traces of wind-blown particles, and bubbles of trapped air, plus a characteristic composition of isotopes that indicates what the climate was like in that particular year.

A big surprise that has emerged from studying these clues to the Quaternary climate is the speed of many of the changes. Sudden jumps in climate were capable of taking mid-latitude climates from Arctic to temperate, over just a few decades. For instance, the warming event 11,500 years ago at the end of the last ice age seems to have been mostly completed in 70 years—within the span of a human lifetime. Of this change, most was concentrated into only 15 years, and most of that occurred in just 5 years.

The warming 11,500 years ago marked the beginning of the long and fairly stable "interglacial" (a name meaning "between glacials") phase that we are now in, but many other warming events in the Quaternary were short-lived. Previously, every time—after a few millennia, several centuries, or sometimes even just a few decades of warmer, moister climates—the Earth would once again be plunged back into cold and aridity. Once again, the signs are that many of these cooling events took only a few decades to be completed, although they are not as well studied and understood as that last big warming event 11,500 years ago. Even during our own fairly stable interglacial, there were some very significant fluctuations. The environmental record of arid zones such as the Sahara and western China shows that during the last 10,000 years there were rapid, alternating shifts between dry and moist climates. Tropical rainforest areas such as the Amazon Basin also seem to have gone through some relatively arid phases during the last several thousand years (Adams and Faure, 1997).

Repeated large and sudden fluctuations in climate and sea level are just the sort of thing that we would expect to reduce populations, and isolate and prevent species from reaching new areas of suitable environment, eventually driving them extinct. There are signs that at the onset of the Quaternary two and a half million years ago, there was indeed the sort of burst of extinction we would expect (see Chapter 3), with a varied assortment of species of mammals, seashells, and trees going extinct.

Yet after that early stage, the extinctions mostly seem to have stopped. Looking across the whole range of life forms, one striking thing about the rest of the Quaternary is how little went extinct. Through most of the last million years or so in particular, one is hard-pressed to find any species that were lost, even though the fossil record is far better and the total number of known species is far greater during this most recent slice of geological time. We have enough clues to talk of the record of corals, seashells, temperate beetles, and tropical and temperate trees (Adams and Woodward, 1992), and it looks like just about everything survived up to the present.

5.5 YET MAMMALS AND BIRDS HAVE SUFFERED A GREAT WAVE OF EXTINCTIONS

However, there were some exceptions. At least two groups—mammals and birds—have suffered a lot more extinction during the most recent part of the Quaternary. Before 50,000 years ago there were 147 genera of large mammals around the world ("large" being defined as weighing more than 44 kilograms, about the average weight of a human—and known to zoologists as megafauna). Now there are only 97 genera of large mammals surviving. A great wave of extinctions across the continents occurred during the second half of the Last Glacial phase, between about 45,000 and 10,000 years ago, taking with it most of the world's diversity of large mammals (Martin, 1989; Figure 5.3).

5.6 AFRICA, 150,000 YEARS AGO

An initial uptick in extinctions seems to have occurred in Africa. There had always been "background" extinctions, happening one by one through the Quaternary, but this looks like something more. Several species of antelope disappear from the continent around 150,000–100,000 years ago (Martin, 1989). However, the predominant picture from Africa is of survival, not extinction. Africa has the only

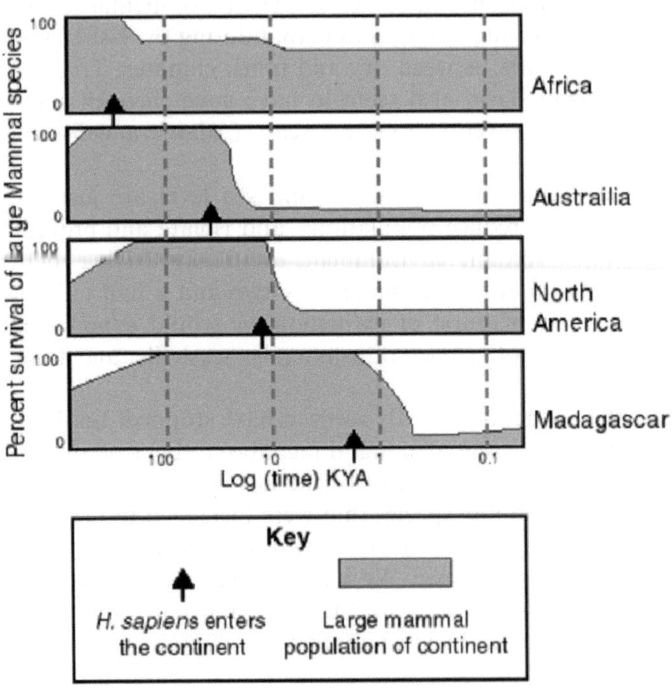

Figure 5.3. The decrease in diversity of large mammals around the world during the most recent part of the Quaternary (source: Wikipedia Commons, after Martin, 1989).

reasonably intact fauna surviving from the late-Quaternary era, and because of this it is often called "a living museum of the Quaternary".

5.7 AUSTRALIA, 45,000 YEARS AGO

The decimation really began in Australia, an isolated landmass inhabited mainly by pouched marsupial mammals (where the young complete the final fetal stages of their development in their mother's pouch). The marsupials had dominated its fauna in the place of placental mammals such as ourselves (which give birth to fully formed young) for tens of millions of years, since the continent first split off from Antarctica and drifted northwards.

In Australia before about 45,000 years ago, several groups of large marsupial herbivores fed upon the vegetation (Brook and Bowman, 2002), including wombat-like creatures such as *Diprotodon* that was about the same size as a hippopotamus. *Palorchestes*, slightly smaller, seems to have been a sort of marsupial analog of a tapir, with a little trunk-like nose. There were giant kangaroos up to 3 m tall, such as *Procoptodon*. Many of the large herbivores would have been preyed upon by extinct carnivores, which included a sort of marsupial analog of a panther (*Thylaceo smilax*). A 400 kg giant monitor lizard (*Megalania*)—resembling the Komodo dragon—also stalked the northern parts of mainland Australia. There was a giant flightless goose (*Dromornis*) and a giant emu-like bird (*Genyornis*) about 3 m tall, which may each have been carnivorous.

All of these are now extinct, that is for sure. But there is a fair amount of uncertainty about when each of these species actually vanished. Some dates obtained on bones have suggested that a few of them lasted until around 30,000 years ago or later (Wroe *et al.*, 2006), but these dates are generally taken to be unreliable. Some studies have averaged the scattered dates that seem to mark the last record of each of the extinct Australian animals, and found that they clustered around 40,000–45,000 years ago, taking this as evidence that it was indeed the "true" date of the extinctions of all these species. However, the statistical methods used to arrive at this date seem rather dubious (Brook and Bowman, 2002).

Some researchers claim that the majority of big marsupials had already gone extinct before 85,000 years ago—not around 45,000 years ago as most archeologists believe (Wroe and Field, 2006). It is certainly striking that only 8 of the 30 or so extinct big marsupial species have actually been firmly dated to around 45,000 years ago. Brook *et al.* (2007) suggest that it is just another case of the Signor–Lipps effect (Chapter 3), where gaps in the record make it look like species went extinct well before they actually did. Certainly, many of the species that look like they dropped out early were not common, with only a few specimens ever having been found (Brook *et al.* 2007). That makes it less likely that they offer a reliable picture of exactly when they died out.

To get the best idea of when things actually went extinct, it is better to concentrate on the commoner species. Evidence from the last records of eggshells of the extinct giant bird *Genyornis* (Miller *et al.*, 1999), and recent findings on the sudden demise of the large marsupials in Tasmania (Turney *et al.*, 2008), seem to make a strong case that most species did go extinct around 45,000 years ago.

5.8 A SECOND WAVE OF EXTINCTION: THE AMERICAS AND EURASIA

What happened in Australia was only the beginning. Just 15,000 years ago, a fantastic assortment of wildlife still extended across both the Old and New Worlds (Martin, 1989). Several species of elephants inhabited temperate and Arctic zones. These included the famous woolly mammoths that were equipped to survive extreme cold with their long shaggy fur; one species living in Eurasia (*Mammuthus primigenius*) and another in North America (*Mammuthus columbi*). There was also a woolly rhinoceros with a similar shaggy coat in Europe and northern Asia (*Coelodonta antiquitatis*), and the large saber-toothed cat (*Smilodon* sp.) in North America. Vanished species of horses inhabited both North and South America, and an extinct genus of llamas (*Hemiauchenia*) lived in North America. Also in the Americas were giant beavers (*Castoroides ohioensis*) the size of Cadillac cars, three species of extinct bison including one (*Bison antiquus*) that was larger than the modern bison. Also in central and South America and on Caribbean islands there were various species of giant ground-living sloths (*Megalonyx* spp.), some of them reaching 5 meters in height.

It is interesting though that we do have one very large native animal surviving in the Americas: the bison (*Bison bison*). However, its survival of the Quaternary extinctions is not what it seems. When humans first entered the Americas, there were the three endemic species of North American bison—but the modern "American" bison was absent. It came across from Siberia after the time when humans spread in, about 10,000 years ago. This further emphasizes how devastating the extinctions in the Americas were.

The final disappearance of most of these animals in Europe, Asia, and the Americas seems to have been concentrated between 12,500 to 9,000 years ago (Barnosky *et al.*, 2004), around the time when the ice age ended and the present warm period—the Holocene—began. In Europe, Siberia, and Alaska, there also seems to have been an earlier pulse of extinctions between 45,000 and 20,000 years ago, mainly of animals adapted to somewhat milder climates (Barnosky *et al.*, 2004). However, it is important to emphasize that there is often uncertainty regarding the exact time at which the last population of any particular species finally went extinct. Because of the Signor–Lipps effect, the last known fossil record of a species does not necessary mark the moment at which it died out, because it might have become gradually less common and survived in undiscovered obscurity somewhere for thousands of years (the example of the last mammoths surviving on islands north of Siberia illustrates this point—see below). Radiocarbon dating can also prove unreliable, especially when the specimens are contaminated by younger or older material (see Box 5.1).

Box 5.1 The radiocarbon time scale

When archeologists or paleontologists find preserved material, they want to know what age it is, and there are various techniques for dating it. Most information on the past 30,000 years or so is from sites or specimens that have been dated using radiocarbon (^{14}C). New ^{14}C is continually made at the top of the atmosphere as nitrogen is bombarded by cosmic rays. The ^{14}C forms CO_2, which enters the carbon cycle by getting incorporated into living organisms, first by photosynthesis. When a plant or animal dies, it stops taking up ^{14}C, and if it is buried and preserved the ^{14}C that was already within it decays away at a very predictable rate.

However, the radiocarbon age scale that would be calculated from first principles (based on the decay rate of the ^{14}C isotope, assuming that ^{14}C was at the same level of abundance as it is at present) is not completely reliable, because there have actually been fluctuations in the rate of production in ^{14}C at the top of the atmosphere. This gives a slow divergence of perceived and actual ages going back in time (see below). The problems are particularly great at around 10,000 ^{14}C years ago, which includes the time of the Younger Dryas (see Box 5.2) when a large influx of old ^{14}C-depleted carbon from the oceans, combined with a decrease in the rate of ^{14}C production at the top of the atmosphere, gives an "age plateau" such that the same ^{14}C age covers a wide span of real time, about 1,000 years.

The reason we know about the errors in ^{14}C dating is that other dating methods (e.g., uranium/thorium, and amino acid racemization) can be used to attempt to check the "true" age of specimens or sediment layers dated by ^{14}C—although these all have substantial error margins of their own. The most convincing way to check the ^{14}C age scale is through biological or sedimentological features which build up annual layers over long periods of time (e.g., tree rings, and annual layers of sediment building up on lake beds); counting back the annual layers will reveal the true number of years before the present, and comparing the ^{14}C age of each tree ring or sediment layer will give an age scale for how ^{14}C age can be converted into "real" age. However, even this method is not completely reliable; "false" double rings can sometimes appear, and occasionally a year may not appear in the record. Because of these problems, individual ring or layer-counting studies often suggest "real" ages differing from one another by several percent, though they all suggest that the "real" age is older than the ^{14}C age before about 3,000 years ago. A recent working consensus ^{14}C-to-real-age conversion scale is given below, but opinions on the appropriate age conversion are likely to change somewhat as more data come in. Useful sources on the current understanding of the radiocarbon time scale include Kitagawa and van der Plicht (1998). Overall, by checking the results of each dating technique against the combined weight of all the others, we can reach an overall consensus on the "true" time scale and the errors inherent in the dating technique. It is necessary to bear in mind that quite apart from all the problems of calibration, a significant proportion of radiocarbon dates are not reliable for any purposes, because they have been contaminated with older or younger carbon that changes the apparent age of the sample. Many radiocarbon-dating specialists still refer to their field as "more an art than a science"! Published radiocarbon dates from sites and layers of fossils and sediments are quite often rescinded, when the materials are found to have been naturally contaminated. Most often the contamination is from older (less ^{14}C-rich) calcium carbonate, coal, or charcoal washed in from other layers, making a sample or layer seem older than it actually is. Although radiocarbon dating is a very useful tool for studying the Quaternary, it must always be interpreted with caution.

Here is an approximate age scale, comparing the "apparent" date from simply assuming ^{14}C has been in constant amounts in the atmosphere, with the best estimate we have so far of the "true" age. Each time a ^{14}C date is obtained, it has to be corrected against a scale such as this:

^{14}C years ago ⇒ Calibrated ("real") years ago

1,000 ⇒ 1,000
2,000 ⇒ 2,000
2,500–2,800 ⇒ 2,600 (sudden shift in atmospheric ^{14}C content)
3,000 ⇒ 3,200
4,000 ⇒ 4,500
5,000 ⇒ 5,900
6,000 ⇒ 6,950
7,000 ⇒ 7,900
8,000 ⇒ 8,900
9,000 ⇒ 10,000
10,000 ⇒ 11,200–12,200 ("radiocarbon plateau", during and just after Younger Dryas)
11,000 ⇒ 12,900
12,000 ⇒ 14,000
13,000 ⇒ 14,500
15,000 ⇒ 17,000
16,000 ⇒ 19,500
17,000 ⇒ 21,000
18,000 ⇒ 22,500
20,000 ⇒ 24,500
25,000 ⇒ 28,000
30,000 ⇒ 35,000
40,000 ⇒ 45,000

5.9 DID CLIMATE CHANGE CAUSE THE EXTINCTIONS ON CONTINENTS BETWEEN 45,000 AND 10,000 YEARS AGO?

The Quaternary was a time of frequent and extreme changes in climate, so it is reasonable to suspect that the extinctions which occurred so widely during this time were caused by climate. If we look at the climatic record of sediments and ice cores, we find that there were indeed changes in temperature and rainfall around the times when the wave of extinction occurred in each region. The question is, whether these changes were different enough from what had been occurring before to offer any sort of convincing explanation. It is important to bear in mind that exactly the same sorts of climate changes had already been occurring frequently over the past several hundred thousand years, well before the extinctions. It is hard to explain why a recent climate jolt would be enough to push so many species to extinction, when they had previously survived many other similar changes.

5.9.1 An Australian drought

In Australia, the extinctions seem centered on a dry phase which lasted several centuries, pushing back the woody vegetation and allowing grassland to expand. Shortly before the large Australian animals went extinct around 45,000 years ago, there was a change in the carbon and oxygen isotope composition in their teeth which shows that they were being forced to eat a different diet. Different plants have different isotope compositions, and a shift in the vegetation showed up in the teeth and bones of the animals.

However, on the scale of the later Quaternary in general this shift in the climate and vegetation was not an exceptional event—indeed it was fairly routine and small (Adams and Faure, 1997). There must have been many events like this, perhaps hundreds of them in the last two and a half million years—but none of these resulted in extinction on such a scale.

5.9.2 A thaw, then a freeze

It was, in a sense, a similar story in North America. The diverse mammal fauna had recently survived the Last Glacial Maximum, an extreme cold phase that brought mile-thick ice sheets down across Canada and into northern parts of the U.S.A. It had also brought cooling and aridity to most of the land that remained uncovered by ice (Figure 5.4). A sudden jolt of warming dragged North America and the rest of the world out of its cold phase around 16,000 years ago, and yet it seems that no species were lost at that stage.

However, even though the most extreme phase of the ice age had ended, there was one last big climate shock in store. After a few thousand years of warm, temperate climate, at about 12,500 years ago temperatures plunged half-way back to ice age conditions. Ice sheets began to grow again, and forests around the northern latitudes died back from the cold and aridity.

This cold phase—known as the Younger Dryas—lasted for some 1,500 years before temperatures again suddenly shot up to levels similar to the present. This final warming event about 11,500 years ago seems mostly to have occurred over just a few decades (Adams *et al.*, 1999). At least 15 species of North American mammals are last found near the start of the Younger Dryas or during it, the period between about 12,400 and 10,800 years ago (Barnosky *et al.* 2004). This seems to suggest that the cold phase had something to do with their extinction. We must bear in mind the limitations of the data we have: exactly when the extinctions actually occurred could

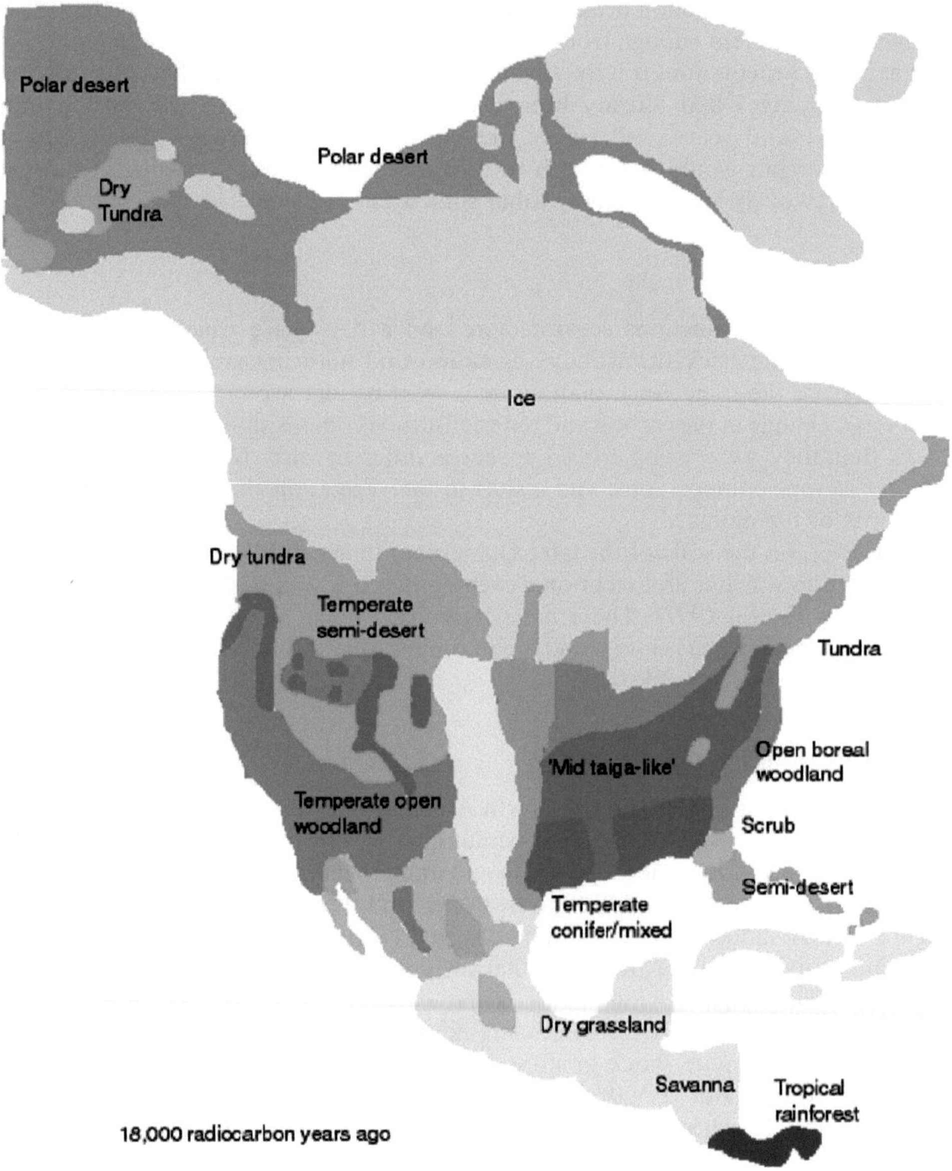

be confused by lack of fossil data—which often gives the impression things died out earlier than they really did (the Signor–Lipps effect, again). There was also a confounding fluctuation in carbon-14 concentrations in the atmosphere (known as the "radiocarbon plateau") which prevents very accurate dating around this time (Box 5.1). Nevertheless, the coincidence in time of so many extinctions with the Younger Dryas is too striking to ignore. Many researchers have suggested that this particular

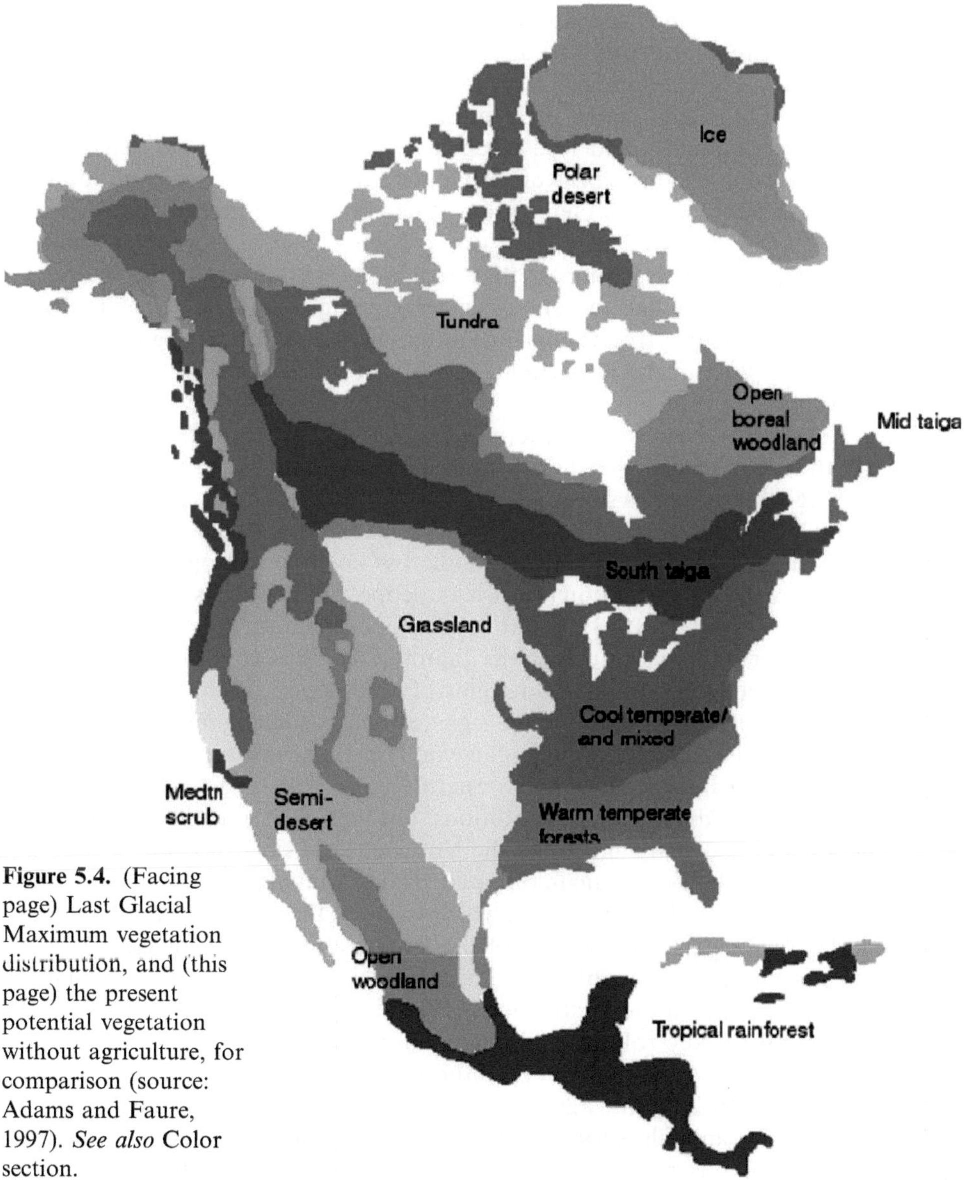

Figure 5.4. (Facing page) Last Glacial Maximum vegetation distribution, and (this page) the present potential vegetation without agriculture, for comparison (source: Adams and Faure, 1997). *See also* Color section.

climate change played an important role in the demise of the late-Quaternary fauna in North America.

The timing of events was much the same in Central and South America. The Last Glacial Maximum (between around 30,000 and 16,000 years ago) had been cooler and noticeably drier through most of the region, with less forest and more open grassland and scrub. Towards the southern end of the South American continent, what are grassland areas nowadays were occupied by desert and semi-desert during the glacial, and a great ice cap spilled down from the southern Andes out across the surrounding lowlands. About 16,000 years ago, the global warming event that marked the end of the main ice age affected this region just as it did elsewhere. Rains also increased substantially, and forest began to spread out over grasslands. The elephants, horses, and ground sloths of the Quaternary fauna seem to have adjusted without much difficulty to this change, and their bones continue to be found in abundance for a few thousand years afterwards. Then 12,500 years ago the Younger Dryas cold event hit. The drier climate that went along with it was severe enough to cause a big reduction in the amount of water coming down the Amazon River (Maslin and Burns, 2000). Around this time, after about 12,000 years ago and before 10,000 years ago, most of the large animals in South and Central America seem to have disappeared.

Europe conforms to the same pattern of extinction. The large animal fauna had survived both the onset and the end of the last big glaciation (between 30,000 and 16,000 years ago), with species shifting their ranges to cope with the changed climate. The time of the glacial maximum had been both cold and dry across Europe, with an extensive ice sheet flowing from Scandinavia across northern Europe, and a mostly treeless open herbaceous vegetation (known as steppe–tundra) across central Europe (Adams and Faure, 1997, 1998). Southern Europe had milder winters than the steppe–tundra, but it too was much colder than the region is at present, with a very dry climate resembling the grassland margins of semi-desert. Then with the sudden warming event 16,000 years ago, we see signs of animals changing their ranges to adjust to the new climate. For example, the woolly mammoths that had lived across central Europe moved northwards into western Russia. The large animal species of the last ice age survived the warmer conditions, but then they suddenly disappeared, some time after 12,500 years ago when the Younger Dryas began (Martin, 1989).

In eastern Asia, the story is again repeated: a large animal fauna (the mammoths, forest elephants, and woolly rhinoceros), which survived the glacial maximum almost intact, went extinct at around the time of the Younger Dryas cold phase.

It seems intuitively reasonable that climate shocks such as the Younger Dryas could in themselves caused these extinctions. The zoologist Dale Guthrie (1984, 1991) is one advocate for a key role of environmental change in causing the extinctions across North America, Asia, and Europe, although he emphasizes the initial warming out of the glacial at 16,000 years ago as the key change that wiped out the animals. It now looks like the animals survived longer than this, with the main extinctions happening a few thousand years later—although that initial change could have reduced populations and set the scene for trouble later on. Guthrie suggests that the important factor was a change away from the vegetation that had predominated across the northern latitudes during the Last Glacial Maximum. During the glacial, a semi-arid vegetation known as "steppe–tundra" formed a patchwork of different plant communities across the landscape. Guthrie suggests that this vegetation had a high growth rate, providing plenty of food for large grazing herbivores, and that its patchwork quality provided diverse niches for many species of herbivores to live off without some of them driving others extinct through competition. The final warming into moist interglacial conditions, Guthrie suggests, meant that the steppe–tundra was replaced by less productive and more uniform forms of vegetation—the moist tundra and conifer forests of the present-day north. And so with the steppe–tundra went the diverse mammal fauna that depended on it (e.g., the mammoths, woolly rhinos, and the American horse). All this may have made life harder for the animals, even if it was not in itself the last blow that finished them off.

Guilday (1984) has likewise suggested that in the eastern U.S.A. changes in the vegetation that were linked to climate change helped to produce extinction. The warming at the end of the glacial caused a shift in vegetation away from open coniferous parkland towards deciduous forest, which would have made the environment unsuitable for certain species.

The case that climate change caused the extinctions looks fairly convincing, overall. But if we zoom out to look at these events against a longer time scale, it seems very puzzling that the extinctions all occurred when they did. Deep ice cores and ocean sediments give us a long-term, detailed picture of what the Earth's climate has been doing during these last couple of million years. It turns out that change has been a constant, and each of the animal lineages which went extinct at the end of the Quaternary had already lived through many large climate jumps. Most of those big animals had clear lines of ancestry in the fossil record going back to the earliest parts of the Quaternary (2.6 million years), or even into the Pliocene period that came before. Many of the earlier climate changes—which did not apparently cause extinctions—were as big or bigger than the ones which occurred in the last 45,000 years (Figure 5.5). For example, the Earth had been through at least six previous glacial maxima similar in scale to the Last Glacial Maximum, during the last 700,000 years.

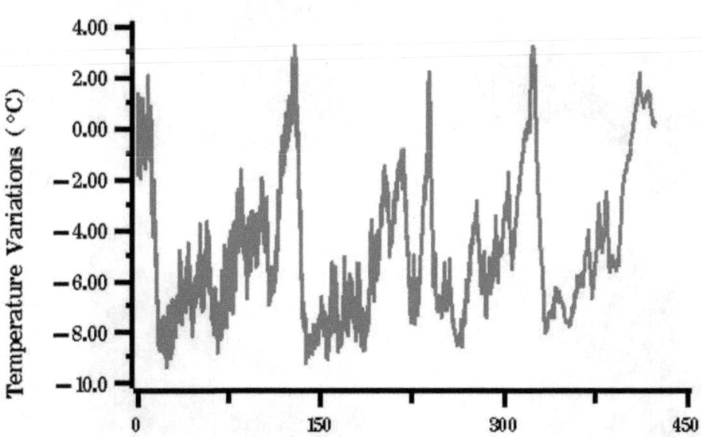

Figure 5.5. The record of the Earth's temperature extending back over the past 500,000 years or so, deduced from Antarctic ice cores. The recent ice age and warm period of the last 20,000 years are only the most recent part of a dramatically unstable history. Most of the large mammals had survived many of these fluctuations in climate, but suddenly went extinct as the last ice age ended (after CDIAC).

It also went through six warm interglacials, each of them rather similar to the present-day world. Mixed in with these broad climate cycles were scores of climate "jolts", many of them on a similar scale to the Younger Dryas. Extending further back during the previous 1.8 million years, before the broadest scale climate cycles had started, there were also numerous large climate oscillations. These were smaller in overall amplitude than the broadest cycles, but often still bigger than the Younger Dryas event that was associated with so many extinctions.

Seen in the context of all this, the sudden, simultaneous extinction of so many species right around the time of the Younger Dryas seems very puzzling.

Box 5.2 The Younger Dryas

As the world emerged from the last ice age, it was hit by a sudden and relatively short-lived cooling event. Coming like an unexpected late spring frost, this cold phase knocked back the forests that had already begun to spread out from their ice age refuges. The reversal in climate was first detected in lake sediments in Denmark, where it shows up as a burst of pollen of *Dryas octopetala*, the Arctic rose—a plant indicative of cold climates (Figure 5.6). Hence this is called the "Younger Dryas" cold phase (the "Oldest Dryas" being the earlier Glacial Maximum when the same plant was also abundant in northern Europe, and the "Older Dyas" being an earlier small cold stage that seems to presage the Younger Dryas). It appears that the Younger Dryas was a global event, producing cooler or drier climates in most parts of the world. Glaciers in mountains around the world

Figure 5.6. The Arctic rose, *Dryas octopetala*. A plant that thrives in the cold climates of tundra and high mountains, its pollen in lake beds in the European lowlands provides a sign of cooler conditions in the past. *See also* Color section.

expanded, and the great northern ice sheets that had been in retreat began to spread outwards again. The forests that had begun to spread across Siberia were now killed and their place taken by a treeless steppe–tundra. In the tropical lowlands, the climate may have turned more arid—suggested by a major decline in the amount of water coming down the Amazon River, for example.

The Younger Dryas seems to have lasted about 1,500 years, interrupted by a brief warming in the middle of it. The cold of the Younger Dryas initially came on suddenly in geological terms, a few hundred years at the most, and it ended with stunning speed. Detailed ice core, lake sediment, and tree ring records show that in the northern latitudes at least, the Younger Dryas finished over just a few decades, with most of the warming occurring within a space of 15 years. Its ending, 11,500 years ago, marked the start our own present interglacial, known as the Holocene.

What triggered the Younger Dryas is still uncertain, but it seems most likely to have been the sudden emptying of a vast meltwater lake (Lake Agassiz) from the edge of the North American ice sheet into the north Atlantic. Apparently this disrupted the circulation pattern and heat flow of the Atlantic, with repercussions all around the world.

5.10 COINCIDENCE OF EXTINCTIONS WITH HUMAN ARRIVAL—DID THE HUMANS DO IT?

Beyond all the evidence of climate changes, there is an extra dimension to this story which seems far too much of a coincidence to overlook. In each of the regions where so many large animal species were lost in the late-Quaternary, the timing of extinctions broadly coincides with the first arrival of modern humans. Perhaps then it was really humans that were in some way the cause of all these extinctions?

5.10.1 When exactly did humans arrive?

The question of when exactly modern humans turned up in each region is key to understanding what role they might have played in the extinctions. In Europe and eastern Asia, the timing of their arrival seems to be fairly well settled. But in both Australia and the Americas, the question of timing has caused a lot of controversy among archeologists. Uncertainties in dating and differences in interpretation have led some archeologists to hold the view that humans turned up tens of thousands of years earlier than most of their colleagues would suggest.

The most generally agreed dates in each of these regions have humans arriving in Australia about 45,000 years ago, and in the Americas by around 13,000 years ago (e.g., Adams *et al.*, 2001). However, some archeologists claim that humans go back as far as 71,000 years ago in Australia, and to 35,000 years ago in North America: it is all a matter of how one interprets the very sparse archeological evidence. In Australia, whenever it was that people actually arrived, they must have made it across by water craft from Southeast Asia. Most likely they then crossed to New Guinea at a time when it was linked by land to Australia, because of low ice age sea levels. In North America, people probably came from eastern Asia by walking across a land bridge that linked Siberia to Alaska, spreading down southwards as the ice sheet in western Canada melted back to expose land on the coastline—or perhaps farther inland along an "ice-free corridor" that opened up between the melting ice sheets. Archeological dates suggest that when humans reached south of the Canadian ice sheets, they quickly spread down throughout North and then South America, turning up at the southern tip of the continent within 3,000 years or less (Steele *et al.*, 1998). I tend to go with the prevailing view among archaeologists, that people arrived around 45,000 years ago in Australia and 15,000–13,000 years ago in the Americas (Adams *et al.*, 2001).

In Europe and Asia, the generally agreed picture is that the modern human presence goes back a lot further than in the Americas: 60,000 years ago is the date that is most frequently mentioned for southeastern Asia, based mainly on archeological evidence. In Europe, colonization by modern humans began some 45,000 years ago—according to both archeology and modern DNA—and was completed by 30,000 years ago. Although on an evolutionary timescale, modern humans are very recent arrivals in each region, in Europe and Asia they do still appear tens of thousands of years before the wave of large mammal extinctions.

The DNA evidence from present-day human populations around the world broadly confirms the direction of spread of the first modern humans in each of the regions where large animal extinctions occurred. Although we cannot precisely date their time of arrival, the DNA also broadly agrees with the general consensus view among archaeologists of when human colonization of each region occurred.

5.10.2 A "blitzkrieg" on animals in the Americas and Australia?

In the Americas and Australia, the coincidence in timing between the arrival of humans and so many extinctions is something too striking to ignore. Humans are one factor that had been missing in both Australia and the Americas throughout the Quaternary, while all those animal species were quite happily surviving so many climate shocks. The humans turned up, and within a few thousand years most of the animals went extinct. To a lesser extent it seems to be a similar story in Europe, which had apparently also lost a few species of mammals around the time modern humans first turned up 45,000 years ago. So what could the humans have done to cause so many extinctions?

One widely accepted and compelling explanation is that the newly arrived humans simply hunted all these animals to extinction. The idea was first put forward in detail by Paul Martin (presented in an updated form in Martin, 1989), based on the North American archeological record. He suggested that the large mammals of the Americas were particularly susceptible to over-hunting because of the peculiar behavior of herbivores that are exposed to a new predator. It has been reported many times that populations of deer or other herbivores that are left isolated from predators tend to lose their fear of them within a few generations. When a predator—including humans—reappears, the herbivores will at first show little in the way of "flight or fight'' behavior, and because of this they are easy to kill. This lack of fear of dangerous predators is called ''naivety'' by zoologists. Only after they have witnessed the predators hunting and killing some of their own kind do the herbivores start to fear them, passing this fear on to their own offspring by example, showing fear at the sight or smell of the predator that the young pick up on. It is only then that most of the animals in the population begin to run away from the predator or fight back against it.

An example that illustrates the effect of naivety is the case of moose (*Alces alces*) in Yellowstone National Park in the U.S.A. (Brook and Bowman, 2002). When wolves—which had long been hunted out of the park—began to spread back in the 1960s, moose were exposed to a predator that they not seen within living memory. In the first years after wolves arrived, moose suffered particularly severe mortality, particularly of the young. Only afterwards did the moose learn to fear the sight and sound of wolves, greatly reducing mortality with their defensive reactions. A similar story occurred in Scandinavia as brown bears (*Ursus arctos*) recolonized areas where they had previously been hunted out (Berger *et al.*, 2001). There were very high kill rates of adult moose which had never encountered this predator before, followed by a large reduction in mortality as the moose learned to become wary of wolves. Of course the species we see today are ones that did learn in time to save themselves from humans. One can perhaps imagine that with a particularly efficient and lethal predator—such as humans—and animals that were a little slower to learn, naivety could have led to overhunting and extinction. The scenario is plausible at least, even if we do not stand much chance of testing it because the animals themselves no longer exist!

In the case of North America, Paul Martin (1989) implicated a particular group of early American Indians who seem to have been widespread just before the extinctions started. These people left behind characteristically shaped spear points, which have a groove (known as fluting) on either side where they would probably have slotted into the end of a cleft stick. Because these spear points were first described from near the town of Clovis, New Mexico, archeologists call them the ''Clovis'' culture. Clovis points turn up in their thousands all across North America, some of them having been brought or traded over hundreds of miles from where they were made. They first seem to appear around 13,000 years ago, and stopped being made and used some time during the Younger Dryas cold event (see Box 5.1) that ended about 11,500 years ago.

Martin suggested that the people who made the Clovis spear points were the first humans to enter the Americas, and that they encountered a continent filled with types of animals that were unable to defend themselves against this new predator. The Clovis people would have been able to slaughter the animals easily, living off their meat and following a sort of primordial Atkins diet. With an expanding population, and so many animals available to find just by moving on further across the landscape, these first arrivals had no incentive to conserve their prey, and no practical impediment to stop them from harvesting their prey unsustainably. When all the herbivore populations in an area had been wiped out, the hunters would simply move on to new areas.

The human population, then, would have spread across North America rather like a wildfire burning through grassland, with a "kill front" at its leading edge. Behind this, almost nothing in terms of large animal life would be left, except for their scattered bones. Eventually, when the kill front reached the far edges of the continent, most of the species of large animals that had existed in North America were extinct. Martin has termed this fast-moving wave of killing a "blitzkrieg", a fast-moving attack that overwhelmed its victims by speed and surprise.

5.10.2.1 Putting the overkill hypothesis to the test for North America

If Martin's hypothesis is correct, we might expect to see certain things in the North American archeological record. For example, we could see archeological evidence of humans killing and eating the ice age fauna. There is in fact some very direct evidence that they were doing this, from one Clovis age site in Texas. Embedded in the skeleton of an extinct elephant—a mastodont (*Mammut americanum*)—is an unmistakable Clovis spear point. Furthermore, on the bones of the elephant are scrape marks that suggest humans were peeling the flesh off with stone tools. It is hard to imagine more convincing evidence of humans both killing and then eating these extinct herbivores.

However, one apparent problem for Martin's scenario is that there is now a fair amount of evidence showing humans were already in North America before the Clovis culture existed. Widely respected radiocarbon dates at several sites suggest that the humans were already present perhaps 1,000 to 1,500 years before the first Clovis spear points (Adams *et al.*, 2001). This means that the animals would already have encountered humans well before the extinctions finally took place, which would surely have given the animals the chance to learn to fear humans. This is puzzling if we are trying to explain the extinctions in terms of an initial wave of hunters killing inexperienced animals—surely given a couple of thousand years the animals would have learned that humans were their enemy, and started to defend themselves.

So, a 14,000-years-ago date presents problems as it means that humans turned up in the Americas a couple of thousand years earlier than when the animals disappeared (which was the Younger Dryas around 12,500–11,500 years ago). Perhaps the somewhat earlier dates of around 14,000 or 15,000 years ago for humans in the Americas are wrong—although most archeologists now seem to feel that they are accurate? Or perhaps it was not naivety as such which made the animals susceptible to extinction, but something else more fundamental about their ecology or behavior?

Having evolved without hominids, the animals of North America may have been in some way genetically incapable of acquiring a behavior pattern that could keep modern human hunters at bay—though exactly what type of behavior was lacking is a matter for speculation. If the naivety was innate rather than learned, the humans could have had a more leisurely period in which to wipe out the animals through over-hunting, without the animals learning to defend themselves.

Can all the animals in one region be dumb, and all the animals in another be smart about humans? There is in fact some circumstantial evidence that the tendency of animals to defend themselves varies by region according to how long they have had to evolve to human hunting.

It is interesting that many large African animals such as elephants, rhinos, and buffalo have a reputation for being extremely aggressive towards humans, or very nervous and flighty as in zebras, compared with their nearest relatives on other continents. We can compare, for example, their behavior with that of animals in Asia, where humans have not been around quite so long. Asian wild elephants, Asian wild rhinos, Asian wild water buffalo, and horses are all generally more docile towards humans than their African counterparts, allowing some of these species to be domesticated. The aggression and nervousness of African animals may have evolved as a necessary response to 150,000 years of hunting by modern humans, and following on from earlier hunting by hominids over hundreds of thousands of years before. Looking back, perhaps that first wave of several mammal extinctions in Africa back around 150,000 years ago was produced as modern humans first honed their hunting skills in the savannas and woodlands of Africa?

Generally speaking, opinion about how humans could have driven the megafauna extinct has now shifted away from a sudden intense wipeout to a longer period of coexistence, but one that ultimately led to the animals' demise. In some regions of the world, the best explanation seems to involve multiple factors interacting: for example, a climate change coming in after a long period of hunting, and finishing the animals off (see below).

5.10.2.2 The plausibility of overkill in South and Central America

In South and Central America, humans making rather similar stone spear points spread southwards around the same time as Clovis. Extending his hypothesis from North America, Martin suggested that it was a similar story of hunters turning up and slaughtering naive animals that caused the widespread extinctions there. In general there is more controversy about the time when humans first turned up in South America, with some archeological dates having been claimed back to 35,000 years ago. However, there seems to have been a pattern of dates before about 13,000 years ago being discredited or widely disputed by archeologists (Adams *et al.*, 2001).

If the earliest dates for humans in the Americas—extending back to 20,000 or even 35,000 years ago—*are* correct, then it is more difficult to explain the simultaneous wave of extinctions right at the end of the last ice age. After all, why would the animals wait for many thousands of years of coexisting with humans before going extinct?

Another potential problem with extending the idea of a wave of slaughter beyond North America is that there is no direct archeological evidence that humans ever hunted and killed the various South and Central American animals that went extinct at the end of the ice age. However, this lack of evidence from bones is not an especially difficult thing to explain: only a very small proportion of animal carcasses ever end up being fossilized, and the chances of one of them being preserved from a short-lived "blitzkrieg" phase are tiny. The slaughter of millions of African elephants during the last few decades has left almost no trace in terms of bones, which quickly rot at the surface. We should not then be surprised that in South America nothing has so far been found from this geological "instant". In fact we should be especially surprised that a slaughtered carcass has actually been found from North America.

5.10.3 Modeling humans driving Quaternary megafauna extinct

So was it really humans or climate? We can make a general case for it being humans, but it would be best to add some rigor to the argument. One approach to the question of whether humans could have driven so many large animals extinct is to "model" their populations mathematically, setting them off interacting as hunter and hunted. Making some apparently reasonable assumptions about the proportion of the prey population that was killed by humans each year, how would this have affected the abundance of prey? The food supply from eating the prey population would also have affected the rate at which the human population built up, and this in turn would have affected the numbers of prey. It turns out that it is not hard, given some plausible assumptions, to have humans alone driving most of the North American megafauna extinct.

Models show that one especially effective way to have widespread extinction would be if the hunters were "prey switching" (Barnosky *et al.*, 2004). Prey switching involves abandoning the hunting of a particular type of animal when it becomes rarer. So when a big, slow-breeding animal like a ground sloth or giant bison becomes rare, the hunters instead concentrate on pursuing another species that remains abundant—such as a small animal like a rabbit or squirrel that can breed fast and maintain its numbers. In this way hunters will maintain their food supply and their own population level, ready to hammer back the big animals if they start to recover their numbers. Kept forever at low levels of abundance, the beasts are much more likely to go extinct eventually, though what eventually finished the last ones off could have been chance dips in their numbers or the effects of inbreeding.

Furthermore, and perhaps even worse for the big animals, is the possibility that humans who are maintaining their numbers on small, abundant prey will tend of pick off the last survivors of the big animal species wherever they happen to encounter them by chance, even if they are not particularly looking for them. For this reason there is no escape from hunting, not even by being rare. And given what we know of human societies, ancient and modern, killing a big rare animal carries special prestige for hunters. The rarer it gets, the greater the kudos involved in being able to find and kill it, thus ensuring its spiral down towards extinction.

One model that included "prey switching" by hunters between large slow-breeding animals and fast-breeding small animals was used to try to predict the fates of 41 different species of large mammals that existed in North America at the end of the Quaternary—whether towards survival or extinction (Barnosky *et al.*, 2004). This particular model did very well generally, correctly predicting which would go extinct and which would survive to the present in 34 out of those 41 species (an impressive 83% success rate). The large mammals that tended to survive—such as the elk or the pronghorn—had sufficiently large background populations and a fast enough rate of reproduction to overcome the effects of persistent hunting. The ones that went extinct, because they could not breed fast enough and were always rarer anyway, tended to last about 900 years from their first encounter with humans to their final extinction.

Another similar model has been applied to the Australian megafauna (Brook and Bowman, 2004), set up with growth and reproductive traits based on similar-looking African mammals. Even though we do not have convincing evidence of a delay between human arrival and megafauna extinction in Australia, it is still interesting to see if it could have worked this way. Brook and Bowman's (2004) model suggests that with a combination of environmental change and hunting, the big Australian animals could have died out in around 750 years (more about such combined effects below).

Brook and Bowman's model seems interesting and persuasive, but since the model was set up around the life history characteristics of African mammals, can it successfully be used to explain why the animals in Africa *survived* the climate changes at the end of the glacial? Apparently in fact it also predicts extinction for the African mammals that made it through to the present (Wroe and Field, 2006)—so obviously something is missing!

Box 5.3 Repopulating the lost North American megafauna

North America has lost most of the large animals that it had as recently as 12,000 years ago. The ecological vacuum caused by the extinction of so many species must have significantly altered the ecology of large parts of the continent. We can get some clues from how North American ecology must have changed by looking at Africa, which still retains many of the same general types of large animals that America has lost. In Africa, for example, we can see elephants causing a great deal of disturbance knocking down and damaging trees and bushes, and scuffing and trampling grasslands. In doing so, they can totally alter the structure of vegetation, the balance of plant species, and the food supply and land cover available to other animals. Elephants and other large herbivores such as rhinos have been termed "bulldozer herbivores" because of their effect on vegetation (Kortlandt, 1984). When elephants and other large animals such as the giant beaver or giant bison lived in North America, they too might have had wide-ranging effects on the vegetation.

It has been suggested that in the desert environments of the American southwest, the extinct North American llamas grazed back the vegetation, and now that they are gone the desert shrubs are much leafier and more abundant. The ferocious spines of cacti may originally have evolved as a defense against the tough mouths of these camel-like animals; nowadays cacti seem much pricklier than is necessary to keep away the existing desert herbivores.

In slightly moister scrubland and grassland environments in the western U.S.A., the wild horses or "mustangs"—reintroduced to the Americas by European colonists and now running wild—tend to become very abundant and graze the vegetation back heavily. Continuous culling seems necessary to keep their populations down to reasonable levels. In Africa, zebras—another form of wild horse—seem limited by lions and other predators. Perhaps when there were lions and other large predators in the Americas, these kept horses from becoming so very abundant?

Pronghorn antelope (*Antilocapra americana*), which live in the dry plains and southwestern deserts of the U.S.A., are capable of running at 95 km/h, far faster than any living predator in North America can manage. They may have evolved to run away from is an extinct species of cheetah-like cat (*Miracinonyx* spp.), which went extinct around 11,500 years ago—cheetahs in Africa can reach 110 km/h.

Thus it is likely that there are many "missing pieces" nowadays in the ecological jigsaw of the North America. For many years Paul Martin—the originator of the overkill hypothesis—has been advocating repairing these gaps, and the ideas have been expressed by Martin and his colleagues in a recent article (e.g., Donlan, 2007) and in a book (Martin, 2007). These zoologists suggest bringing in surviving species of mammals from other continents which offer the closest ecological analogs to what has been lost. Most of all, the fauna of Africa seems to offer the widest range of species to replace the extinct megafauna. For example, they advocate releasing lions into the western mountains to control horse populations, plus elephants to restore the natural structure of vegetation. Releasing camels into the southwestern deserts is also suggested. One obvious question that springs to mind is whether African animals would be able to survive the winters in North America. Martin answers that many African animals in fact do occur in parts of Africa that have a colder season (e.g., in South Africa, or in the Sahara Desert). As an example of the ability of African animals to survive harsher winters than one might expect, he points to a thriving population of North African camels on a ranch in Kansas, where winters can certainly get bitterly cold.

However, the biggest obstacles to reintroducing the "lost" megafauna are human. Ranchers and the general public would probably not respond at all well to the idea of something like African lions or elephants being introduced to the U.S.A., even if was only in sparsely inhabited areas. There has been enough resistance to reintroduction of wolves to the Rocky Mountains from Canada, and they rarely attack humans (unlike lions or African elephants). Many of the wolves released into the wild have ended up being shot by those who oppose the plan.

Also, ecologists argue that African species are just not the same as the native North American ones that were lost, and that they may behave in unexpected ways as part of the ecological system. There are certainly many examples of disasters caused by deliberate introduction of foreign species—some of which are mentioned in this chapter—and that seems good reason to be suspicious of the idea.

Given all the logistical, conservation, and political problems that reintroducing these animals would present, it seems unlikely that Paul Martin's interesting idea will ever come true.

5.10.4 Problems in the dating: Did Australian animals really die out just as humans arrived?

To really make the case that humans really "did it", it would be nice to have some examples where species were clearly present and abundant right up to the point when humans turn up, and then vanished. In the Americas and Europe, there is no doubt among archeologists that the main extinctions occurred after humans arrived. But in Australia this has not always been accepted.

Part of the problem is that the record of human arrival and megafaunal extinction in Australia is less clear. The dry, erosive environment of most of Australia is not so conducive to preservation of bones and artifacts, and in any case events generally seem to have occurred further back in time which also gives a less clear picture. As I mentioned earlier, there has been a fair amount of controversy over the idea that the megafauna actually went extinct around the time humans turned up 45,000 years ago. Some have claimed that many of the large animals survived with humans around, then died out one by one between around 40,000 and 20,000 years ago. This would tend to open up the possibility that something other than human hunting was to blame, perhaps climate changes. However, the dates that have been used to make this claim are now widely disputed, and the idea that the megafauna stayed around that long is not now generally believed.

Others have suggested that the megafauna mostly died out a long time *before* humans arrived, maybe back before 80,000 years ago. For example, Wroe and colleagues (see above) have strongly advocated this viewpoint.

However, in the last few years two significant studies have made quite a strong case that the extinctions really did occur just around 45,000 years ago, the time when humans turn up—not later, or earlier. There is the work I mentioned above by Miller and colleagues that includes an extensive set of dates on the eggshells of the extinct giant flightless bird *Genyornis*. Abundant finds of eggshells continue right up until 45,000 years ago, and then suddenly there are no more. As in the case of the end-Cretaceous extinction (Chapter 3), the strongest case that lots of species went extinct at once can be made using a particular species that was abundant and widespread in the fossil record, and then suddenly vanished. We can use this to suggest that the same actually happened to all the less commonly fossilized species that disappeared around the same general time. So *Genyornis*, through its abundant eggshells, fits this requirement as a "mass extinction indicator" species. *Genyornis* was abundant, right until approximately the time when humans appeared in Australia, and then went extinct. That looks like fairly good evidence that the extinctions were the work of humans.

The other study, which came out just as I was about to send this book to the publisher, was the one by Turney *et al.* (2008) that I have also mentioned above. They showed several different large marsupials survived in Tasmania right up to the time when humans arrived there, and then suddenly vanished. This also clearly implicates the humans.

For anyone who still suspects that climate, and not humans, really caused the extinctions in Australia, it is important to emphasize that nothing like this wave of extinctions happened there ever before during the Quaternary. The extinctions in Australia ended evolutionary lineages that had continued for many millions of years—and they waited all that time up until just about the moment when humans turned up! While the case that humans did it was for a long time ambiguous, it is still astounding to me that anyone could ever have seriously doubted the idea. Nevertheless, as always in science the skeptics have fulfilled an important role, for they have concentrated minds rigorously on the evidence, and also motivated many to go out and look for evidence that could either prove or disprove the hypothesis. Thanks to all the controversy, it now looks like we have a very good case that humans drove the Australian megafauna extinct, rather than just the vague belief that would have prevailed otherwise.

And what of the survivors? The small marsupials such as wallabies and possums that now make up much of the Australian fauna are merely remnants of what previously existed. Even some of the species which now survive are apparently pygmy versions of their ice age ancestors; for example, the modern-day koala (*Phascolarctos cinereus*) is a third smaller than its likely ancestor (*Palorchestes stirtoni*), and some kangaroos seem to be pygmy forms of their ancestors which existed tens of thousands of years ago. This again is what we might expect if human hunting was a key force. Smaller animals are less desirable to hunt, less likely to be found, and are more likely to escape. They are also more likely to mature and reproduce before having the bad luck to be caught. Similar evolutionary changes have occurred in ocean fish populations in recent years, as a result of overfishing.

5.10.5 An explanation for the delay in Europe and the Americas: A "double-whammy" combination of climate change and over-hunting?

In Europe and Asia there is no doubt that modern humans were present for tens of thousands of years before the extinctions. In these places, it seems especially hard to explain why humans refrained from hunting the animals to extinction for so long. And why, by the time the animals did actually go extinct, hadn't they already gained the defensive measures to avoid being over-hunted? Even the delay of just a few thousand years in the extinctions in the Americas is rather hard to explain. If hunters were capable of killing off the North and South American large mammal fauna, why would the extinctions wait for thousands of years?

And there remains the nagging question of why the main waves of extinction outside Australia—in the Americas, Europe and Asia—waited until just about the time of the Younger Dryas climate shock (above).

One rather elegant explanation is that all these animals were actually eliminated by a *combination* of climate change and hunting. So according to this idea, when modern humans arrived in Europe, Asia, or the Americas, there was a step-up in predator pressure on the animal populations. Although there would be fewer of the animals around because of this hunting, they would still be capable of maintaining a steady (but lower) population level.

However, eventually a sudden big climate change would come along, and this would provide an extra "stress" on the animal populations. When climate changes, vegetation will respond. Some plants will grow better, and others more poorly: this can alter and even cut off the food supply for herbivorous animals. A warmer climate could also cause heat stress on animals adapted to cold climates, while cooling could make life more difficult for others. Eventually, given a few centuries, the animal population might be able to shift its distribution to suit the new climate. But in the meantime in many parts of its range, it might reproduce less and die younger.

It is possible that by themselves, the effects of climate changes on animal populations were not usually drastic enough to send them to extinction. After all, the big animals around the world had managed to survive so many large climate changes before during the Quaternary. Yet the combination—of a climate change against the background of continual depletion of each population by hunting—may have been enough to put many species into a steep decline. To use a boxing term, the animals were hit by a "double whammy": a one–two punch that finally floored them.

This idea could then explain why the extinctions waited for some time after initial contact with humans. The naivety of the animals when humans first encountered them might not after all have been the one all-important problem. Instead it could have been the interactions of multiple factors on their population biology. While the idea of the hunting and climate "double whammy" is plausible, in science it is always best to follow an idea through with mathematics to help show that the reasoning is solid. The representation of an idea in mathematical form is called a "model", and nowadays models are usually run on computers.

The question of whether the Quaternary extinctions were likely to have been caused by the "double whammy" effect has been investigated using a different model from the one outlined previously. Mithren (1993) decided to concentrate on the mammoth in eastern Europe at around the end of the last ice age. He modeled the population starting just before the end of the ice age 16,000 years ago, and predicted an extinction following on around the time of the Younger Dryas cold phase several thousand years later.

All-important in a population model are the parameters it is based on, detailing such things as generation time, potential life span, and potential reproductive rate. There are no mammoths around nowadays of course, so it was necessary to come up with some reasonable guesses. Mithren chose to base the population parameters of the mammoth on present-day African elephants. This is a reasonable choice as African elephants are about the same size as mammoths, although in evolutionary terms mammoths were actually more closely related to Indian elephants.

A key aspect of the model was the population density of mammoths that could be supported by the past vegetation. Given the available reconstructions of ice age vegetation, which envisage a rather open dry grassland and low scrub across much of Eurasia, the mammoths were assumed to have had a fairly high population density grazing off this material.

According to this model, mammoth densities would suddenly have decreased across Europe with the Younger Dryas cooling, with further climate fluctuations adding additional problems. This sort of thing had happened many times before during the long span during which mammoths had existed on Earth. But all the previous times, the only humans around in Europe had been the relatively unsophisticated *Homo heidelbergensis* and Neanderthals, and the challenge of trapping and killing such a huge beast is likely to have been too much for them. Now the difference was that modern humans were present in abundance in the region (having arrived in eastern Europe around 45,000 years ago, they were assumed in the model to have built up their population in the region slowly over many thousands of years), and all the ingredients for a disaster were in place. Mithren estimated what level of hunting that the humans could plausibly have subjected the mammoths to. During the conditions after the first warming 16,000 years ago, which actually gave an increase in plants that mammoths could eat, his model suggested that this hunting was not especially a problem. Calculations showed that the mammoth population could replace lost individuals fast enough to keep itself going, even with occasional losses of mammoths to the hunters. Then the big climate shock of the Younger Dryas hit—temperatures decreased drastically and conditions became less favorable to mammoths. In particular, the supply of food from vegetation would have decreased. The mammoths reproduced less often, and grew and reached maturity more slowly. Suddenly, the relatively small "tax" on the mammoth population from human hunting became much more important, because they could no longer make up for this depletion of their numbers. The model finds that the mammoth population would have gone into decline, each generation not quite replacing itself, until after a few hundred years there were none left. This fits nicely with an extinction several thousand years after the warming at the end of the main glacial cold phase.

This case study using a model provides some reassurance that the "double whammy" effect could really have operated in causing extinctions in the late-Quaternary. However, there were many other species and genera lost from across the Old and New Worlds, and Australia too. How can we know that it could really work for a range of shapes and sizes of animals, each with its own separate ecology? At present, we do not have a separate model for every extinct animal, and it is difficult anyway to know if our assumptions about their life cycle are correct (since they are no longer alive to study in detail). Nevertheless, the details are not so important as making the general point that interaction of factors with hunting could produce a relentless decline to extinction. I am persuaded by the general argument that it was a special and new *combination* of circumstances that caused such widespread extinction around the time of the Younger Dryas.

5.10.6 Smaller animals tended to survive

One very striking general pattern in all these waves of extinction that swept the continents in the late-Quaternary, is that smaller animals were far more likely to survive than larger ones. Small creatures such as rabbits, mice, and shrews on the northern continents and South America survived almost unscathed. Similarly on Australia, while the bigger marsupials were lost, the smaller ones such as marsupial mice, hare wallabies, and koalas survived up into recent times (although a few of the smaller species have been lost since European settlement). This selective pattern of survival makes sense if we think of a situation where animals all around are being hunted to extinction. A species is more likely to survive is if it is hiding under rocks or bushes or in burrows where humans cannot see it or reach it. Even if it is not initially scared enough of humans to run away and try to save itself, a small animal is less likely to be noticed. By contrast every last mammoth or *Diprotodon* was an unavoidable feature in the landscape—and perhaps an irresistible target to hunters. Another thing that smaller species have in their favor is that their total numbers tend to be far greater; so even with a very high kill rate it is more likely for some to be left in the end to continue the species.

An additional interesting aspect of the extinctions is that (as I mentioned above) some species which survived the late-Quaternary ended up smaller (e.g., koalas now are about a third smaller than they were when humans first appeared in Australia). It is possible that only the smallest individuals were able to avoid the hunters, so as a result of this selection the populations now are dominated by genes for small size.

In general, species such as the koala, the tree sloth, or the orangutan that are arboreal seem to have survived better those that lived on the ground, which is hardly surprising given that humans are relatively awkward in climbing and hunting in trees (Barnosky *et al.*, 2004). Among the larger ground-living animals, the ones that survived the extinctions often lived in very barren or rugged high-mountain environments, where human populations would have been smaller and where escape from hunters would have been easier. Examples include the llama (*Lama glama*) and vicuna (*Vicugna vicugna*) of the Andes of South America. Again this pattern fits in well with what we would expect if human hunting had played a critical role in the extinctions.

5.10.7 Fire-setting by humans in the Australian extinctions?

In Australia, as in South America, there is no direct evidence of anyone hunting or eating the extinct fauna. The only evidence we have is circumstantial: the earliest respectable dates for a human presence is Australia are around 40,000–45,000 years ago, and this is where the last records of many extinct species cluster.

Something that humans did plenty of in Australia was burning the vegetation. It could be that this too was part of the cause for the extinctions. At the time when they were first encountered by Europeans, the aborigines were very adept at manipulating their environment through fire, for a variety of purposes. Fire was used to drive animals such as kangaroos into a trap where they would be killed, or to encourage green regrowth that would attract grazers in. It was used to clear away scrub to give a clear run for hunting, and it was also used as a defense in warfare. There are signs of an increase in burning of vegetation in Australia around 45,000 years ago (Adams and Faure, 1997), at around the time when humans seem to have first turned up there, and when the large marsupials seem to have become extinct. Among the signs that fire was having effects on the environment are a shift in the carbon isotope composition of the eggshells of the extinct bird *Genyornis*, suggesting a change away from bunchgrass vegetation towards more fire-tolerant *Eucalyptus* scrub. By depriving grazing animals of their usual diet, humans may inadvertently have impacted them in ways that extended beyond hunting itself.

Of course, it is also possible that the change in vegetation from burning combined with hunting pressure—and perhaps climate change too—to drive animals extinct. The debate about what caused the late-Quaternary extinctions has tended to become polarized between different academic camps, their members sometimes barely on speaking terms with one another. Experiences of extinction in more recent history (e.g., the heath hen, and the great auk—see below) suggest that often it can result from combinations of unlucky events combined with hunting. It may be worth reminding ourselves that nature frequently works through multiple causes.

5.10.8 Or was it a disease? Or meteorites?

Some have suggested that the first humans to arrive in the Eurasia and Americas not only hunted the large mammals, but also brought with them a lethal new disease which killed off a range of species. Perhaps it was something related to rinderpest, a virus that kills many types of hoofed animals? Some scientists have looked in frozen carcasses of mammoths from the permafrost for evidence, but found nothing. But even if there were such a disease, it is unlikely that we could find a carcass from the brief time of the plague itself.

All considered, it seems implausible to me that a few small groups of colonizing humans could have been so successful in introducing diseases that killed large animals to so many places. For one thing people do not normally carry diseases of large herbivores: the strains of microbes are different. Short-lived viral diseases should tend to die out quickly in small groups of hunter-gatherers, which explains why such maladies as influenza, smallpox, and measles never reached the New World (leaving its inhabitants unusually susceptible to them once they did finally reach there after contact with Europeans). Even if modern humans did carry some sort of disease that also affected animals, it is very unlikely that it could have infected mammals from the diverse range of orders that were hit by extinctions in the late-Quaternary. It is especially far-fetched that the distantly related marsupials of Australia—as well as the birds that went extinct—were killed by a disease that was suited to living on humans.

Another idea, something that has been around for a while but was recently revived, is that many of the extinctions occurred because of a comet or meteor shower that hit the Earth (Firestone *et al.*, 2007). There is now some evidence that around 13,000 years ago, impact craters formed in various places across North America. These include the strange elliptical ponds known as "Carolina bays" which are scattered down the coastal plain of the eastern U.S.A. Recently published analyses of these sediments in these ponds reportedly show high concentrations of iridium, an element that is abundant in meteorites and which was first used to identify the impact at the end of the Cretaceous (Chapter 3). Mammoth bones and tusks from animals living out on the Siberian tundra at the same time also seem to have been peppered with tiny fragments of meteorite (Firestone *et al.*, 2007). However, this work needs to be checked and corroborated. Many geologists who work on past extinctions due to meteorites are incredulous about the idea that there was any major impact event at this time.

In any case, what we do not see at the time of these end-Quaternary impacts is any sign of wholesale ecological collapse, on the scale of other widespread extinction events further back in time (Chapter 3). We do not see any "fern spike", for example. It also seems to be a bit too early in time for the main peak of extinctions, which followed maybe 1,000 years later. If impacts from space 13,000 years ago did contribute to the extinctions, it might have been in terms of providing an additional stress on populations that reduced their ability to recover from hunting, or from humans burning the vegetation.

One question that nags in my mind is why the "big one" would have waited and waited for more than two million years to strike the Earth in the geological instant after modern humans had spread around the continents? This is another reason I have problems with accepting this as a pre-eminent cause of the extinctions. If there really was an impact at that time, my guess is that these meteorite bombardments are a lot more common that we have tended to realize, and the one picked up 13,000 years ago is only one of many extending back into the Quaternary—and that none of them was large enough to drive the megafauna extinct. The last fairly substantial impact was only 100 years ago, in the Tunguska region of Siberia in 1908, where it flattened hundreds of square kilometers of forest. It is only a matter of time before another of these comes along: just hope that you are not underneath it.

5.11 THE WAVE OF EXTINCTION SPREADS TO ISLANDS

After the extinctions on the continents had mostly been completed, a second wave began on remote islands around the world. In most of these island extinctions, it is beyond any reasonable doubt that humans were the cause. Humans were by now trying out their skills as ocean navigators, reaching a high point within the past 3,000 years with the Polynesians—who managed to reach virtually every isolated speck of land in the Pacific.

Over millions of years, distant islands had been loading up with endemic species. These were descended from waifs that had blown or floated across to turn up on their shores. Perhaps sometimes small groups of males and females turned up at once, but even a single fertilized female could have been capable of establishing a new population. Island animals tended to be bizarre: often dwarfs, sometimes giants, and occasionally morphed out of all recognition compared with their nearest relatives on the continents. Time after time, the archeological record shows humans appearing on an island, and the indigenous fauna vanishing, often within an archeological instant—a span of time so small that it cannot be distinguished by archeological dating.

5.11.1 The last mammoths: Wrangel and St. Paul Island

Off the coast of northern Siberia there is a large island covered in boggy, treeless tundra. In this desolate landscape, one last population of mammoths survived long after their mainland cousins had gone extinct. Probably they were already on the island when it was cut off from the continent by sea level rise 9,000 years ago. The Wrangel Island mammoths were about a third smaller than those that had previously existed on the Siberian mainland, and are termed pygmy rather than dwarf forms because they are not quite as small as the ones found on many other islands.

The Wrangel Island mammoths seem to have survived up until about 3,600 years ago, then suddenly they vanished. It seems too much of a coincidence that the first archeological remains of humans are found on Wrangel Island around the same time. Levy (2006) has suggested that it was a combination of climate change and hunting that finished the population off, much as it had done for the mainland mammoths. In this case the cooling of the regional climate out of the warmer early part of the interglacial might have been the trigger, although the precise coincidence with the time when humans turn up is certainly quite suggestive that hunting alone was enough to wipe out the pygmy mammoth.

There was a similar pygmy form of mammoth on St. Paul's Island off Alaska, although there it did not survive as long as its cousin on Wrangel Island. The St. Paul's mammoth died out around 8,000 years ago: so far we have no archeological evidence from that time suggesting colonization by humans, although the ancestors of today's American Indians were certainly in Alaska by this time.

The island mammoths were not the only late-Quaternary animal to linger for a while. The "Irish elk" (*Megaloceros giganteus*) likewise managed to survive (fully sized) in western Siberia until about 7,000 years ago, after having lost most of its previous range across Europe and Asia at the end of the last ice age (Stuart *et al.*, 2004).

5.11.2 Mediterranean islands

Scattered around the Mediterranean Sea are various islands, each several tens of kilometers away from the mainland shores. After the end of the last ice age, many of these islands retained a strange mixture of giant and dwarf animals, descended from species from the European or North African mainland. For example, on Cyprus there was a dwarf elephant that is estimated to have weighed only 2% as much as its full-size ancestors on the mainland. On Crete there were eight species of dwarf deer (some of these no more than 40 cm at the shoulder). On the Balearic Islands there was a giant dormouse (*Hypnomys*), along with a dwarf elephant. On Malta, there was yet another dwarf elephant, and a giant swan (*Cygnus falconeri*) half as big again as any present-day species.

Dwarf elephants seem to have occurred on most of the islands scattered around the Mediterranean Basin, including at least seven of the Greek islands. Only around 1 m high at the shoulder when fully mature, these tiny elephants were mostly descended from the European straight-tusked elephant (*Elephas antiquus*), a large elephant—now extinct—that lived in southern Europe up until around 11,500 years ago and perished as part of the main wave of extinctions on the northern continents. The dwarf elephant on Sardinia, however, seems to have been a type of mammoth (*Mammuthus*). It is thought that each of these Mediterranean island dwarf forms originated independently from normal-sized elephants which were washed out to sea and ended up stranded on an island. Each type of island elephant is classified as a distinct species, reflecting its own separate origin. It is possible that the legend of the ancient Greek cyclops—a single eyed humanoid giant originated from the skulls of dwarf elephants found on Greek islands. The large nasal cavity of the elephant's skull could easily have been mistaken for a single eye socket in a misshapen human skull.

It is not known why elephants produce dwarf island forms so easily, but it has certainly happened several times, from different species of large ancestral elephant. In Indonesia, another type of elephant (*Loxodon*) produced several island dwarf species, although only one of these survived into the last 50,000 years. This was the dwarf elephant of Flores, which was probably hunted by the dwarf human *Homo floriensis*—and they may both have been wiped out by the arrival of modern humans around 30,000 years ago. Also there was a pygmy mammoth on the California Channel Islands just off Los Angeles at the end of the last ice age, and there were also those pygmy mammoths on Wrangel and St Paul's islands in the Arctic (see above).

Why did size reduction in island elephants occur so many times? Presumably, there must have been some consistent selection pressure for size reduction—perhaps it is the absence of predators on such islands which allows elephants to decrease in size because that way they are more efficient at finding food. It may be that an elephant isolated on an island will often shift its niche to take the place of pigs, which do not commonly manage to reach islands. Combined with this it may be some poorly understood plasticity in the elephant genome that allows it to shift the course of its growth and development so drastically.

The various endemic island animals in the Mediterranean mostly disappeared between about 9,000 and 6,000 years ago, around the time when archeological evidence shows humans beginning to turn up on most of these islands. However, on the Greek island of Tilos, the tiny elephants may have survived up until 4,300 years ago, perhaps showing up in an Egyptian tomb painting (Theodorou, 1988).

5.11.3 Madagascar: lemurs and elephant birds

Madagascar broke off from the side of Africa some 17 million years ago, forming a great raft of rock that has drifted out into the Indian Ocean. As it became an island it would presumably have carried along some of the life forms that were living on it as part of the African continent. Other species may have reached it by chance dispersal across water before Madagascar had drawn too far away from the mainland.

In some ways Madagascar has been a museum of life, where certain groups of animals that had once been widespread and important around the world managed to survive, at least up until recently. In other senses it has also been a laboratory of evolution, as its fauna has diversified into a wide range of niches—taking on similar roles to other animal groups on the mainland. Illustrating Madagascar's role as a living "museum" is a group of primitive primates—the Strepsirrhini—which have survived on Madagascar even though they went extinct on the mainland of Africa before 20 million years ago. They include the lemurs and the aye-aye (*Daubentonia madagascariensis*), which are only found on Madagascar. Yet also illustrating how

Madagascar has been a "laboratory" of evolution, the lemurs there diversified into four families, ranging in size from the mouse-sized (*Microcebus*) to large cow-like forms such as *Archaeoindrus* which reached 240 kg—several times the average weight of a human. All were herbivorous, often vaguely resembling other groups of plant-eaters in different parts of the world—probably as a result of natural selection having pushed unrelated creatures towards similar forms, because of the demands of a similar lifestyle. For instance, one large lemur (*Palaeopropithecus*) had taken on a remarkable resemblance to the tree sloths of South and Central America, which hang from branches and spend most of their time munching on leaves. Although we cannot be entirely sure how *Palaeopropithecus* lived because it is now extinct, the resemblance to the sloth is too striking to ignore.

Also locally evolved and isolated on Madagascar was a pygmy species of hippopotamus, and several species of primitive mammalian carnivores—all of them distant relatives of the mongoose, known as fossas (*Cryptoprocta*). Another very notable feature of Madagascar was the elephant bird (*Aepyornis maximus*), an enormous relative of the ostrich which reached 3 m in height. Its huge eggs—several times larger than an ostrich egg—are still dug up around the island where they were once laid, buried, and forgotten.

The turning point for Madagascar came about 1,500 years ago, when the first humans turned up on its shores. Surprisingly these people arrived not from the African mainland, as one would expect, but from the islands of Indonesia in far-away Southeast Asia. Both their language and their DNA reveal their Asian origin, overlain by more recent blood and dialect words from the African mainland. It seems that their boats had been caught by the broad gyre current that sweeps west across the Indian Ocean and then curls down past Madagascar. In an archeological instant, the elephant bird, the pygmy hippo, the largest fossa, and all the biggest lemurs vanished—just when we see the first traces of humans turning up. The unavoidable conclusion is that humans wiped them out, probably by hunting, and perhaps also by changing their habitat. Today, only the smaller lemurs weighing less than about 22 kg survive on Madagascar.

5.11.4 Several thousand islands: the story of the Pacific

Islands lie scattered across the Pacific Ocean like stars in the sky. Most of them were formed as undersea volcanoes that broke the surface and then died. Some of these islands are scattered in lines or clusters—archipelagos such as the Bismarcks, the Tonga, or Marquis Islands. Others are way out on their own, thousands of kilometers from anywhere, like tiny Easter Island, only 15 km in diameter and 3,500 km from the nearest landmass. Just about all of these islands, no matter how small or how isolated, were reached by seabirds which can live for weeks or months out at sea. Quite often they were also colonized by land or freshwater birds that had been blown far off course—reaching land against all the odds before they could die of exhaustion. On many of the Pacific islands with enough resources for stable, viable populations, evolution took hold and produced new species of birds. Many of these endemics were flightless, for there were no places to fly to and no predators to try to escape from. They nested on the ground even where there were trees, and had no capacity for fear of humans or any other form of predator. Many islands had a dozen or more species of endemic birds—often flightless, and often members of the rail family Ralidae that for some reason seem to be especially able to establish isolated populations, and evolve into new forms once they are isolated. As well as birds, there were frequently endemic animals that had by chance managed to make it across from distant landmasses. For example, endemic species of iguanas evolved on many Pacific islands.

This then was the world of islands that awaited the first humans who ventured out across the Pacific. Starting around 3,000 years ago, a thin wave of human population began to fan out across the Ocean. Many of the islands around and eastwards of New Guinea were reached by the Melanesians, who had once been part of the first wave of modern humans into the region. The majority of the most remote Pacific islands were first reached by Polynesians, a people speaking languages that are related to those of far-away Madagascar, reflecting their common origin in Southeast Asia.

The Polynesians had mastered ocean navigation like no-one else in the ancient world. They combined robust ocean-going canoes with navigation techniques that enabled them to detect the presence of land far beyond the horizon. Changes in ocean waves, in cloud patterns, and fragments of floating vegetation were all used as indicators of distant land. Where they reached land, Polynesians established a characteristic farming culture that combined root crops with domesticated chickens (originally an import from Africa) and rats kept for meat. Many other Pacific islands, especially those closer to New Guinea, were reached instead by the Melanesian peoples who shared some common ancestry with the Polynesians, though they spoke quite different languages.

The story of the disappearance of these endemic birds after human arrival has been seen in the archeological record of nearly 70 islands in 19 different island groups in the Pacific (Steadman and Martin, 2003). Time and again on Pacific Islands, the bones of extinct birds stop at around the point where the first archeological evidence of humans appears. For example, on Lifuka Island, in the Tonga group of the east-central Pacific, at least 16 endemic species of birds vanish abruptly around 2,800 years ago: right where the characteristic "Lapita" pottery of Polynesians and the bones of their domesticated chickens turn up (Steadman *et al.*, 2002). The stratigraphic precision on this particular island is enough to show that the extinctions occurred in an archeological "instant"—certainly less than a century—of humans settling there. It is suspected that extinctions followed this rapidly on many Pacific islands, although no other island has been found to have as detailed an archeological record as Lifuka. However, in other cases the process of extinction took longer. On at least some islands, there is reasonable evidence from the scatter of dates that endemic birds survived for several centuries before finally succumbing. In many cases, not all of the endemic birds went extinct, with a few fortunate survivors persisting to the present—or at least long enough to be wiped out after European colonization.

Some Pacific islands seem to have suffered much more complete extinction of their endemic birds than others (Steadman and Martin, 2003). What favors extinction or survival? Low-lying islands have tended to fare the worst: there were few places for the endemic birds to remain hidden or out of reach of the invaders. Islands with steep topography that is difficult to farm or hunt in, and also difficult for pigs or rats to scale, tended to do much better in terms of survival of endemic birds and other animals. Among the birds themselves, flightless forms on islands have gone extinct far more often than those that could still fly.

It is thought, then, that the extinctions on Pacific islands resulted from a combination of factors that were all associated with humans turning up there. As well as direct overkill by human hunting there was predation of birds and their eggs by the rats and pigs that Polynesians kept for food. The humans also cleared the natural vegetation for farming, depriving endemic birds and reptiles of their habitat. Exactly how many different species of birds went extinct as humans spread across the Pacific cannot be known, but some estimates put it at as high as 2,000 (Steadman and Martin, 2003).

5.11.5 New Zealand and the moas

One large but grimly representative example of the pattern of extinctions around the Pacific is the group of islands that make up New Zealand. It was formed in a similar way to Madagascar, in this instance breaking away as a huge slab from the flank of Australia some 70 million years ago. It is not clear how much of New Zealand's original fauna survived the initial violent volcanic phase that marked its first separation, and how much dispersed across from Australia before it had moved too far away. Either way, New Zealand ended up with a rag-tag fauna—lacking terrestrial mammals but dominated by certain groups of flightless birds. The most abundant and diverse of these were the moas, flightless long-legged birds distantly related to the emu and cassowary of Australia. There were at least 15 species of moas, which varied between the turkey-sized *Euryapteryx* sp. and the 3.6 m tall *Dinornis maximus*. All were herbivores—large land predators being absent from New Zealand. The only predator for the moas to fear seems to have been an enormous extinct eagle, Haast's eagle (*Harpagornis moorei*). It has been suggested that it killed moas by swooping down and pulling their head off in its talons—which would explain why many moa skeletons are found mysteriously missing the head.

To the great misfortune of the moas, Polynesians reached New Zealand around 1250 AD. The moas seem to have hung on till the 1400s AD, but there is no doubt that the Maori ate them. Large roasting pits—a traditional Polynesian form of barbecue—have been found filled with moa bones. Eventually, it seems, the moa populations could not sustain the onslaught against their numbers and all of their various species disappeared. Archeologists have asked themselves, could human numbers really have built up fast enough in a couple of centuries to extinguish the moas so completely? Population models seem to provide confirmation that in a favorable environment with plenty of food, the Maori could have expanded in numbers fast enough to eliminate the moas in such a short space of time. All that is left of the flightless bird fauna now on New Zealand are a few small distant relatives of the moas, such as the kiwi, joined by a menagerie of introduced animals such as pigs, Polynesian rats, and Australian possums.

5.11.6 The Hawaiian islands and their birds

An especially diverse and impressive Pacific island fauna existed on the Hawaiian islands in the northern tropical Pacific. When the Hawaiian islands were first reached by Polynesians around 1000 AD, they found an island biota that had evolved over millions of years from rare long-distance dispersals, derived from continents around the edges of the Pacific. In terms of building up diversity, the Hawaiian islands had in their favor that they were part of a much older group of islands that had each risen and fallen in turn over millions of years. This had given a long time frame for accumulating endemic species, which would have been able to spread to new islands in the group before their old homelands sank below the waves. The indigenous bird fauna included an array of both flightless and flying ground-nesting forms. There were no land predators on the Hawaiian islands, and the birds lacked any form of defenses. It is still uncertain just how many endemic bird species were lost in the century or so after the Polynesians first arrived there; the bones of extinct and previously unknown species are still being discovered. The way it looks so far, dozens of species vanished around this time. The elimination must have been carried out partly by the Polynesians themselves hunting for food, but probably also to a large extent by the animals they introduced. Ground nesting was common among the birds of the Hawaiian islands, which had evolved in the absence of predators. This turned out to be the undoing of many of them when the domesticated rats and pigs of the islanders escaped out into the wild and ate the eggs. The victims of this first wave of extinctions included the nene-nui (*Branta hylobadistes*), a close cousin of the endemic Hawaiian goose or nene (see below), which was evolving towards flightlessness on the Big Island of Hawaii. On some of the other islands there was a large flightless duck—the moa-nalo (*Chelychelynechen*, and other genera). The various moa-nalo species were the largest land herbivores on the islands, at up to 7.5 kg. Other flightless groups on the Hawaiian islands included the ubiquitous rails.

The troubles of Hawaii continued when European colonists turned up with goats and rabbits that ate the vegetation in which birds fed and made their nests. The dogs, cats, and mongooses that escaped into the wild killed many of the island birds. Moreover, the Europeans hunted ducks and geese for sport, using rifles. More species became extinct, and others teetered on the brink. The flightless Lasan rail went extinct as a result of competition with rabbits that were introduced to the only island it occurred on, despite some last-ditch efforts to conserve it. The Hawaiian goose or nene (*Branta sandvicensis*) was thought to be extinct for many years until a small colony was found in the 1940s. Since then, it has fortunately rebounded to several thousand individuals. For now, the future of the nene looks secure with the benefit of protection from hunting. Even in an ecological disaster on this scale, there are some stories of hope that demonstrate the benefits of trying to save what is left.

Box 5.4 The Polynesian rat: agent of extinction

The Polynesian rat (*Rattus exulans*) has been blamed for more extinctions than any other species apart from humans. It is a smallish rat, dark brown in color, originating in Southeast Asia. When the Polynesians began to spread out across the Pacific starting 3,000 years ago, they took the rat with them as a domesticated animal which was kept caged for its meat. Being an undemanding and unselective eater, it was perfect for recycling scraps into protein.

On each new island, it would not take long for some of the caged rats to escape and establish their own population. The rats would feed on the undefended eggs and nestlings of ground-nesting birds, and being unable to reproduce, each species of bird would generally go extinct. Time and again, the archeological record shows that appearance of the Polynesian rat on a Pacific island correlates with the last records of numerous flightless endemic species of birds.

5.11.7 Mauritius and the dodo

The dodo (*Raphus cucullatus*) has in itself become a symbol of extinction: "Dead as a dodo" used to be a common phrase in English, and it appears in that role (extinct, yet paradoxically alive) in *Alice in Wonderland.* Essentially a big flightless pigeon, the dodo had followed familiar paths of island evolution towards gigantism and loss of the ability to fly. Isolated on Mauritius, a small speck of land in the Indian Ocean, the dodo was first encountered by humans around the early 1500s when the island was discovered by accident by Portuguese ships. Also present on Mauritius were several other species of flightless birds, and lush forests composed of endemic species of trees. Rodrigues Island, about 500 km to the east of Maurius, had a vaugely similar-looking bird, known as the Rodrigues solitaire (*Pezophaps solitaria*). DNA evidence from the remains of the dodo and the Rodrigues solitaire show that both were descended from a pigeon (something closely related to the Nicobar pigeon—*Caloenas nicobarica*—of the northeast Indian Ocean) that must have been blown across the ocean millions of years before.

The name "dodo" was apparently given to the bird by Portugese-speaking sailors on the ships that stopped by at Mauritius—the word coming from a slang term meaning "dozy" or "stupid". To the amazement of the sailors, the dodo seemed to lack all common sense; it was completely unafraid of humans, making no effort to save itself when attacked. Although the flesh of the dodo was oily and rather unpalatable, after months at sea it made a welcome addition to the diet. Over the next century, trading ships made frequent stops at Mauritius and Reunion to harvest dodos. At some stage European rats were introduced accidentally from the ships to both islands. It was probably the rats, more than any other factor, that sealed the fate of the dodo and the other flightless birds of Mauritius. By eating the eggs that were left in nests undefended on the ground, the rats would have prevented the dodo population from replacing itself. The last definite sighting of a dodo was in 1662. A few years later, another visitor reported that the dodo was nowhere to be seen—and no subsequent voyage ever reported finding it. The dodo population probably became too rare to find but lingered on for a while before it finally went extinct (a sort of above-ground Signor–Lipps effect). A statistical modeling study by Roberts and Solow (2003), which allows for this, suggests that the dodo probably died out sometime between 1688 and 1715.

To the layman, the dodo is an example of a creature that simply deserved to go extinct, because of its stupidity and refusal to change. But from the point of view of natural selection, the dodo was simply well adapted to its island environment. In the absence of predators, and with nowhere to fly to, functional wings would have been useless. Nesting in trees would have been a waste of effort, and fear of predators was unnecessary. The arrival of humans and all their associated animals was a bolt from the blue, as unexpected and unpredictable as the asteroid that wiped out the dinosaurs.

When full-time colonists (rather than just sailors passing by) turned up on Mauritius, more extinctions followed as dogs, pigs, and cats joined the rats that were already established there. The native forests were felled and cleared for farmland, and even the forest that survived became filled with invading plants from other parts of the world. Many of the endemic plants of Mauritius are now rare, and only preserved against the tide of introduced foreign species of trees and shrubs by clearing of small plots in the forests that have been specially weeded of introduced plants, to make some room for them. It has been suggested that one of the native types of trees of Mauritius (the so-called "dodo tree", *Sideroxylon grandiflorum*) is now rare and in danger of extinction because it depended on its seeds passing through the digestive system of the dodo to prepare it for germination (Temple, 1977). Temple claimed that the tree was confined to just a few very old individuals. However, searches of the forests of Mauritius have turned up quite a number of young individuals of the "dodo tree", so it is not certain that it is in such immediate danger, nor that it was ever really so dependent on the dodo in order to germinate its seeds.

5.11.8 St. Helena and its daisy trees

St. Helena sits thousands of miles from anywhere, on top of a deep sunken ridge in the middle of the Atlantic. Hilly and only a few kilometers in diameter, it was formed by undersea volcanoes that built up to the sea surface several million years ago. No mammal had reached St. Helena before humans, and the largest animals to inhabit the island were seabirds. Very few plants ever managed to reach St. Helena too, and those that did were mostly weedy herbaceous types which had small, light, easily dispersed seeds. When St. Helena first formed out of volcanic ash and lava, it would have had no trees. Yet the mild, moist subtropical climate of the island was quite suitable for it to become covered in forest. Nature abhors a vacuum, and an ecological vacuum is no exception to this rule. With only a few types of woody plants having already arrived in the form of seeds, most of the trees of St. Helena evolved in place from small herbaceous species that may have crossed the ocean as seeds clinging to the feathers of birds. Many of them were derived from one original ancestor, a small weed belonging to the daisy family. The island became covered in dozens of types of odd-looking daisy trees, many of them with fat soft stems rather like an overgrown cabbage, and each one bearing sprays of daisy-like flowers.

The island was discovered in 1502, and some goats were deliberately released there as a source of meat for future passing ships. Within a few decades, the island was reportedly populated by vast herds of goats, which had multiplied without any predators to keep their numbers in check. Isolated on a land without large herbivores, the trees of St. Helena did not possess the tough leaves, spines, or the astringent poisons that could put off something as determined and unselective as a goat. The island lost most of its forest cover, and much of the ground was left bare and eroded.

It is difficult to guess just how many species of trees went extinct on St. Helena before the first botanists reached there, but we can be reasonably sure that there were many. By the 1800s, most of the endemic trees of St. Helena were confined to just a few individuals clinging to rocky slopes where it was difficult for goats to reach them (Cronk, 1989). The renowned Victorian botanist Benjamin Hooker suggested that—given how many species he found just on the edge of extinction—there must have been dozens of types of trees that had not quite been so lucky. The extinctions continued one by one, as the last individuals of several species died during the 20th century. Many of the endemic trees of St. Helena remain on the brink of extinction, with just one or a handful of individuals surviving. Despite the efforts of many who are concerned about the island's flora, the extinctions are still happening. The latest victim was the St. Helena olive (not actually an olive, but a member of the buckthorn family Rhamnaceae), *Nesiota elliptica*. The last wild tree of this species was lost in 1994, and the last one in cultivation died in 2003 while conservationists looked on helplessly.

5.11.9 Guam and its ground-nesting birds

Bursts of extinction on Pacific islands have continued to within living memory. Guam in the northwestern tropical Pacific had evolved its own diverse set of flightless bird species, that by luck had managed to avoid the onslaught of humans right up to the 1940s. Then the island was occupied as a U.S. military base in World War II. Shortly after the end of the war, there arrived a stowaway, riding on one of the supply ships. This was a new species to Guam—the brown tree snake (*Boiga irregularis*), a native of the islands and landmasses near New Guinea. The brown tree snake is a very adventurous and effective predator, hunting in the branches of trees to feed on birds. The native birds of Guam had never before been exposed to any predator, and so they were completely defenseless against it. Within a few years, the majority of the endemic birds of Guam were extinct.

Another recent burst of extinction occurred on Tahiti, with its snails of the genus *Partula*. After reaching Tahiti, it had proliferated into 76 endemic species. This all came to an end when a predatory snail from the Caribbean was introduced to control an introduced African snail, that had become a pest in the fields. The predator instead turned on the native *Partula*, and wiped out all but 5 of the species in the wild. A handful of species survive in captivity, but most of the endemic *Partula* snails of Tahiti are now completely extinct.

5.12 WHY WERE ISLAND SPECIES SO SUSCEPTIBLE TO EXTINCTION?

In general, island species have been particularly susceptible to extinction. On the continents, human-induced extinction was almost entirely confined to large mammals, but on islands it could eliminate almost any life form (e.g., plants, insects, snails, small birds, and small mammals, in addition to any large herbivores that happened to be present). Some 75% of known animal extinctions since 1600 have been on islands—including 90% of the bird extinctions, 58% of the mammals, and 80% of the known mollusk extinctions (Groombridge, 1993).

Part of the reason that so many island extinctions are recorded may be an illusion caused by taxonomy—the practice of classifying and naming organisms. Forms of island life which would otherwise only be classified as subspecies or races of mainland species tend to end up being designated as species in their own right, just because they are out on an island (Groombridge, 1993). If every race or subspecies that has ever gone extinct on the continents in the last few centuries had been labeled as a species, that would certainly make for a lot more extinctions. This tendency of biologists to make distinctive-looking island populations out to be species may tend to inflate the overall estimate of island extinctions, but it cannot explain such a large difference between mainlands and islands—many of the lost island creatures were clearly unambiguously different species from their nearest relatives on the mainland or on other islands.

The very real vulnerability of island life forms to extinction in the last few thousand years can be put down to two main causes. Both relate to the fact that distant islands were colonized in a rag-tag way, by rare chance events, which ensured that they were only ever reached by a tiny minority of the life forms that lived on the nearest mainland.

The first result of this chancy process is that many islands had far simpler food chains than would normally be found on the mainland. For example, many of them had never been reached by predators of any sort, even though herbivores were present in abundance. As a result, island species of mammals and birds tended to lose the defenses that would have prepared them to cope with humans and all the other predatory species that humans tended to introduce. Reptiles likewise tended to become vulnerable through evolution. For example, some races of giant tortoises on the Galapagos Islands in the Pacific evolved an open "saddleback" neck to their shell, which enabled them to reach up to browse vegetation. When dogs were introduced to the islands in the past few centuries, the saddleback forms were vulnerable to being attacked by their neck, and went extinct (only one specimen of saddleback tortoise still lives, hanging on with incredible longevity since the time of Darwin). It seems to have been the same general story with many island plants. With large and effective herbivores lacking, many island species lost the thorns or chemical defenses necessary to keep mammalian herbivores at bay. The daisy trees of St. Helena were devastated by goats, partly because they did not have the chemical or physical deterrents necessary to keep themselves from being eaten. Similarly, the defenseless little *Partula* snails in French Polynesia were wiped out by introduced predatory snails.

Second, many island species adapted in an *ad hoc* way to the opportunities presented by empty niches, where the normal occupiers on the continents were lacking. The giant duck that was the only large land herbivore on Hawaii was one example of a species that had jumped niches, as is also the giant tortoise on the Galapagos. Also on the Galapagos there is a finch that uses a cactus thorn to pry grubs out of rotting wood: it is essentially living in the niche of a woodpecker, in a place that woodpeckers have never reached. On Hawaii there is currently a moth caterpillar that ambushes its prey like a mantis, presumably because the preying mantises of the continents never happened to reach there. While they represent fascinating examples of the adaptability of life forms under natural selection, there can be little doubt that many of these island creatures would not survive in open competition with mainland species. Continental predators and herbivores represent the honed product of many millions of years of evolution, with many different lineages jostling for the same niches. It is harder to point to examples where this factor has likely been important in the actual extinction of an island species. One can speculate that this is because these species have already gone, so they cannot be closely compared with the invaders that drove them extinct! One possible example though would be the losses of marsupials on the "island continent" of Australia as a result of competition with placental dogs, rabbits, and cats from the Old World.

The overall result of these factors has been that islands have suffered terribly from the combined onslaught of humans, plus the animals and plants they introduced. Islands are known to have lost twice as many species around the world as mainlands since 1600 (Groombridge, 1993). Extending back to the past several thousand years before historical records began, the number of island extinctions has certainly far exceeded the total of large animal extinctions on the continents that occurred towards the end of the last ice age.

Are island extinctions becoming any more frequent, or has the main phase of extinction passed? The record is not detailed enough to be able to say for most groups of organisms. Island birds have the clearest record, left by generations of ornithologists and casual observers—which is hardly surprising because birds are highly visible and naturally tend to attract attention. In the case of birds, there is no clear trend in extinctions over the last several centuries. There are peaks in the 17th and mid-18th centuries, but also a general increase in extinction rate between the 1800s and the 1930s (Groombridge, 1993). The majority of recent extinctions seem to have been due to habitat change, which has been taking over from hunting and introduced animals as the major cause of extinction (Groombridge, 1993). Since then, extinctions among island birds seem to have declined, perhaps partly because of recent conservation efforts, but also surely because the majority of what was most vulnerable to humans has already gone extinct.

5.13 BACK TO THE MAINLAND

While humans were spreading to islands around the globe, their numbers were also building up back on the continents. The invention of agriculture some 13,000 years ago had given a massive boost in food supply that began a long, slow rise in human population. By about 300 years ago our species numbered about 450 million, at least 40 times what it had been in hunter-gatherer days before agriculture. Then suddenly around 1800 AD, population exploded across Europe and North America, as a result of modern medicine that greatly reduced mortality, plus modern agriculture which was able to feed all the extra mouths. The population explosion spread out around the world in the 1940s and 1950s, as medical knowledge and agricultural science reached Asia, Africa, and South America. As humans expanded their numbers, the onslaught against nature began with renewed vigor. Ancient forest was cleared and grassland ploughed up, and the remaining wildernesses were scoured by hunters and loggers. Wildlife began to suffer, and a new burst of extinctions on the continents began.

5.13.1 The great auk

Some of the first to go in this new burst of human population were species that had long been in decline as a result of hunting and persecution. One example was the great auk (*Pinguinus impennis*), a flightless diving bird some 75 cm in height that bore a remarkable resemblance to a penguin. But whereas penguins are only found in the southern hemisphere, the great auk belonged to an unrelated group, living on the shores around the North Atlantic. It was the northern hemisphere analog of a penguin, a remarkable example of convergent evolution which shapes creatures of different ancestries to a similar form, to suit the needs of the same way of life. Yet in a sense, the great auk really was a penguin. The name penguin was an old name for the great auk, probably coming from the Welsh words *pyn gwyn*, meaning "white head": presumably a reference to the white patch on the top of the bird's head. When European explorers visited the southern hemisphere, they found a bird that reminded them of their *pyn gwyn* back home, and they labeled this southern bird too as a "penguin".

The great auk had once had a range that stretched all around the rocky edges of the north Atlantic, from the British Isles, through Iceland, Greenland, to the east coast of North America. As recently as 1400 AD, it even seems to have been present as far south as Florida. During the last ice age 30,000 years ago, it also extended its range into the Mediterranean, where it was recorded in cave drawings.

The range and abundance of the great auk declined over the centuries as it was hunted for its skins, its blubber, and its eggs. Though agile in the water, it was clumsy and easy to catch on land, and it had a slow rate of reproduction—producing only one egg a year. Eventually the great auk was only known to survive on Iceland, and then only as one small colony on a small tower of rock on the island of Geirfuglaskir just off the coast of the Iceland. By sheer misfortune, this last refuge sank during a volcanic eruption and the remaining auks were forced to relocate to a more vulnerable location on the coastal rocks of Eldey, a nearby small island, where they were picked off by local fishermen. The last remaining pair were killed on the morning of July 3, 1844 by two Icelandic sailors, Jón Brandsson and Sigurður Ísleifsson, who had been sent by a Danish natural history collector to collect specimens. There was a fairly reliable sighting of a single great auk off the Grand Banks of Newfoundland in 1852, but no more were seen after that, and we must assume that the species finally went extinct around this time.

5.13.2 The passenger pigeon

Several centuries ago, the passenger pigeon (*Ecopistes migratorius*) was probably the most abundant bird in the world. It has been estimated that there were five billion of them in the forests of eastern North America, where they ate seeds, especially acorns. During their annual migration—north in spring and south in autumn—the sky was reportedly darkened for days by the sheer size of the flocks that passed overhead. These flocks could be up to 1.6 km wide and 500 km long, consisting of up to 2 billion birds. When passenger pigeons nested, they did so in huge aggregations of millions, taking over a large area of forest. It was said that trees literally creaked under the weight of the birds on their branches, and that the over-burdened boughs would sometimes come crashing down taking thousands of adults and their nests with them. One mass nesting in Wisconsin reportedly occupied 850 square miles, and contained an estimated 136 million birds.

European settlers found that the meat of the passenger pigeon was good to eat, and it was considered an especially useful cheap food for slaves and the poor. Commercial hunters set about harvesting passenger pigeons on an ever-increasing scale using weapons and traps. By the mid-1800s the harvesting was being done almost on an industrial scale, shipping box cars full of the slaughtered pigeons to the cities. For example, one hunter reported shipping 3 million birds in 1878 alone. Some early conservationists were concerned by an apparent decline in the bird's numbers, and wondered how long it could continue to bear such an onslaught. In 1857, a select committee of Congress rejected the idea of protecting the passenger pigeon by law, arguing that it was far too common. Yet already by the late 1870s, it was noticed that the passenger pigeon was becoming much less common. By the 1890s it was rare, and laws were passed belatedly to protect it from hunting. The last wild sighting was in 1900 in Ohio, and the last passenger pigeon of all (named "Martha") died in captivity in 1914.

How could a species go from several billion to zero, in the space of a little over a century? The passenger pigeon was clearly suffering huge losses from hunting, and at the same time the forests that it depended on for its food supply were being cleared rapidly. Starting from an almost unbroken forest cover when the first settlers arrived, by the mid-1800s the forest cover was only 15% to 20% of the landscape in many eastern states. It is possible that the extra energy expended in moving from one small forest patch to another in search of food was too much for the pigeons—that their dietary intake could not now meet their needs.

Another possibility is some sort of disease accidentally introduced by the settlers, perhaps a bird virus carried by domestic fowl. Newcastle disease, a virus brought from the Old World, has been suggested as a possible candidate, although it was not discovered in North America until 1926, well after the passenger pigeon had gone extinct.

Perhaps the most convincing explanation is that something peculiar to the passenger pigeon's biology eventually doomed it. Having existed in such numbers, and nested in vast aggregations, each passenger pigeon may have needed to be surrounded by millions of other birds in order to feel comfortable enough to breed. Huge numbers offer safety against predators, and so natural selection could have favored this as a survival trait. There are many species of birds that strongly prefer to nest in aggregations, and the passenger pigeon may just have been an extreme example of this. However, behavior only made good evolutionary sense under conditions of vast, stable populations. It was not equipped to cope with a world in which humans were harvesting large numbers of pigeons, and breaking up their habitat by clearing forest. Once the population density of passenger pigeons declined by a certain amount, the birds would become more reluctant to breed. With less breeding and raising of young, the population density of the pigeon would decline further, causing more difficulties in reproduction, and so on. The passenger pigeon would thus have entered a spiral towards extinction, and once it started there was no stopping it. In a sense it was a victim of its own success; a trait produced by it having existed in its billions eventually ensured that it went extinct as a result of the changes brought about by humans.

5.13.3 The Carolina parakeet

It is hard to believe that in the eastern United States there once lived a native parrot, the Carolina parakeet (*Conuropsis carolinensis*; Figure 5.7). These little green and red birds managed to live all year round as far north as New Jersey, despite the hard winters, feeding off a range of seeds and fruits. But although they could survive the northeastern winters, they could not survive humans.

Formerly ranging across most of the eastern U.S.A., the Carolina parakeet's numbers declined as a result of destruction of the old forest wildernesses that it used to inhabit, and hunting for its colorful feathers, which had become fashionable for ladies' hats. By the time laws were passed for its protection, it was already too late. The last one died in a zoo in 1918.

Figure 5.7. The Carolina parakeet, extinct in 1918 (source: painting by Audubon). *See also* Color section.

5.13.4 The thylacine

Nothing illustrates the principle of convergent evolution more perfectly than the thylacine (*Thylacinus cynocephalus*). This marsupial carnivore was Australia's own native analog to a dog or jackal—evolving from some possum-like ancestor, entirely independently from the placental dog of the Old World and the Americas. The matching of form was so complete that when the skull of a dog and a thylacine are placed side by side, it is nearly impossible to tell them apart. One of the few straightforward ways to distinguish the skulls is in the hinge mechanism of the jaw—in the thylacine like all marsupials the lower jaw nests up into the upper jaw, while in dogs it sits outside the hinge of the upper jaw. In the animal when it was alive there was also one very striking visual difference: the thylacine had stripes along its back and hindquarters, hence the name "Tasmanian tiger" that has often been given to it.

Apparently the ecological similarity between the thylacine and the dogs of the Old World was too close for its own good. Around 4,000 years ago, the Australian aborigines received an Asian dog—which we now know as the dingo—by sea trade from New Guinea. Kept as a pet, the dingo soon escaped into the wild and spread throughout mainland Australia. Just at the point that the dingo appears in the archeological record, the thylacine suddenly vanishes. The coincidence is too striking to attribute to anything other than competition between the thylacine and the dingo. This is a classic example of the principle of "niche exclusion"; if two different species try to live in exactly the same way, one will eventually drive the other extinct. What actually gave the dingo an advantage over the thylacine is something of a mystery. The thylacine seems to have had a stronger bite, but the dingo may have been better engineered to gripping on to larger, struggling prey (Wroe *et al.*, 2007).

If the timing of the thylacine's extinction on the mainland were not evidence enough that the dingo "did it", what clinches the case is that the one place where the dingo did not reach provided the last refuge of the thylacine. The island of Tasmania, at the southeastern tip of Australia, had been cut off from the mainland by rising sea levels around 8,000 years ago, isolating its population of thylacines. With a water barrier in place, there was no way for the dingo to reach Tasmania. It is a similar story with the Tasmanian native hen (*Gallinula mortierii*), a small flightless bird which disappeared from the Australian mainland at the same time that the dingo appeared, and continued to survive on the island of Tasmania. But unlike the thylacine, this species is fortunately still with us in the present-day world.

It has been proposed that other things may have contributed to the extinction of the thylacine and the Tasmanian native hen on the mainland (Johnson and Wroe, 2003). For example, there is evidence of an ongoing increase in aboriginal populations around the time the thylacine disappeared. However, invoking this factor all seems a bit unnecessary given the timing of the dingo appearing, and the fact that both species survived on Tasmania where the dingo was absent but humans were present. I get the impression that there is quite a tendency in the archeological world to "make noise" for the sake of it, when existing explanations are perfectly satisfactory and make more elegant use of the facts.

The thylacine survived 4,000 more years on Tasmania, thanks to its protective barrier of seawater. That enabled it to be found by European invaders from the other side of the world, and then finally finished off by them. The last thylacine died in Hobart Zoo in Tasmania in 1936, apparently of neglect from being left out in the winter cold.

5.13.5 Yangtze River dolphin

We should not think that extinctions of iconic species are confined to the past. Far from it. When a country's development is at stake, species are still expendable, even if they are large and lovable. The endemic dolphin (*Lipotes vexillifer*) of the Yangtze River in China is one of several aberrant freshwater species of dolphin that have evolved in major river systems around the world. In the thick silty waters of the Yangtze, the river dolphin had little use for sight, and its eyes became small and ineffective. Instead it relied on sonar, sound waves generated under water, to catch the fish that it fed off. This probably proved to be its undoing, for as the boat traffic on the Yangtze grew and grew, the incessant noise of propellers would have made it difficult for the dolphin to hear its own sonar. A survey in 2003 found a handful of river dolphins still surviving in the Yangtze. When the same group of researchers returned in 2006, they found none, despite months of searching. It is fairly likely then that Yangtze river dolphin is now extinct.

5.13.6 Cichlids in African lakes

There are some things that just seemed a good idea at the time. One of these was the introduction of the Nile perch (*Lates niloticus*) and Nile tilapia (*Oreochromis niloticus*) to Lake Victoria. Even though Victoria's abundant cichlids had sustained a thriving fishing industry for centuries, it was hoped that the Nile perch and tilapia would increase it further. So in the 1950s these fish were released into the lake. The Nile perch is an effective predator, and in Victoria it has fed voraciously on cichlids. The Nile tilapia feeds off plankton, competing with the plankton-eating species of cichlids. Added to this, there has been a big increase in human activity around the lake, resulting in a lot of pollution and clouding of the water. The result of it all is that the cichlids have become far less abundant, and many of their couple of hundred species can no longer be found. It has been the same fate for Victoria's various other native fish, including its own two endemic species of tilapia. We can only assume that many of the species that have vanished from Victoria are now extinct, though a number of them do survive in small isolated lakes around the edges of Victoria. It is not clear how long the extinctions will continue, or how much will be left at the end of it. Ironically, the Nile perch and tilapia turn out not even to be particularly tasty, or easy to catch, or easy to prepare, by comparison with the cichlids that they are driving extinct.

5.14 CURRENT EXTINCTION, SEEN AND UNSEEN

Around the world, humans are altering habitat—to make money or just to survive. In at least some areas that are very species-rich, rapid habitat change must be causing extinctions each year. But we do not know each time a species goes extinct, and we may not even know that it ever existed in the first place.

In tropical rainforests, the majority of species have certainly not been recorded and named by science, let alone censused in enough detail to know if they were confined to places that have now disappeared (Chapter 6). But there is really no doubt that rainforests are already losing species all the time, as loggers and settlers cut into them at the rate of millions of hectares per year. After all, there are so very many species in the rainforest, and so much rainforest being cleared. However, there is a great deal of uncertainty about the actual *numbers* of species going extinct. Estimates of how many rainforest species are being lost each year depend mainly on two different parameters, and both of them can only be estimated in a rather shaky and very indirect sort of way.

First, the figure will depend upon the total number of species in the rainforests. Estimates which put the number of species in the world at 30 million or 100 million, concentrated in the tropical rainforests, would of course lead us to expect that many more extinctions are occurring than we would get from estimates putting the total at 4 million or 6 million species. For if there are greater numbers of species around in the first place, cutting down a given area of rainforest is likely to finish off more of them.

Another thing that affects estimates of the rate at which species are being lost, is how localized each species is on average. If most species in tropical rainforest occupy only a few square kilometers, it is much easier to drive a species extinct by cutting down habitat than if they are evenly spread across many thousands of square kilometers. How do we get a reliable picture on this aspect? The estimates of how large range sizes of tropical species tend to be are based on surveys in pristine tropical forest. In these surveys, researchers take a series of expanding sample sizes, noting how many extra species appear in each sample compared with the smaller previous sample. The faster that new species accumulate in such a study, the more localized they must on average be, as well as there being more of them out there in total. The equation of the slope on the line of accumulating species (which incorporates two "constants" c and A that summarize the starting height and angle of the line of species richness per sample against area) is drawn out ("extrapolated") to cover the whole vast area of the tropics. Zooming back down the scale to smaller and smaller areas can provide an estimate of how much would be lost if a given area of tropical forest was removed by deforestation. If the slope with which species accumulate with ever-expanding sample sizes is a steep one, then many more species can be expected to be lost if the amount of habitat is reduced, compared with if the slope is relatively flat. Estimates of the slope of species accumulation across the whole tropics differ between different studies which have each started in a different place and expanded the sample size up to a certain point—beyond which it became impractical to continue sampling. It generally looks like most species are fairly widespread, which decreases the likelihood of them going extinct. One widely quoted generalization, based on averaging the species–area lines of many individual studies, is that reducing the total area of tropical forest in the world by 90% would cause half the species to go extinct (Reid and Miller, 1989).

There are various things that could complicate this. One is how patchy the diversity is. If different species are strongly clustered together in localized species richness "hotspots", this will also affect the rates of extinction, depending on whether the deforestation is concentrated in these hotspots.

Let's then try making some estimates. Say we assume that something like 1.8% of the world's pristine tropical forest is being lost each year, either completely cleared or badly damaged such that it loses most of its species (this is about where UNEP estimates put the rate, at present). A fairly conservative estimate based on slopes of species accumulation with sample size suggests that around 2% or 3 % of rainforest species are being lost per decade (Groombridge, 1993). The question then becomes what number this 2% or 3% is of. For instance, if we assume that there are 20 million species in the world's rainforests, this would make for the extinction of 100,000 species each year. Lower and more generally believed estimates of species richness in tropical rainforests—such as 6 million species—would give correspondingly lower rates of extinction, but still many thousands of species per year. In any case, most of these will be forest insects, and mostly still unknown to science.

Other areas of high species richness, such as the fynbos of the Cape of South Africa, are almost certainly losing species as land is taken away by expanding development, and as introduced trees and shrubs from other continents push into the vegetation. At least 26 species of fynbos plants are now thought to have gone extinct in recent years. In Lake Victoria (above), extinctions are likely still occurring as a result of introduced fish species such as the Nile perch.

In each of these cases, extinctions are still happening unseen. In a big lake in Africa, it is hard to census fish thoroughly enough to be reasonably sure of what no longer exists—all we can say is that it looks like many species of Lake Victoria have already gone, and many more are close to extinction. In the Cape of South Africa, many plant species occur only over a few hundred square meters and will disappear very rapidly if that area is ploughed up for a field. Even building a single house and its yard could potentially wipe out the entire range of a species.

And also scattered around the world outside hotspots, species are still going extinct all the time. In the southeastern U.S.A., several endemic species of freshwater mussels seem likely to become extinct as a result of the invading zebra mussel (*Dreissena polymorpha*), an aggressive competitor that was accidentally introduced from Asia. In Australia, a colleague told me of what was probably the last population of an endangered snail he found living in a power station intake pipe. That population went when they cleaned the pipe. And there is also the probable example of the Yangtze river dolphin, sacrificed for progress in a rapidly developing economy.

Some species which have not been seen for decades are almost certainly now extinct, but cannot yet officially be declared as such. Current convention in the world of conservation dictates that a species can only be declared dead if it has not been seen for at least 50 years. It is understandable why there should have to be such a rule, as there are certainly examples of species that had been thought extinct but were rediscovered alive decades later—some even after the 50-year limit had passed (see Chapter 8). But it is interesting that technically there is a built-in lag period to extinction, no matter how strong our suspicion that a species has been lost. This obligatory waiting time is one reason why the extinction rate of island birds may have seemed to decline over the last several decades (Groombridge, 1993). It also delays the recognition of any upturn in extinction that may result from increased habitat change in any part of the world.

How many species will join the ones that we have already lost depends largely on what happens over the next few decades. In Chapters 7 and 8, I will turn more to the threats to species diversity which loom on the horizon, and what we might do about them.

6

Knowing what is out there

So far we have mostly considered the species richness that is already known to exist, and the mechanisms which may control this. However, it is clear that not all the world's species have so far been discovered. Every year, biologists find thousands of "new" species (i.e., new to science), and describe and name them. Most of the new species that are presently being discovered are either insects, or their relatives the mites that are universally present in forests and soils, or the little nematode worms that burrow into sediment everywhere. Additionally, some of each year's new species are plants and a few of them are birds, mammals, or fishes.

Given that we do not yet know every species out there, there are some questions which have to be asked. For instance, how much uncertainty is there about species richness patterns? How can we estimate the number of species that are still awaiting discovery? And how were the existing totals for global biological diversity arrived at? Are they reliable in the first place?

6.1 NATURE'S CURRENT TOTALS

Centuries of hard work by thousands of biologists have led to somewhere over a million living species being formally described, named, and listed. This is the business of taxonomy, of finding and describing the diversity of life. Most members of the general public (and most ecologists, too) assume that there must be some central database of all the species so far described in the world, so that it will be possible to obtain an unambiguous figure on the known number of species in every group of organisms. The real situation is far more chaotic.

A few brave souls have attempted to pull together information, to come up with more precise totals for the numbers of species found in each of the broad groups of life. Although the estimates in these different studies are mostly of the same order of magnitude, they do vary quite a lot. For example, one estimate published in 1990 has it that there are 45,000 known species of mollusks, and another published the same year says there are 100,000 of them (Groombridge, 1993). This says something about the difficulties that are inherent in the business of species counting.

What is evident from published figures, and Tables 6.1 and 6.2, is that the estimates for the most diverse groups are rounded, because they tend to be given as even thousands. That is the best anyone is able to do for the time being. There are just too many species in these groups for it to be worth the effort of counting every last one that has ever been recorded by science (especially as the number always changes from one year to the next).

Something that makes the task harder is that for most groups of organisms, there is no central list where each new species is added as it is discovered. The raw information on the discovery of new species remains dispersed through hundreds of specialized scientific journals. For example, in the case of insects there are several learned societies and institutions around the world that provide lists and guidebooks (known as floras and faunas), and new descriptions of species in their journals. One example is the American Entomological Society, which publishes *Entomological News*, published several times a year and packed full of new insect species descriptions. It is the same for plants, for mollusks, and so on. Because the information is so voluminous and so scattered, anyone trying to compile an overall total for global species richness must in effect fit together a lot of jigsaw pieces from different sources. When Brian Groombridge, Martin Jenkins, and their colleagues (Groombridge, 1993; Groombridge and Jenkins, 2002) came up with their totals for the world's current known biological diversity (Tables 6.1 and 6.2) it was partly a result of phoning around or writing to a lot of experts in different fields of taxonomy who could put a plausible number down. Each of these experts must in turn have relied upon combinations of other, smaller jigsaw pieces of information. These were the estimates of total known species within particular narrow subgroups, obtained from the specialists who work on them.

Adding to the problems involved in reaching an overall total is the fact that not all species discovered and named in the scientific literature survive scrutiny. Months or years after a species has been named, other taxonomists may take a closer look at it and decide that this species is really just the same as one that had already been described by someone else. The problem of "double description" was especially bad in the recent past when taxonomists were working independently because they were separated by linguistic and political barriers (e.g., the Soviet or Chinese taxonomic literature vs. Western taxonomic literature) and the mess is still being sorted out.

Even with good communications, the problems of double description are huge. Among insects, for example, somewhere between one-fourth and one-third of the new species described each year turn out to be ones already named by others—so they are removed from the lists (Groombridge, 1993; Groombridge and Jenkins, 2002). It is difficult to keep pace with such turnover in names, and any list made up of the new species just announced within the scientific literature will contain many which eventually turn out not to be accepted by the taxonomic community, so they in effect vanish. The best estimates of the world's known species try to take this attrition rate

Table 6.1. Current totals of known species for certain prominent groups of organisms.

Colloquial name of group	*No. of species in group*
Vertebrates	
Mammals	4,630
Birds	9,750
Reptiles	8,002
Amphibians	4,950
Fishes	25,000
Invertebrates	
Insects	950,000
Mollusks	70,000
Crustaceans	400,000
Others	—
Plants	
Gymnosperms (Coniferophyta, Cycadophyta, Ginkgophyta	876
Angiosperms (Anthophyta, flowering plants)	250,000

Source: Groombridge and Jenkins (2002).

Table 6.2. Current known species for major groups of life, left-hand column. Some plausible estimates for the total number of species "out there" (including those as yet undiscovered) are in the right-hand column.

Domain	*Eukaryote kingdoms*	*No. of described species*	*Estimated total*
Archaea		175	?
Bacteria		10,000	?
Eukarya			
	Animalia		1,320,000
	Craniata (vertebrates), total	52,500	55,000
	Mammals	4,630	
	Birds	9,750	
	Reptiles	8,002	
	Amphibians	4,950	
	Fishes	25,000	
	Mandibulata (insects and myriapods)	963,000	8,000,000
	Chelicerata (arachnids, etc.)	75,000	750,000
	Mollusca	70,000	200,000
	Crustacea	40,000	150,000
	Nematoda	25,000	400,000
	Fungi	72,000	1,500,000
	Plantae	270,000	320,000
	Protoctista	80,000	600,000
Total		*1,750,000*	*14,000,000*

Source: Groombridge and Jenkins (2002).

into account, lowering the total by a certain correction factor to allow for the ones that will eventually be removed.

So, accepting all the limitations in any estimate, what is the overall picture so far for global species richness? The compilation put together in 2002 by Martin Jenkins and Brian Groombridge of the World Conservation Monitoring Center (Table 6.1) seems likely to be as good as any.

All in all, something like 1.75 million species have been identified so far. One thing that leaps out of the totals in Table 6.1 is how many types of insects there are: somewhere around 800,000 to 950,000 species of them have been found so far (Groombridge, 1993; Groombridge and Jenkins, 2002). The next most diverse group is the flowering plants—the angiosperms—with more than 250,000 species. Although vertebrates such as mammals, reptiles, and fish tend to get a lot of attention, far fewer species of these backboned animals have been discovered so far—several tens of thousands. The majority of known vertebrates are fishes (around 25,000 species), and next most diverse are the birds (around 10,000 species). For mammals we know of around 4,000 species, the majority of these being rodents (mice, squirrels, and suchlike) and bats. There is a similar number of known amphibian species, most of these being frogs.

6.2 IDENTIFYING NEW LIFE FORMS—TAXONOMY AND ITS CHALLENGES

We have seen the overall totals of known species, but how exactly do the raw data come about? How is a new species found and identified, and officially announced to the world? This business of discovery and description has its own craft, its own long traditions and even its own language (Taxonomic Latin). The name of this craft is taxonomy—and the guilds which its craftsmen and women belong to are the taxonomic unions.

In a sense we are all taxonomists. Humans have probably always wanted to enumerate and classify the organisms around them—there is basic survival value in knowing what is useful and what is harmful out in nature, and this needs some sort of classification, identification, and naming (even if it is very rough and ready, and unwritten). Written accounts of the diversity of life go back as far as the ancient Chinese and Greeks, and there are others from the time of the ancient Romans. In the 1600s, a shaking up of European academia by the Enlightenment led to a fresh burst of interest in nature, and that included going out into the world and classifying whatever could be found. For instance, the English botanist John Ray and his students traveled extensively in Europe, and collected, described, and named many hundreds of plant species in a series of books. However, the detailed traditions of modern-day taxonomy have their origins more with a later generation of natural historians who worked in the 1700s. This began when the process was formalized and greatly advanced by a Swedish biologist who is nowadays known by the Latinized version of his name—Carolus Linneaus.

Linneaus came up with a simple way of naming each species, using two Latin names. In Linneaus' system, the first name is used for the cluster of closely related species that it belonged to, known as the genus. The second name is reserved only for that unique species. So, for example, the North American white oak tree I see outside my window as I write is known formally in biology as *Quercus alba*. The first name—*Quercus*—means "oak" in Latin, and is the genus or generic name used for all oaks. The second name *alba* means "white" in Latin, and is the specific name—reserved for that species in the oak genus. Generic and specific names are many and varied. Some are straightforwardly descriptive, such as the name *Quercus alba*, but others are named after the colleagues, friends, or family of the taxonomist who discovered or first described the species or genus. So, for example, the flowering plant genus *Stewartia* was named by Linneaus in honor of a friend of his who was a respected Scottish botanist.

Before Linneaus, only long meandering Latin names involving 10 or 20 words had been used to describe species—without any short snappy name to use. Linneaus' two-name (or binomial) system proved to be a good practical way to label a species; not too long, but with the genus name putting each species in the context of other things it resembled. Linneaus also went further, suggesting a hierarchy of categories that further placed each species into the wider framework of life. In the opinion of my eclectic and ever-curious former tutor at Oxford, Barry Juniper, the system that Linneaus came up with was based loosely on the way librarians in medieval Europe categorized the books within their collections: they too used a "binomial" system to classify them—first a word for the general category, and then a word that specifically described that particular book.

Linneaus was a fanatical worker, and he named and classified thousands of species in his lifetime. His students voyaged around the world aboard trading vessels in search of new forms of life to bring back to him, often losing their own life in the process. The tradition that Linneaus started and passed on to others continues today, with various modifications to his techniques

It is interesting that Linneaus himself had earlier in his life been made to adopt a Latin binomial to describe himself. In Sweden in those days, people traditionally did not have a surname as such. But to enroll in a university, each student had to adopt a single snappy surname that administrators could handle, and Linneaus chose a Latinized version of a name that described the farm he came from. The word *Linneaus* means "a big linden tree", something which grew close to the family farmhouse.

6.3 THE STAGES IN DISCOVERY OF A NEW SPECIES

The story of the discovery of each species is unique, but there are certain common patterns to these stories. In general, nowadays the discovery and naming of a new species goes through several different stages.

6.3.1 Collection

First, a specimen of a previously unknown species is collected by biologists on fieldwork or by interested amateurs. It may already have been noticed as something unusual and unknown at the time it was collected. Often, instead of having immediately been recognized as new, it will just have been collected incidentally by ecologists, as part of a general inventory being made to try to summarize the species composition of a particular place. For example, it might be just another plant in a bag full of specimens taken back for identification, or just another insect or mite among many in a bottle of preserving alcohol.

Sometimes, taxonomists themselves go out collecting in the hope of finding new species—although nowadays there does not tend to be very much money available for such fieldwork. More often they must rely on material that was collected as a part of other studies, or collected specially for them on request by friends who are ecologists or conservationists.

There are many different stories of how new species were discovered serendipitously. Some species have first been recognized as new by chance after they sat in museum collections under another name for decades or even centuries. One recently discovered mammal known as the rat-squirrel (*Laonastes aenigmamus*) was first identified from a killed specimen on sale for food in a village market in Laos (see below). There is one tale (possibly just apocryphal) that an insect taxonomist collected a new species of beetle when it fell into his gin and tonic as he sat outdoors at a bar. He is said to have given it the very apt name of *ginandtonicus*. I tried to verify this story, but I can find no reference to a species of this name in the academic literature—so it looks to be a sort of taxonomists' urban legend.

6.3.2 Identification

Let's suppose that a taxonomist or an ecologist collects a range of material out in the rainforest: for example, the shoots of plants that are put in bags, or insects trapped in a net then kept preserved in alcohol. The next stage is to get the collected specimens to somewhere where they can be identified. Most often, ecologists will try identifying the material they collect themselves back at the field station or lab, using "keys". These keys are books (or computer software) which guide biologists through the identification process until they eventually arrive at the name of the species that they are looking at. When the collectors cannot successfully "key out" a specimen to something that is in the books, it is possible that this is a new species, previously unknown to science. The specimens may then be sent to a specialist in that particular group of organisms, whether it be beetles, birds, bivalves, or whatever. Such experts tend to be employed at museums and zoos or botanical gardens, or sometimes in university departments. Taxonomists who are prepared to take on the task of trying to describe and identify new species are known in the trade as "alpha taxonomists" (somehow this term makes me think of "alpha male" silverback gorillas, though I am not sure if that is how alpha taxonomists think of themselves).

So looking carefully at a specimen, and comparing it with published descriptions of other species—or actual specimens preserved held in the collections—the expert can decide whether it is a new species or not. Increasingly nowadays, techniques that focus on DNA are also used to help figure out whether a specimen belongs to a previously known species, and if so what species it is. A small part of the organism, such as the leg of an insect or a leaf, is sent to the lab and analyzed to produce a "DNA barcode". This is a profile of genetic information which can be compared against other known species. If the DNA barcode looks very different from anything already known, the specimen is worth a second look by a taxonomist.

Another new technique currently under development is to use an "automated recognition system": a computer program that has been taught to recognize specimens of a range of species from color patterns and shapes. A digital image of the specimen is examined by the program—and if it is in the database it is identified. If it is not in the database, it could be a new species. This may turn out to be the way of the future, but at present the development of this technique is in its very earliest stages, and not reliable enough to use for anything other than cursory identifications that help narrow down what group the specimen belongs to.

In the end, the decision whether to name something as a new species, or just as a slightly different variety of an existing species, is often rather subjective. We do not usually have any sure way to know that the new species doesn't intergrade with some other described "species" somewhere out there in nature. Or perhaps it would hybridize easily with some other species it doesn't normally encounter, it if was given the chance. In either case both "species" should probably just be relegated to "subspecies", if we actually knew enough about them.

Taxonomists who are reluctant to name new forms of organisms they find as separate species are called "lumpers": they lump things together into the same species. Those who tend to label each variety they find as a new species are called "splitters": they tend to work by splitting up the world into a greater range of species. The overall species richness we find in the world depends upon the balance between the activities of these splitters and lumpers, and how many species they each feel like describing. Across the broad field of taxonomy, there are local traditions that tend more towards splitting or lumping—often partly depending on the influence of one particularly prominent taxonomist and his or her former students. Scores of species can sometimes vanish overnight when someone takes a new look at the classification of a group and decides that the "species" in it are not really separate species at all, but just local varieties. At other times swarms of species can suddenly appear after a taxonomist attempts a revision of another's work and finds reason to split varieties or subspecies into fully-fledged species. This is what ecologists such as myself must contend with in the study of species richness—we deal with the raw material that taxonomists choose to hand us, and which they sometimes take away again later.

6.3.3 Description and naming

So a new species has been discovered. What happens next? It is necessary to announce the discovery of this species to the world, just as the birth of a baby is announced by sending out cards. The taxonomist writes a careful description of a specimen (known as the "type specimen"), typically consisting of several pages written in English with Latin-based technical terms that precisely convey what it looks like—and what makes it differ from all other species known. Usually a precise drawing of the specimen is also provided, produced by a specialized zoological or botanical artist. This description is published in a scientific journal (e.g., *Entomological News*, which I mentioned above), often the in-house journal of a zoo or botanic garden (e.g., the *Kew Bulletin*, in the U.K.) which may publish hundreds of new species descriptions each year.

It used to be possible to name a new species after just about anyone or anything. Taxonomists would name a new species after themselves, their latest sweetheart, or their children ... or (perhaps) give it a funny name like "*ginandtonicus*". Then the taxonomic unions decided this was getting too much and clamped down. Nowadays it is not possible to name a species after oneself, nor just any old friend or family member: only serious scientists, or just occasionally famous media people.[1] Any names like "*ginandtonicus*" are definitely out.

When the description of a new species has been made, the original specimen is carefully preserved, as a reference point for any future taxonomist who wants to come back and check on the identification. If it is an insect, it will probably be pinned out in a specimen drawer next to rows of other insects. If it is a plant, it will be pressed and dried between sheets of paper and filed away rather like a book. And if it is a mammal, it will probably be preserved floating in a bottle of formalin. This preserved individual is known as the "type specimen". There is only one fully described living species, however, for which no type specimen has been preserved, and that is humans. When producing his first work classifying animal life, Linneaus described himself for the published description of humans. However, he did not leave his body to science to be preserved!

6.4 THE UNCERTAINTIES IN CURRENT ESTIMATES OF SPECIES RICHNESS

Everyone working in the field of biological diversity agrees that there must be many unknown species still out there, remaining to be discovered by science. The ease with which new species can be discovered in certain parts of the world makes it clear that there are a lot more to find. I am told that a single day's fieldwork, collecting thousands of insect specimens in a remote area of tropical rainforest, is sure to yield several new species of insects. A day spent collecting hundreds of plant specimens in a previously unsampled locality is also quite likely to yield a new species of rainforest plant. A single deep-ocean trawl from a research ship is likely to yield new species of invertebrates from the sea bed. It just makes sense then that there must be a lot more species out there, still undiscovered.

[1] The American comedian Richard Colbert recently had a newly discovered species of spider from California named after him, after his alter-ego character on *The Colbert Report* stated that someone ought to name a spider in his honor.

However, it is not quite as easy to discover new species of certain other forms of life. For example, new mammals and birds are relatively difficult to come across. At the time of writing in 2008, typically a dozen or so new species of mammal are found each year—most of them rodents, the most diverse group of mammals in the world. This figure is down from around 37 new species a year in the 1970s and 1980s (Groombridge, 1993), suggesting that the supply of undiscovered mammal species is running out. Around half a dozen species of birds are discovered in the world each year, suggesting that there are fewer birds than mammals still left to be found.

Various biologists have tried to put a figure on how many species still exist undiscovered in nature, something I will talk about later. It seems strange, actually, that we could ever try to quantify things which we have not yet found. Yet this is just an extension of the same commonsense principle that if we can find several new species in a single day's collecting in the rainforest, there must surely be a lot more species there in the forest.

If we just had a limited geographical perspective and only considered the showiest groups of animals and plants, it would be easy to get it completely wrong about how many species there must be in the world, including the undiscovered ones. Taxonomists tend to be specialized in one particular group of animals and plants, which must bias their own perspective, and they are also limited by the state of knowledge of the time that they are living in. For example, in the late 1600s the pioneering taxonomist John Ray suggested that there must be about 10,000 to 20,000 species of insects in the world. Even just in Europe, there are actually a lot more types of insects than that! Ray's perspective on the world's diversity would have been biased by most of the work having been done on larger, showier insects and then mostly in Europe, without much knowledge of all the small drab ones and also the vast diversity of some other parts of the world. In fact, we now already know of about 900,000 insect species—and still counting.

Despite the difficulties of coming up with a reliable estimate, biologists are still fascinated by the question of how much is really out there in nature. Using some mathematics, plus a more careful global perspective based on the much greater amount of collecting that has been done these past three centuries, we can just maybe pin down the numbers a bit better. One way to try to make an estimate of how many undiscovered species are still out there is to plot a graph, showing how many known species have accumulated over time. Because of the different histories of study of different groups, each main group of organisms (e.g., fishes, beetles, mammals, etc.) shows its own individual curve, in terms of the cumulative total of species found (Figure 6.1). Some groups were very thoroughly sampled and studied even a hundred and fifty years ago. For example, half of the 10,000 bird species we now know of in the world had already been discovered by 1843—and most of the present-day total was known by the early 1900s. This general pattern of discovery is typical of organisms that are visually appealing, large, and easy to spot. The same goes for trees (large and imposing), butterflies (pretty and slow-moving), and mammals (generally furry and appealing, fairly large, and often relatively easy to spot).

Using a simple method, it is possible to make a rough estimate of how many species there are still out there remaining undiscovered. By following the slope of each

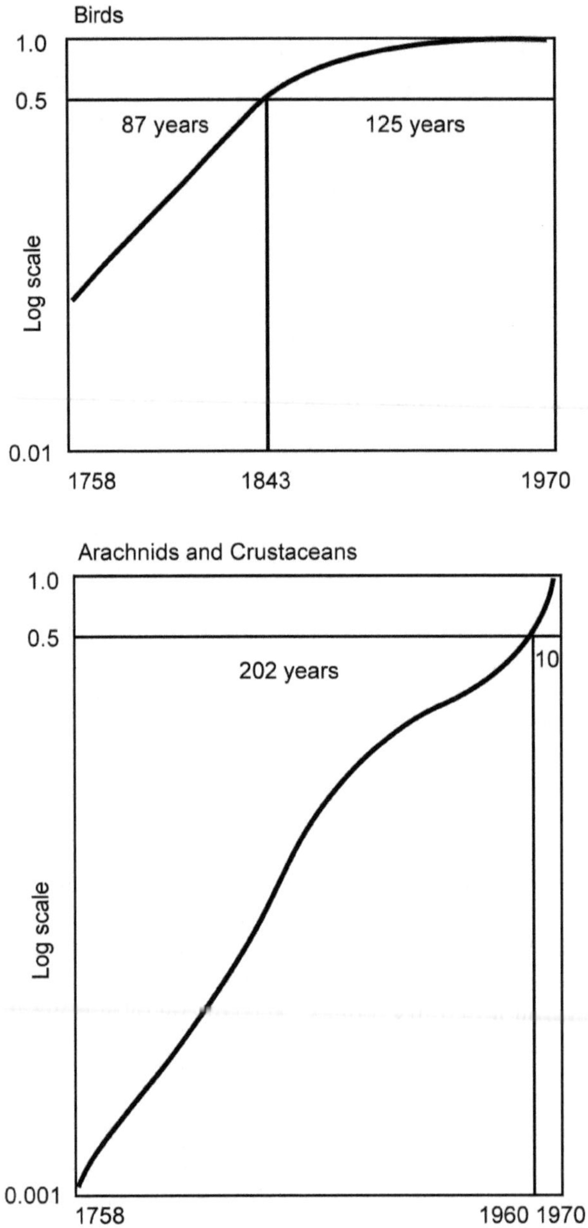

Figure 6.1. Cumulative known species reached by 1970 since the 1700s for two broad groups of organisms (top) birds and (bottom) arachnids and crustaceans, which include most of the known non-insect arthropods. Note the logarithmic scale. In birds, half the species were known by 1843 and the trend is leveling off. In the arachnids and crustaceans, the half-way mark was only reached in 1960, with a very steep rise in new species discoveries since then (after Groombridge, 1993, after May, 1990).

graph showing the number of known species for each group of organisms, and then continuing it onwards and upwards (a process known as "extrapolation"). If new discoveries of species in that group have started to level off, this probably means that we are already approaching the true total number of species out there: there is not much left to find then. The tell-tale sign is a slight bend in the line of the graph, called the "inflection point". Once we have seen the inflection point, it may be possible to extrapolate the graph to the point where it would most likely reach a plateau, once just about all the species have been discovered. This can be done for particular groups of organisms, or in theory if we averaged enough different graphs it could be done for all the species on Earth. But because there is no central global database of species and because each group has own its individual trend, we cannot reliably plot one huge curve for all species on Earth. The best we can do is to look at groups individually, and try to make some reasonable guesses for each one about where the overall "true" species richness might be.

What this graph-plotting technique suggests is that for many groups, such as mammals, birds, fishes, and reptiles, we are close to knowing all of the species out there. This is even with all the best efforts of DNA technology and suchlike in finding new "cryptic" species (see below). With birds, for example, the main phase of discovery of species by science was in the late 1700s and 1800s as European naturalists first voyaged around the world. The cumulative curve of bird species found has now flattened off almost completely. For some other groups such as insects, new species are still being found as fast as ever, suggesting that we are nowhere near to finding all of them. One can also draw curves of accumulation of new species at a narrower taxonomic scale (e.g., in particular families). In the crow family Corvidae (Figure 6.2), the curve for the accumulation of known species leveled off many decades ago, suggesting that we are not likely to find any new members of this family.

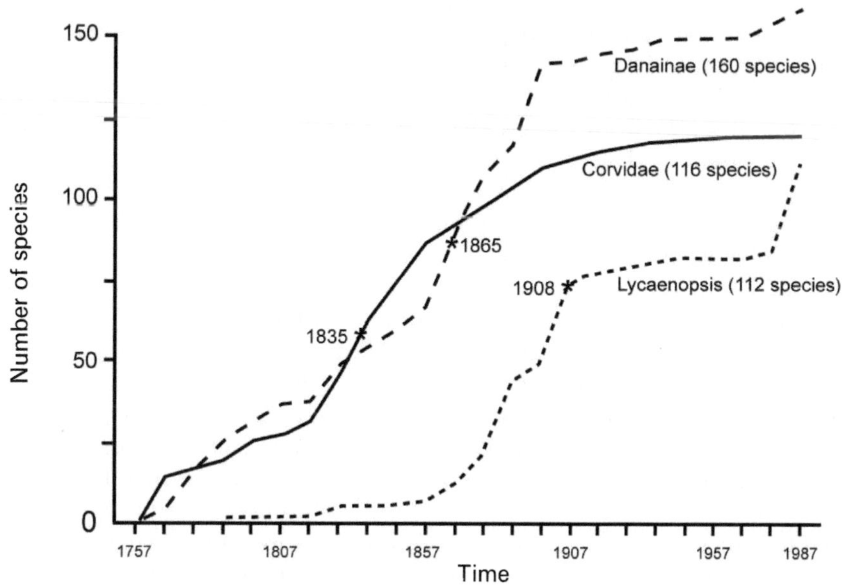

Figure 6.2. Cumulative known species since the 1700s in three taxonomic families of animals: the crow family (corvids), and the butterfly danids and lycaenids (after Groombridge, 1993). The corvids and danids leveled off early on, so apparently most species are now known. Lycaenids suddenly jumped up in richness in the 1960s as many more species were found in Asia.

However, even if the rate of discovery of species in a group of organisms has leveled off, there is still always a chance that it could undergo a renewed burst. The known diversity of the butterfly family Lycaenidae (the "blues") plateaued through the 20th century, but suddenly many more species were discovered after 1967 due to more extensive collecting in the mountains and islands of Southeast Asia, an area that happens to be especially rich in lycaenids. As a result there are now 33% more species of lycaenid butterflies that are known compared with 1967 (Figure 6.2).

Another approach to judging how many more species might be out there undiscovered is to do a much more localized study, with just a few samples of (say) insects in rainforest trees, or the invertebrates found in a series of trawls of the deep-sea bed. In these samples all the species are carefully identified, and the numbers of new species (new to that sample, though not necessarily new to science) in those samples are counted up. The rate at which new species are added with each extra sample is then extrapolated, to the total number of species—both named and unamed—in that general environment all across the world.

6.5 THE DEEP OCEANS: A BIG UNKNOWN

Deep oceans more than 3,000 meters deep cover slightly more than half of the Earth's surface. They are the main environment on Earth, yet they are a big unknown in terms of species richness. Visiting the deep-sea floor requires expensive technology, and back-up ships. For this reason, just a relative handful of biologists have ever seen this world directly and taken samples while they were down there. The only alternative is to trail scoops on long ropes from a research ship and then see what they bring up—a fairly haphazard way of sampling this vastness. Nevertheless, for the relatively few samples we do have, there are tantalizing signs of great biological diversity down there in the deep oceans. When Grassle (1989) tried trawling a scoop known as an epibenthic sled along the sea floor, he found that around half the species in each successive sample of sediment were new to science! These included crustaceans such as amphipods (shrimps and suchlike), and large numbers of different types of polychaete worms, as well as many tiny animals such as nematode worms. Grassle and Maciolek (1992) found that in their studies of the sea bed in deep water (1,500–2,500 meters down) off the east coast of the U.S.A., as a successively larger area of sea floor was sampled the rate of discovery of new species showed no sign of leveling off—again an indicator of a lot more out there waiting to be found. In their study they found a total of 798 species in just 21 square meters of sea floor (both known and new)—many only represented by a single individual. From the ease with which new species could be found on the deep-sea floor, Grassle and Maciolek suggested that there are at least 1 million deep-sea species, and perhaps more than 10 million. If they do exist, most of these species are still unknown to science. However, Robert May of Oxford University has poured some doubt on the idea that deep-sea diversity is really this high, pointing out that from the fact that half the species in these samples were already known, this should scale up to "only" 500,000 deep-sea species globally (still a very large number, and most of these still unknown). A later study by Poore and Wilson (1993) that sampled at greater depths in the South Pacific suggested that diversity is higher there than in the sea off the eastern U.S.A. If this greater diversity in the Pacific is more representative of the oceans in general, then overall diversity of the oceans may be a lot higher than May suggested—perhaps pushing the figure back up into the millions for all the world's deep oceans. But for all we know, future estimates elsewhere on the ocean floors might counterbalance this with lower levels of species richness.

Estimates of the global diversity of deep oceans depend very much on certain other assumptions too. One is the assumption of how uniform they are in terms of species composition. It might turn out that many species are in fact very widespread if we sample broadly across a region of ocean floor. In a sense there is only one "climate" on the world's deep ocean floors: chilly (around 2–4°C) and dark all year round. This might mean then that the same sets of species can occur just about everywhere in the world, and so perhaps when one has thoroughly sampled just a few areas of the ocean floor, then most of what is out there has already been found? On the other hand, some newer studies have suggested that there is actually a lot of variation in species composition from one place to another in the deep oceans (e.g., between different ocean basins, and different depths within those basins, and even between different deep-canyon systems). It also seems that species composition and diversity in the deep ocean varies with latitude, in response to differences in the seasonal "pulse" of food from above (Chapter 2). It is difficult to try to average out all these different factors to get any reliable overall estimates of the world's deep-ocean diversity.

Given how much of the world is actually deep ocean, it is frustrating that we know so little about its biological diversity. There are just a few cursory samples, from which it is hard to scale up to the whole world.

6.6 OTHER TRICKS FOR ESTIMATING UNKNOWN SPECIES RICHNESS

A range of other tricks have been used to estimate species richness in particular groups, and some of these tricks are quite ingenious. However, ingenuity is not necessarily any assurance that a study in correct, because each set of sweeping predictions of global species richness rests on certain assumptions, which might or might not be justified.

One example was a clever attempt to estimate global species richness of fungi, based on looking at just one part of the world which is relatively well studied (Hawksworth, 1991). Hawksworth started by noting that 12,000 fungal species have been found in the U.K. If we compare this number with the total number of vascular plant species in the U.K., there are about 6 fungal species for each plant species (even though not all of them live off plants). Hawksworth then proposed an assumption: how about we assume that this ratio between plant and fungal species richness remains the same everywhere in the world? Since about 250,000 plant species have been found in the world, we need to multiply this by 6 to get the total number of fungal species. This gives a huge global total of 1.5 million species of fungi, almost all of them still undiscovered.

While it is an interesting starting point for discussion, there are many possible weak links in this estimate. For instance, how do we know that the ratio of plant to fungal species is necessarily the same everywhere? Many of the 12,000 species of fungi in the U.K. do not attack living plants and rot down dead leaves and roots in soil, or attack living or dead animals. Thus it is possible that their diversity could depend on factors that do not so strongly parallel plant species richness, so their global diversity trend might not go in the same direction as plant diversity. Furthermore, particular plant species may occur over smaller areas than fungal species—many fungi appear to be extremely widespread, across thousands of miles or even globally. So whereas we may find a totally different set of plants if we move to another part of the world, many of the fungi could be the same—not adding much to their species richness. However, there seems to be a general feeling that the global species richness total for fungi is far greater than the current total of 70,000 species. Tropical fungi in particular are very poorly studied, and it does seem very plausible that in the damp, warm, and generally species-rich rainforests there could be many hundreds of thousands of fungal species out there—many of them adapted to attacking particular species of plants or insects.

6.7 A BOUNTY OF NEMATODES?

Nematodes are little, unglamorous worms that belong to their own phylum (Nematoda), quite distinct from the many other worm-like life forms in the world. They are anatomically fairly simple, with no legs or feelers, and many of them are very small—no more than a few millimeters.

Yet they are everywhere, in soils, in sea floor sediment, in the guts of animals, on the roots of plants ... So far no more than 15,000 species of nematodes have been found, but it is likely that there are still many more out there undiscovered. When taxonomists go over a sample of good nematode habitat with real determination, they find high species richness and many new species. For example, in one study a few cubic centimeters of coastal mud yielded 200 species (Groombridge, 1993). It is difficult to know how this diversity scales up if one samples more widely. Would we find more or less the same set of species in coastal muds everywhere, or a different set of species on every coastline? It seems quite reasonable to suggest that there might be at least several hundred thousand species out there (Groombridge, 1993), but this is still only a guess.

6.8 A PLETHORA OF MITES?

Mites are also low down on the scale of charisma. Most of them are little more than a speck to the naked eye, and many are too small to be seen at all without the aid of a microscope. They live in abundance in soil and leaf litter, playing a role as decomposers in food chains, but many others suck the juices of plants or live as parasites burrowing into the skins of animals.

Some 30,000 species have so far been described (Groombridge, 1993), but there is no doubt that the mites as a group have been neglected because they are harder to find and less immediately appealing than insects. If we assume that the differences in species composition between one place and another, and one habitat to another, are much the same as for insects then this would mean at least several hundred thousand species of mites in the world—nearly all of them so far undiscovered. It is not clear whether mites really should be assumed to follow similar patterns in diversity to insects, for they have their own biology and habits, but this is at least a reasonable basis for a first guess. Another study that indirectly tried to estimate global diversity of mites was Erwin's (1982) study which lumped the mites in with insects, as fellow arthropods.

6.9 ESTIMATING THE UNKNOWN SPECIES RICHNESS OF TROPICAL INSECTS

From what we know of biological diversity to date, the majority of species in the world are insects, and most of them live in the tropics. Tropical forests cover only 7% of the Earth's surface, but they may contain more species than all the rest of the world combined. Recent studies of forest canopies in different parts of the world suggest that tropical forests contain roughly ten times the species richness of insects and other arthropods (such as mites, pillbugs, and millipedes) per unit area, compared with temperate forest. However, within tropical forests there is a lot of local-scale variation in the composition of insect faunas that varies with the conditions in the canopy, such as the abundance of epiphytes (plants growing up on the branches of big trees). It is also uncertain just how widespread each insect species is on average: if you have the same species occurring over large areas this makes for a lot fewer species than if most species have very localized distributions and are replaced by other species in other localities. This patchiness in insect faunas in rainforest canopies makes for a lot of difficulty in figuring out just how many insect species there are. Discussions of the species richness of tropical canopies are in fact central to the whole question of how many species there are in the world. Some estimates put the number of species of arthropods in tropical forest canopies as high as 30 million to 100 million species (most of them still undiscovered).

How were these estimates of total insect diversity arrived at? They depend mostly upon limited local-scale studies where insects in the canopy are killed by spraying insecticide from a machine hoisted high up into the trees. The insects are then collected on sheets or nets as they drop down to the ground. A classic early study using this technique, known as "fogging", was carried out by Erwin (1982) in Panama. He took 19 trees all of one species (*Luehea seemannii*), scattered through a rainforest reserve. Over three seasons, he saturated the canopy of each of his selected trees with a fog of insecticide, and collected the insects that dropped out. He decided to concentrate just on the beetles, and the sample included 1,100 species of beetles! Many of these were already known species, but many were new to science. The important thing here, though, is only how many species there were in total—either new or previously known.

The next key step was to take a guess at how many of those beetles were unique to the type of tree that they were found on. With a certain sleight of hand, Erwin then estimated that 160 of his sample of beetle species were host tree–specific (i.e., found only on that one species of tree).

So now Erwin had a figure for the beetles on that type of tree in the rainforest, but there are a lot more creepy-crawlies up there in each tree than beetles alone. Beetles represent about 40% of all known arthropod species (the group including insects, mites, etc.), therefore Erwin estimated $160 \times 100/40 = 400$ arthropod species that live only on each tree species.

But, another thing that has to be considered is that not all of these creepy-crawlies live up in the canopy. Some spend most of their time on the ground, on the forest floor. Erwin estimated that the rainforest canopy is roughly twice as species-rich as the forest floor, and that the species there are quite a different set from those up in the canopy. Therefore, including the forest floor this brings the total to 600 arthropod species per tree species.

Erwin then multiplied this figure by what he considered to be the likely total number of *tree* species in the tropics—each presumably with its coterie of narrowly adapted insect species. The estimated total number of species of tropical trees in the world is around 50,000. Therefore, the total number of tropical arthropod species is estimated as $600 \times 50{,}000 = 30$ million. So the figure he arrived at was a stunning 30 million arthropod species—far in excess of the million or so species so far discovered.

Elegant though it is, each step of Erwin's argument has a fair amount of uncertainty surrounding it. For example, if the fraction of arthropod species specific to that tree species were only half the number he had estimated, then the total estimate of species in the world would be reduced by half. Another thing is that the species on the forest floor might not be particularly specific to that one type of host tree, for example. After all, these are mostly just feeding off the junk that falls off trees all around: dead leaves, and fruits all mixed in together, plus the dead bodies and waste material of other creatures that feed off these on the trees; so it is hard to believe that they would be that choosy about exactly which tree they live under. Another thing that adds to the uncertainty is that the diversity of arthropods in the canopy might have either more than or fewer than 40% beetle species. Furthermore, for all we know, the one type of tree Erwin examined might not be typical of tropical rainforest trees in general.

What many ecologists regard as a more reliable estimate of the total number of species of insects in the world comes from a study on tropical rainforest bugs (the group known as hemipterans) on the island of Sulawesi in Indonesia by Hodkinson and Casson (1991). They sampled hemipteran bugs over a one-year period using several sampling methods at several sites including a variety of host trees—not just fogging with insecticide but also using various sorts of traps set to catch the insects.

Hodkinson and Casson's study was based on how easy it was to find new species never before described by science, so it is along similar lines to the studies I referred to above for the deep sea. They found that the rate of accumulation of previously unrecorded hemipteran species leveled off towards the end of the study period, indicating that they had by now identified a substantial proportion of the species in that area. They found a total of 1,690 hemipteran species of which only 37.5% were previously known to science. Globally, at that time the total number of described species of hemipterans was 78,656. Therefore, a simple estimate for the real global total of hemipteran species (including those not yet discovered) is:

$$\begin{pmatrix}\text{total known hemipterans}\\ \text{in world}\end{pmatrix} \times \begin{pmatrix}\text{percentage of new species findable}\\ \text{in an area thoroughly searched}\end{pmatrix} = \begin{pmatrix}\text{total estimated hemipterans}\\ \text{both known and unknown in the world}\end{pmatrix}.$$

So this calculation goes:

$$78{,}656 \times 100/37.5 = 209{,}000.$$

Hodkinson and Casson suggested hemipterans represent about 10% of all described insect species; therefore, the estimate for the total number of hemipteran species is about 2.1 million, giving an estimate for the total species number for insects of about 5 million—consistent with earlier intuitive estimates from taxonomists. However, unlike Erwin's study this is not a figure for all arthropods, and if it were applied to all the rainforest arthropods it would presumably add a few million more.

Always with such local-scale studies of insects in the rainforests, the global total one comes up with for species richness is highly sensitive to assumptions. In Erwin's (1982) study, a key assumption (and potential weakness) is how host-specific insects tend to be to particular species of trees. Some recent studies by Novotny *et al.* (2002, 2006) in New Guinea suggest that tropical forest insects are not nearly as host-specific as Erwin or most other tropical biologists had assumed. Another fairly major uncertainty is what proportion of each group (e.g., beetles, or hemipteran bugs) represents all the insects or arthropods in the world. Hemipteran bugs in particular happen to have been a popular group for taxonomists to work on—being as they are relatively large and often colorful insects. This means that they are probably over-represented in the species lists, and that in turn means that there are really more non-Hemipteran insects than we might think, and thus more insect species in the world than Hodkinson and Casson estimated. Groombridge and his colleagues (1992) estimated that hemipterans may really only be around 5% of all the insect species in the world—and that would put the total global estimate of insect species up to around 6.5 million to 11 million species!

Another important aspect to consider, if one is sampling only canopy insects, is the proportion of insect or arthropod species that live instead on the forest floor and in the understory layer of shrubs and small trees. For example, some studies (data from Hammond's 1990 study in Sulawesi, Indonesia, see below) suggests that there are actually a lot more species of arthropods (maybe twice as many) in the lower parts of the rainforest below the canopy, and this would be typical also of what is found in temperate forests. Having proportionally more species in the lower parts of the rainforest would tend to multiply up the number of species in the world beyond those only in the canopy. But then again we don't know if these species living near the ground are really host-specific or can just occur under any type of tree: if they are not selective about where they live and there are just the same species throughout the forest, this would lower the estimate.

In yet another attempt at trying to estimate global insect species richness, Hammond (1990) tried a third type of number-juggling trick. Based on looking at the beetle (coleopteran) species richness of sites in the tropics in Sulawesi (the same Southeast Asian island coincidentally where Hodkinson and Casson worked, see above), he noted that there seem to be typically about five times as many beetles as one would find at a site in the temperate zone. Now we have 0.2 million species of beetles described from colder regions outside the tropics. Probably a fairly high proportion of all the world's temperate beetles have already been found, given how much work has been done on them by the relatively abundant entomologists of those parts of the world. Hammond suggested that about half the outside-the-tropics (extratropical) species of beetles have been found and named by now, judging by how often new species are found. That gives 0.4 million species of extratropical beetles. If we assume that the ratio found at select sites such as Sulawesi holds true generally for the tropics, this means we have to multiply the 0.4 million species of extratropical beetles by 5 to get the number in the tropics—that brings up the global beetle total to 2.4 million species. Assuming that about a third of all insect species in the world are beetles, we then have to multiply our total of 2.4 million by three to get the true global species richness total—and this gives an estimate of 7.2 million insect species.

Another twist on using Hammond's results comes from using his raw collecting results in a different way. This relies on scaling up from the numbers of beetle species he found in his local-scale tropical study to a global total (Groombridge, 1993). In his intensive collecting, Hammond found 6,000 beetle species at his tropical rainforest sites. Assuming he found all the beetle species in that locality, and making further assumptions about how widespread each species was, this could scale up to 1.8 million species globally. If beetles are about a third of all insect species, this translates into somewhere between 5 million and 6 million insect species. However, this all involves a great deal of assumption about how localized each species is on average. For example, if most of the species that Hammond found are really a lot more widespread than he thought, the total estimate would have to be lower.

There have been several other published studies in the last 15 years or so, and though they use slightly different methods and assumptions they all seem to arrive at a figure of between about 5 million and 10 million arthropod species in the world. For example, Odegard (2000) had a go at sifting through the data that others had gathered, and challenged some of the previous assumptions (e.g., about the proportion of species that live in rainforest canopies and how narrow a range of tree species each insect will be found on). He reached what is nowadays the fairly typical conclusion, that the true figure is around 5 million to 10 million species of arthropods, the majority of them insects.

6.10 SO, HOW MANY TYPES OF ARTHROPODS ARE THERE?

Despite the huge amount of effort that has gone into describing insect species by taxonomists, and the fascination that biologists have with estimating how many more remain undescribed, it is frustrating to have to admit that we still do not have a firm idea of how many insect and arthropod species there are in the world. Looking at the estimates arrived at by the various approaches outlined above, and averaging out the more recent and better received ones, a reasonable guess would be that the true figure is about 4 million to 7 million insect species, and at least a couple of million more arthropods of other kinds—bringing about a total of around 6 million to 9 million arthropod species.

6.11 CRYPTIC DIVERSITY

It is not always necessary to look in remote rainforests or the deep oceans to find new species. Hidden diversity is right under our noses out in nature, wherever we live. Sometimes, even in densely populated and much-studied parts of the world, taxonomists still find things that clearly, obviously belong to new species and unsurprisingly, most of them are arthropods. This even happens from time to time in my native England, which from a natural history viewpoint is one of the most thoroughly picked-over places on Earth.

But there is another sort of hidden diversity that is all around us: the diversity that exists *within* the lists of species that are already known. Over the last several decades, detailed study has sometimes revealed that what was thought to be a single species is actually several different species, which do not interbreed with one another. These hidden species are known to biologists as "cryptic" species.

Cryptic species have been found within groups as different from one another as tropical butterflies, deep-sea clams, freshwater fish, and Arctic plants. Over the past 50 years, more than 3,500 scientific papers have been published on the subject (Bickford *et al.*, 2007). Looking across the range of cryptic species discovered so far, relatively few of them are higher plants or microbes. The majority of those that have been found are temperate zone organisms, but this can only be because temperate-zone life has been closely studied enough to pick up the cryptic species that exist there. It is likely that the tropics hold many times more cryptic species that have not yet been discovered, simply because the vast diversity of the tropics has never been studied in such detail.

Though the different cryptic species in a group often occur separately in different places, in some cases they have overlapping distributions. So while some of the birds or insects we may see in a place belong to one species, other identical-looking ones in the same trees can sometimes belong to another species.

Sometimes cryptic species were first found just by taking a closer look at external color patterns, anatomy, or some other feature—so that what had been thought to be one species turned out to be two or more quite distinct forms with their own behavior and breeding patterns. One example is the two species of thrushes in the genus

Catharus in North America. The gray-cheeked thrush (*Catharus minimus*) and Bicknell's thrush (*Catharus bicknelli*) used to be classified as one species, which is understandable as they are only distinguishable to human eyes by a slight difference in their average size. It is not always the case, though, that we can actually find a simple identifying feature that is recognizable to the human eye.

Some cryptic species were first discovered by looking at the chromosomes—the packages of DNA in the nucleus—that show up within dividing cells stained under the microscope. In certain of these cases, what had previously been thought of as just a single species of animal or plant turned out to be several different ones: their chromosomes were so different that each type would be unable to interbreed with the others. This is because having mismatched chromosomes can cause a whole range of problems for any hybrid organism that is produced between the species. For example, if their chromosomes do not match up properly at the time of meiosis (when sperm or eggs are produced), the hybrid offspring will be sterile.

Other groups of cryptic species have been found using more modern genetic tools that focus on the details within chromosomes, not their overall shape. This requires DNA-sequencing technology, looking for differences in the genetic codes of different populations within a species. Often after cryptic species have been distinguished from one another using chromosomal shapes or DNA sequence evidence, certain subtle diagnostic features for each of these newly found species are found in terms of their body form (e.g., the number of bristle hairs found on the leg of a fly may mark one cryptic species off from another).

Although DNA studies can be very useful in looking for cryptic species, they should not be used as a substitute for careful classification work. It is quite possible to have two populations of the same species that seem very distinctive in DNA sequence terms, yet quite able to interbreed—presumably because the genes which are different do not affect the function of the organism too drastically, while the ones that remain about the same are the really important ones. So, overall DNA sequence difference in itself is not really the key factor in causing organisms to belong to different species. If we are not careful in checking out the biology and behavior of the creatures we are studying, we could easily end up filling the world with imaginary species.

Some well-authenticated complexes of cryptic species have been found in economically or medically important groups of insects (e.g., mosquitoes; these are important because they carry disease). For example, there is the *Anopheles gambii* complex: several species of mosquitoes which look the same to us but do not interbreed with one another, each species having its own distinct behavior pattern. Some of them are important as carriers of human malaria, while others never bite humans. Clearly, if we want to understand mosquito breeding habits, malaria transmission, and control strategies, it is important to know which particular species we are dealing with!

In organisms that have less practical significance to humans, it is generally too time-consuming and expensive to be worth doing the exploratory work that would be required to find out if there are cryptic species around. It is quite possible that this "hidden species diversity" is far more common in nature than we realize. Humans tend to classify the world in visual terms, but for many other groups of organisms this is not the case, and they use non-visual clues such as sound or odor to distinguish themselves from other related species. Thus it is thought that cryptic species (cryptic from our point of view, anyway) are more common in groups that rely heavily on these other cues for choosing their mates. For example, relatively drab birds such as the North American thrushes (*Catharus*) and Old World warblers (family Sylviidae) use details of song to distinguish which of their neighbors belong to their own species. Mosquitoes use the detailed pitch of sound generated by the wings, while many other insects use a precise combination of volatile chemicals (pheromones) produced by the male, the female, or both. Frogs also tend to rely on both calls and pheromones. These are all traits that tend to go unnoticed by humans, likely leading to the existence of undiscovered cryptic species. It can be quite a revelation when every once in a while we discover how much we were missing: a group known as the rhacophorine frogs in Sri Lanka suddenly expanded to more than a hundred species after many of the original 18 named species turned out to consist of groups of cryptic species.

There are also many life forms that do not have much morphological complexity that we could ever use to distinguish them by eye. For example, sponges have a very simple form, so their species are often by nature cryptic.

Another factor tending to lead to cryptic species (that look similar or identical to our eyes) being produced is a certain niche that is very demanding in terms of the exact physical form of the organism. For example, Antarctic bonefish living in an extreme environment of very cold water under floating ice are one example of a group that has produced many species, with little change in physical form.

It used to be thought that all groups or pairs of cryptic species had only recently diverged from one another, perhaps in the last few tens of thousands of years. In fact, DNA studies show that very often they have been separate for millions of years. For example, the cryptic species of bonefishes, plus many cryptic species of amphipod and copepod crustaceans, turn out to have diverged millions of years ago.

6.12 FALSE SPECIES DIVERSITY: SPECIES COMPLEXES

In some other cases, what we find is really the opposite problem: "false species diversity". Sometimes what seemed to be several separate species, each one having quite a distinctive look about it, turn out only to be different populations of a single species. They may all be capable of interbreeding, and if we were to study these populations closely enough they might also be found to intergrade in nature. Interbreeding and intergradation breaks the formal definition of a species in biology, which is essentially "something that it is distinct in form and does not interbreed with anything else". These clusters of false species are known as "species complexes". In the plant world they sometimes also go by the name of "ochlospecies", the Greek word *ochlo* meaning a crowd.

An example of a species complex that is just now being uncovered is the "scarlet oak" complex of eastern North America. Scarlet oak (*Quercus coccinea*) is one of the commonest trees in the eastern forests, and for more than two centuries botanists have thought of them as separate and distinct species. Yet now they are finding a plethora of clues suggesting that this "species"' actually intergrades with several other eastern U.S. oak species (such as willow oak and southern red oak), to form one big "super-species". The evidence comes partly in terms of careful examination of the form of leaves, flowers, and fruits of the trees, which shows that many individuals have mixed combinations of traits that do not seem to fall into any one species. This is confirmed with studies of DNA from the chloroplasts—photosynthetic bodies in the plant cells—which show that indeed the different "species" are often exchanging genes. The chloroplast DNA of the other oak species in one locality is often the same as that of the scarlet oaks nearby (Jeanne Romero, pers. commun., 2008).

A famous species complex in the animal world is the "ring species" of gulls of the genus *Larus*, which links populations around the Arctic spanning Scandinavea, Siberia, Canada, and Alaska. Several "species" of gulls across this broad span have been found to merge into their neighbors. All except, that is, the two species where the ends of the ring join, which are in some way too distinct to interbreed. There is a similar ring species in the warblers of the genus *Phylloscopus* whose range loops round the Himalayan Mountains in Asia. To the north of the Himalayas, the various "species" merge. Yet the two forms that hook round to the south from either side do not interbreed where they meet there, even though that are really part of the same complex of species that is busy exchanging genes elsewhere.

In the sea, clamshells (bivalves) seem to show a lot of geographical variation within particular species. In the past, taxonomists often classified these variants as separate species in their own right. DNA studies are showing that in fact many "species" are just different populations that have grown to look somewhat different by being isolated from one another (Bickford *et al.*, 2007).

Where else might species complexes be found? I remember one of my professors from undergraduate days, Frank White, speculating that the tropical rainforests might be full of them. He suggested that many of the tens of thousands of tree species in the tropics are actually just a part of complex species (or "ochlospecies" as he called them), that might vary geographically or perhaps even on a local scale—looking like different species when in fact they are one. This complexity could be fooling us into thinking that there are many more species than is actually the case. So far I have heard no evidence that rainforest trees tend to form ochlospecies. But few tropical species have been really intensively studied—so if they were in fact ochlospecies, would we really know?

6.13 THE HIDDEN WORLD OF MICROBIAL DIVERSITY

Bacteria, viruses, protozoa, and the smaller fungi are abundant everywhere, but they represent the greatest challenge in terms of species richness. Microbes have always been difficult to classify in terms of species, because they have such a simple physical form. One species of bacterium, protozoan, or fungus tends to look a lot like many others under the microscope, and since taxonomy has traditionally been based on morphology, there was not much basis to distinguish them on. Could there be a vast hidden world of species richness in the microbial world, unseen and all around us? The existence of enormous hidden microbial diversity is now becoming apparent, with the aid of new molecular techniques. So far the work has mainly focused on bacteria, but other groups will no doubt soon become subject to similar scrutiny.

It is becoming increasingly obvious that the traditional ways of counting up numbers of bacterial species in samples from nature (taken from soil, water, or whatever) miss most of the diversity that is really out there. For more than 150 years, bacteria have traditionally been identified by culturing them out from very diluted samples, spread onto a Petri dish or floating in a bottle of a growth medium—essentially a soup full of nutrients. Each colony that forms in culture is identified under a microscope or using special techniques for staining and biochemical activity. For this to be a reliable way of finding what species were present in the initial sample, one requirement would be that most species actually grow in a culture medium. It now seems that all along, this was not the case: most species were being completely missed. A new set of techniques based on picking out the DNA sequences present in raw samples from nature shows that many, many forms that are present in samples simply do not grow in culture. Hence they had always been missed, and gone uncounted. As an example of how much more can be found, Gans *et al.* 2005 found around a million quite distinct forms of genes coding for ribosomal RNA sequences in a few grams of soil. Within each of those bacterial "species" that have been studied in detail and seem to correspond to what we would normally call a species, the gene for the rRNA, as it is called, does not tend to vary—so this implies that each of these gene types in the samples comes from a separate species. In the bacterial world, the equivalent of the plant and animal diversity we might find in a whole region of tropical rainforest is concentrated into a single pinch of soil!

The initial studies set off a scramble to catalog the bacterial diversity of different environments on Earth. Now everywhere they look using these molecular techniques, microbiologists are finding similar levels of diversity: what look like millions of bacterial species, many of them not previously found in the other environments that have been sampled.

Complicating the business of cataloging species richness of bacteria has been the discovery that many different types of bacteria are able to exchange genes directly with others that would not normally be regarded as belonging to the same species—because these have such a distinct biochemical makeup, or most of their DNA is quite different from one another. In this sense, the traditional definition of "a species"—unable to exchange genetic information with others—breaks down for bacteria living out in nature. Perhaps much of the total diversity of bacterial life could better be described as one huge species complex (see above) through which any gene can gradually spread if it carries some advantage for its host.

The problems with identifying species of bacteria are also found to some extent with fungi. Although their cell structure and the form of their fruiting bodies is a lot more complex and variable than in the case of bacteria, they are still hard to discern from one another. Once again, growing things in cultures has been the traditional way of finding and identifying species of fungi—yet we cannot be sure how many types there are in each soil sample that have simply failed to grow. Molecular techniques may eventually show us how many species have so far gone unnoticed.

6.14 NATURE STILL YIELDS SURPRISES

Most new species of animals and plants are fairly similar to forms that have already been found, usually in the same genus as previously named and described species. But every once in a while, a new species is found which stuns those who discover it: something novel enough to be featured in news media around the world. Most dramatic of all are life forms that are in a sense both old and new: "living fossils". Living fossils are life forms that were known from the fossil record but thought to be long extinct. It is as if someone had gone back in a time machine and brought us a real live specimen. In just the last few years, two such living fossils have been found, one of them a plant, the other an animal.

In 1997, a ranger at the Wollemi National Park in Australia abseiled down a steep gulley and found a strange-looking tree. It turned out to belong to a genus of conifer that was last known from 30 million years ago, related to the monkey puzzle tree (*Araucaria*) of Chile and the kauris (*Agathis*) of Australia and New Zealand. The Wollemi pine (*Wollemia*), as it is called, had somehow survived in three tiny populations totaling just a hundred or so individuals, barely clinging to existence in gullies close by one another. Almost as amazing was the fact that its refuge was only 50 km from Sydney, the largest city in Australia with several million people.

In 2002, zoologists on fieldwork in Laos found a strange animal on sale in a village market. It was a rat-squirrel (*Laonastes aenigmamus*), something akin to the common ancestors of the rats and mice on one hand, and squirrels on the other. Forms very much like it had existed 20 million years ago, known from fossils, and yet here it was alive in the present-day world.

Sometimes, species thought to have gone extinct in the much more recent past have been rediscovered, alive and well. For example, the Bermuda petrel (*Pterodroma cahow*) was thought to have gone extinct around 1615, but to the amazement of ornithologists 18 pairs were found alive in 1951. The population has now expanded under protection to about 150 individuals.

Discoveries such as these remind us that we do not know everything that is out there, and that in addition to finding more nematodes or more beetles, there may still be some genuine surprises out there awaiting us.

6.15 THE SHADOWY WORLD OF CRYPTOZOOLOGY

While we can be sure that there are many species out in nature that have not yet been discovered, there are some that seem to exist in a twilight zone of the human imagination—suspected by at least some serious biologists, but never having been formerly captured and described, and dismissed as myths and wishful thinking by many others. The business of investigating claimed but unproven animals is known as "cryptozoology". In some senses crypozoology has a venerable history: many unusual life forms were first known from scattered, fleeting sightings or tribal legends, and treated with skepticism or even derision by scientists at the time. The gorilla (*Gorilla gorilla*), for example, was regarded as a mythical animal until a skull was obtained in 1847. It is the same story with the giraffe-like Okapi (*Okapia johnstoni*), discovered in 1901. However, cryptozoology also tends to attract amateurs and eccentrics, whose zealous enthusiasm has helped ensure that cryptozoology is banished to the sidelines of respectable science.

Often, the animals that are purported to exist by cryptozoologists are "living fossils", resembling species that are known from the fossil record but which are generally agreed to have died out. Among the more widely accepted of these "living fossils" is the yeti of Nepal and Tibet, said to be a large upright-walking ape covered in shaggy reddish hair. It is purported to be a living descendant of a rather similar fossil form *Gigantopithecus*, which survived in western China up until half a million years ago. With a background of many centuries of reported sightings continuing up to the present, the yeti is an integral part of folk culture in the region. Intriguingly, mysterious ape-like footprints have been found in snows high up in the Himalayas, including a set of tracks examined in detail and photographed by a British expedition in the 1920s. It is difficult to explain these away as being simply a hoax or the tracks of some other animal. On the other hand, there is the puzzling lack of any recent physical remains or droppings of these creatures, although in the first half of the 20th century expeditions did claim to find these. In those days, DNA analysis was not available but hair, skin, and feces could be compared with those of other known creatures. Those who gathered and analyzed the samples tended to note that the samples (and the parasites within them) did not resemble known forms that lived off other animals in the region—and then always the samples were lost—conveniently, some might say. Recent news reports (BBC, January 2009) mentioned another "yeti" hair sample, which was analyzed for its DNA. It turned out to belong to a species of wild goat.

Occasionally, respected naturalists have made fools of themselves by declaring too early the existence of something which did not turn out to be supported by the evidence. The most famous example was when Sir Peter Scott declared the existence of the legendary "Loch Ness Monster", on the basis of a sonar photo of a diamond-shaped fin in the loch, which resembled the fins of the plesiosaurs of the Mesozoic period. The photo turned out to have been greatly retouched from the original, which offered only a very vague pattern, and it is now generally regarded as a hoax. Scott had even coined a Latin name for the monster: *Nessiteras rhombopteryx*. Adding to the embarrassment, someone pointed out that this was an anagram of "Monster hoax by Sir Peter S".

Other well-known tales in cryptozoology involve species that are generally thought to have gone extinct in the recent historical past, but may have been sighted since. The most famous of these in recent years has been the ivory-billed woodpecker (*Campephilus principalis*) (Figure 6.3). This very large woodpecker—with a wing span of about 75 cm—was formerly widespread through the southeastern United States,

Figure 6.3. The ivory-billed woodpecker. Extinct in the 1940s—or is it? (source: Audubon). *See also* Color section.

where it nested in holes in large old swamp cypress (*Taxodium*) and pine (*Pinus*) trees. The ivory-billed was reportedly nicknamed the Lord God bird, because people seeing it for the first time would often exclaim "Lord God, what's that?!" Felling of the old growth cypress and pine forests across the South occurred on a large scale between the 1860s and 1920s, and eliminated the ivory-billed from most of its range. The last confirmed specimens were seen in the early 1940s in a tract of cypress swamp in Louisiana which was subsequently cleared. There had been sporadic reports of sightings through the last several decades, but these were generally met with skepticism. Then in 2005, Fitzpatrick *et al.* (2005) announced the probable rediscovery of the ivory-billed, based on sightings and a video of the bird taken in a surviving old-growth cypress swamp in Arkansas. This was at first met with a wave of euphoria, later dampened when another group of ornithologists claimed that the bird seen in the video was more likely a common pileated woodpecker, not an ivory-billed. Listening devices stationed in the forest later seemed to pick up the distinctive "double rap" of ivory-bills, and the scientists who had claimed the video did not show an ivory-billed woodpecker now accepted the audio evidence of the bird existing there. However, there are a number of prominent skeptics remaining who suggest that the double raps have other sources, and there have been no apparent sightings of ivory-bills in the area subsequently. So at the time of writing, the ivory-billed remains basically in limbo—neither fully extinct nor fully alive in the collective consciousness of science.

Another possible example of a "crypto-species" is the thylacine or Tasmanian wolf (*Thylacinus cynocephalus*). This marsupial carnivore was very similar to a dog, representing a striking example of the process known as convergent evolution that molds species from diverse origins into the same niche. The most obvious difference from dogs, however, was a set of stripes along its back, which may have served as camouflage in hunting. Thylacines were once widespread all across Australia and the island of New Guinea, but they died out several thousand years ago—presumably as a result of competition with the dingo, an introduced placental dog from Asia (Chapter 5). By the time Europeans arrived, the thylacine only survived on the island of Tasmania, where the dingo had never been introduced. Deforestation, persecution by farmers, and perhaps competition with introduced dogs and cats ensured that the last known thylacine was dead by 1936. However, there have been many reports of sightings of a thylacine-like animal. These reports were taken seriously enough to warrant the setting up of a thylacine "reserve" in Tasmania, perhaps the world's only reserve designated for an animal which might or might not exist. Reported sightings of the thylacine were not only on Tasmania but also on the Australian mainland, and apparently also from aboriginal people from northwestern Iryan Jaya. Several times, photographic or video evidence has been purported to show the thylacine, but the quality to the images is always poor. The common concerns of zoologists are that they might represent another known species of animal (such as a wallaby), or that they may be outright fakes. Skeptics reasonably ask why—if the animal still exists—no tracks or droppings have been found. Once again, the thylacine exists only in the twilight, with the quality of evidence for its survival being poorer than that for the ivory-billed woodpecker.

6.16 THE TWILIGHT WORLD OF SPECIES RICHNESS

It is not just the thylacine, the yeti, and other semi-mythical forms that exist in the twilight. Much of the diversity of nature, indeed most of it, is unseen and unknown. Almost certainly, the majority of species in the world have not yet been discovered. Science has never seen them, but it feels it knows that they are there. Most of these undiscovered species will be fairly mundane: worms or beetles or rainforest trees that belong to groups that are already well known. Yet these unknown life forms may yet include some weird and wonderful things, perhaps ones thought to have gone extinct long ago. However, we should be wary of the wishful thinking that often drives a belief in such things among enthusiasts and the general public.

In a sense then, most of the present-day world's diversity exists in an unknown state: it exists yet does not fully exist. The situation reminds me of quantum uncertainty in physics—an intermediate state that might only exist in the mind of the observer. It is rather unsettling to have to admit it, but most of the species richness that this book deals with is still more or less imaginary.

7

The current threats

In the 12,000 years or so since agriculture was first invented, our own species *Homo sapiens sapiens* has seized control of the world. The extra food from agriculture has allowed a huge increase in human population, and the need to grow crops, keep farm animals, and build homes gave our ancestors reason to alter the land surface wherever they went. Now there are more than 6.7 billion of us in the world—nearly a thousand times as many as there were before farming. As a species we have reached such levels of abundance and technological sophistication that we are now altering the basic properties of the Earth's system. A large proportion of the world's vegetation cover has been changed: forests have been cleared away, and much of the natural grassland and scrub has been ploughed up. The transformation began on a large scale in the temperate zones of the Middle East, Europe, Asia, and then North America and Australia, after which it moved to the tropics. At present, somewhere around 6 million hectares of tropical forest—an area about the size of the state of Delaware in the U.S., or Wales in the U.K.—are being lost each year (FAO, 2005; Achard *et al.*, 2008). Such changes in the world's vegetation caused by humans are in themselves enough to have far-reaching effects on the climate system. There are good theoretical reasons for thinking that the forest loss that has already occurred must already have altered patterns of temperature and rainfall. There are observations of particular areas both "before" and "after" deforestation that seem to validate this (see my other book *Vegetation–Climate Interaction*, Adams, 1997, for more on this topic).

Most recently, in the last couple of centuries, the influence of humans has extended to altering the fundamental composition and heat balance of the Earth's atmosphere. We have poured out greenhouse gases from our machines, from the vegetation that we have cleared, and from the soils of the croplands we plant. The carbon dioxide content of the Earth's atmosphere is set to reach double its background pre-industrial levels within the next few decades, and even after that it will probably continue to rise.

And in addition to all of this, we have been busy breaking down geographical barriers in the living world that have existed for millions of years. Species are constantly now being transported thousands of miles by modern trade and travel, to places where they were never found before. Biological communities around the world are being altered by the arrival of these newcomers.

In a previous chapter we considered the losses of species richness that humans have already caused. In this chapter we will examine the likelihood of further loss of diversity occurring in the near future, as a result of all the things that humans are doing to the world.

7.1 THE GREENHOUSE EFFECT AND EXTINCTIONS

Much of the world's environmental concern is currently focused on the greenhouse effect—and quite rightly so. It has huge potential to affect human lives as well as the biological diversity of the planet. Contrary to the view pedaled by a few eccentric scientists, amateur enthusiasts, and fringe pressure groups, there is now overwhelming scientific evidence (and overwhelming consensus among scientists) that the increasing greenhouse effect is already having substantial effects on climate, and that these effects are going to intensify over the coming decades (IPCC, 2007).

The greenhouse effect is basically the trapping of heat that occurs when certain gases prevent infrared light from escaping from the Earth's surface back into space. Instead these greenhouse gases bounce some of the infrared they trap downwards, to warm the surface and the air just above it. Carbon dioxide is one of these greenhouse gases, produced when fossil fuel such as oil or coal is burned, or when forests are either burned or felled and allowed to rot. Already the concentration of CO_2 in the atmosphere is some 40% higher than it was in the more or less "natural" state that existed more than 200 years ago before the industrial revolution, and it is increasing every year (Figure 7.1). By the end of this century, it will likely be at more than double that natural concentration.

Another of the most important greenhouse gases is methane (CH_4), produced by bacteria in the soils of rice paddies: as rice cultivation has spread and intensified to support a growing human population, methane output from these rice fields has raised its concentration in the world's atmosphere. CH_4 is also produced in the guts of cattle, and more cattle farming around the world has added to the increase in methane. Currently, CH_4 is about twice its pre-industrial world level (Figure 7.2). Unlike CO_2, its concentration seems to have more or less stopped increasing, though it is staying at its raised levels with no sign of any decline. Just in the last two years or so, it also seems to have been showing a slight renewed upward trend.

Some other greenhouse gases, such as CFCs (which are also notorious for destroying the ozone layer) and nitrous oxide, are also present in the atmosphere at greatly increased concentrations resulting from human activity. Water vapor is another very strong greenhouse gas in the atmosphere, and one that was always present in abundance. However, the concentration and the heat-trapping effect of water vapor are expected to increase because each initial bit of warming from other

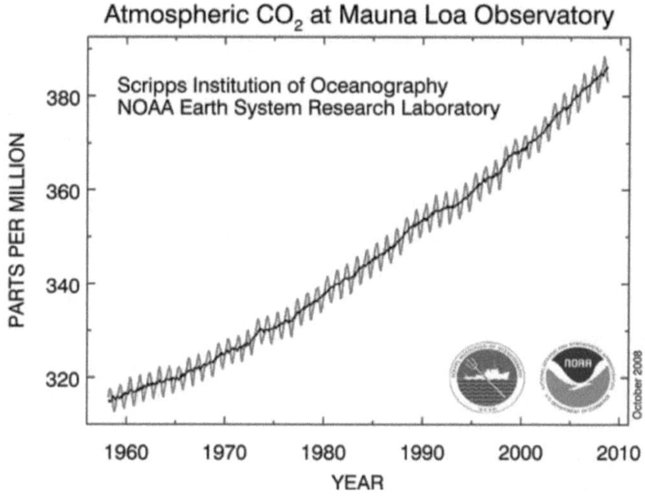

Figure 7.1. Rising trend of CO_2 concentration in the atmosphere (source: NOAA, CDIAC).

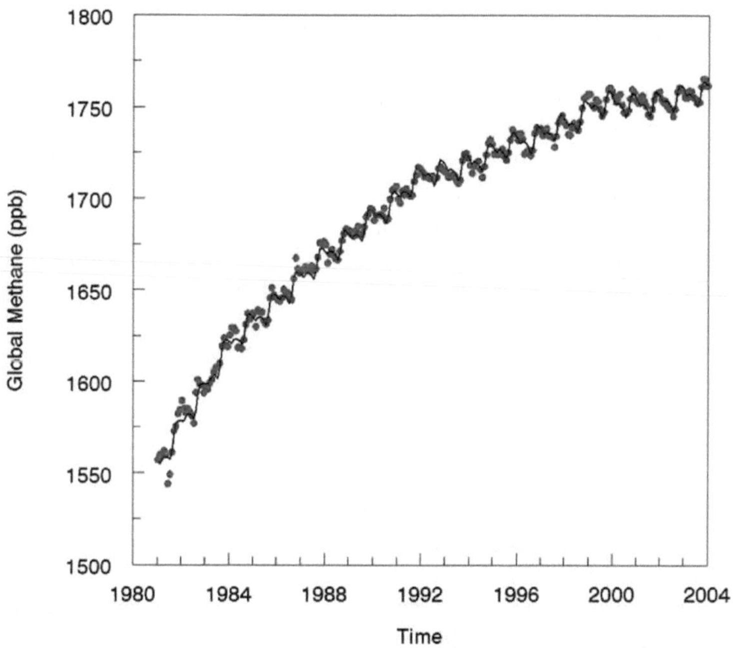

Figure 7.2. Trend in CH_4 concentration in the atmosphere (source: CDIAC).

greenhouse gases such as CO_2 and CH_4 is amplified by the evaporation of more water, adding to the warming. This is a process known as positive feedback, something which amplifies change by "feeding back" on itself. Observations suggest that, indeed, the global amount of water vapor in the atmosphere is increasing in line with expectations (IPCC, 2007).

Knowing that greenhouse gases will produce warming is not enough. We need to be able to forecast *how much warming* there will be in each part of the world for any given increase in greenhouse gases. Only then can we begin to think about the effects on biological diversity, for example. Complex computer models of the climate system are used to forecast the amount of warming which should result from adding a certain quantity of greenhouse gas to the atmosphere. One thing that the models forecast is that the Earth should already be seeing measurable warming effects from the greenhouse gases that have accumulated over the last 200 years or so as a result of human activity, and that this warming will continue to intensify over the next century or so as greenhouse gases build up. Different teams' models differ somewhat in their predictions, according to slight differences in their internal workings and assumptions about how best to simplify such a complex world. Generally, the model-based forecasts suggest that by 2060 the world will have warmed by somewhere between 2.5°C and 5°C, relative to the pre-industrial world of 200 years ago (IPCC, 2007). The warming is expected to be particularly strong close to the poles, but all parts of the world will warm to some extent.

7.2 SPECIES RANGES CHANGING UNDER GLOBAL WARMING

If global warming occurs as the models predict over the next century or so, it could well have serious effects on many species around the planet. The natural distribution ranges of species often seem to be limited by climate: it is frequently possible to find striking correspondences between the edge of a species' distribution and some particular aspect of climate (e.g., summer or winter average temperature, or mean annual rainfall). What we see from these broad climatic relationships is that each species is to some extent specialized to a particular range of climates: no single species is able to occur in every climate around the world, so each has a sort of "envelope" of climate in which it can exist. Many of the clearest relationships between climate and natural distribution ranges are found for plants. This is partly because plants tend to be easy to find and map, but perhaps they also show especially close relationships to climate because they depend so directly on warmth and the water in the soil.

If climate shifts, the potential range in which each species can live will also shift. Plants in particular are especially likely to be susceptible to climate change, because they cannot migrate as easily as many animals in search of more favorable conditions.

As well as taking away areas in which a species can live, warming may open up other areas that are suitable for it if only it is capable of reaching them. For example, the familiar sugar maple *Acer saccharum* (whose leaf is on the Canadian flag) presently occurs across a broad area that extends from Quebec in eastern Canada down to Tennessee in the southeastern U.S.A. (Figure 7.3). It is thought that the limits of its range are mainly defined by climate. As climate warms by several degrees C during the 21st century, the area of territory suitable for sugar maple will tend to move northwards, up towards the Hudson Bay. Meanwhile, at the southern end of its range in Tennessee it will disappear, as the climate there becomes too warm for it.

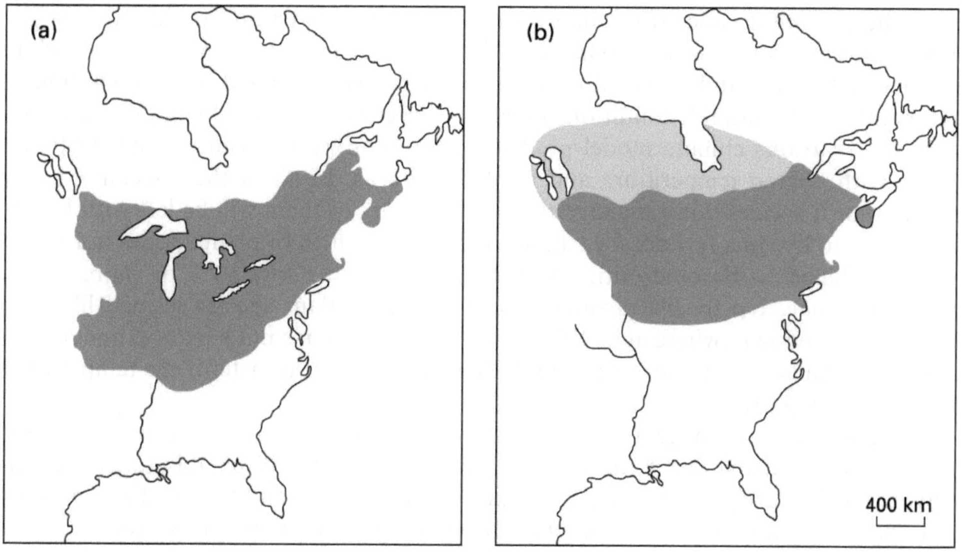

Figure 7.3. Current (a) and future (b) projected ranges of sugar maple (*Acer saccharum*) (after IPCC, 2007).

Eventually, if the greenhouse gases continue to build up, the warming might be so extreme that all of the sugar maple's present range is lost. A new potential range will open up to the north—but this will only be available to sugar maple if it can reach there.

So the opportunities provided by new areas of potential range must be considered against the risks resulting from part of a species' old range disappearing. If populations fail to move quickly enough to catch up with the changes in climate, the whole species may eventually go extinct as the range that it had occupied vanishes under the new climate. For instance, we have no sure guarantee that sugar maple will be able to continue moving up northwards just as fast as climate shifts. The same could well be the case for many other temperate and Arctic species of plants. It stands to reason that the faster and more extreme the warming, the more species are going to be left out of sorts and die out. At the same time as climate warms, the seasonal distribution and total amount of rainfall can be expected to alter too, as rain-bearing winds and storm systems change their tracks.

So as new areas of suitable climate open up, places where a species now lives may become too warm, too dry, or too wet to support it any more. Saxon and colleagues (2005) suggested that by 2100 some 60% of the U.S.A.'s land area will have climates that have nothing really corresponding to them anywhere in North America at present (these "novel" climates are known as no-analog climates). What this will mean for the plants and animals of the U.S.A. is an open question.

To a large extent, how much of a threat global warming is to wild species of animals and plants will depend on their ability to migrate as the climate changes. If we assume a very glum scenario in which all species are unable to move their ranges over the coming decades (for whatever reason), then extinctions are likely to begin. For example, Thuiller *et al.* (2005) took the current distributions of 1,350 wild plant species in Europe, and assumed each has a range that is limited by combination of temperature and rainfall conditions in which it occurs. Then as a "what if" scenario, they used various climate model predictions for Europe between around 2050 and 2080, with altered temperature and rainfall patterns. In all of these scenarios they found that if species don't manage to migrate, many of them will be left with hardly anywhere to live in a few decades' time. As many as a fifth of plant species in Europe were predicted to become critically endangered, having lost 80% or more of the previous range. Up to 2% (around 30) of Europe's plant species seemed likely to go extinct, having nowhere at all left to live. This is of course not a mass extinction for Europe's plants, but it does seem to set the scene for later disaster if the temperature keeps on rising after 2080.

If species do manage to migrate in response to climate, there will be fewer extinctions—and the faster they migrate, the better they will be able to save themselves. But in European landscapes broken up by farming and towns, it is especially uncertain how fast these wild plants will be able to shift themselves around as the climate changes (Thuiller *et al.*, 2005).

One group of organisms that many conservationists are particularly worried about are the "old-growth" forest species: these are animals and plants which only seem to be able to live in forest that has been uncut and mostly undisturbed for hundreds of years. They include, for example, the bluebell (*Hyacinthoides non-scripta*), a flower which forms beautiful blue carpets on the forest floor of English and Welsh woodlands, before the tree leaves come out in spring. If it cannot move north between the isolated fragments of forest that it inhabits in a mainly agricultural landscape, the bluebell may ultimately go extinct. The ivory-billed woodpecker (*Campephilus principalis*) has already paid the ultimate price for being a specialist in old-growth swamp cypress forests in the southern U.S.A. It was lost just as a result of humans destroying its habitat, even before modern climate change could get to it. The northern spotted owl (*Strix occidentalis caurina*) of the northwestern U.S.A.—also dependent on old-growth forests—is already in danger from habitat destruction. It may go the same way as the ivory-billed woodpecker if climate change eliminates the remaining areas of old-growth forest it lives in, without providing adjacent areas that it can colonize.

It is not just in the high latitudes that warming could eliminate species. Although the amount of warming in the tropics is expected to be less than elsewhere, they may lose species too—and in total there are many more species there to lose.

For instance, Hughes and colleagues (1996) studied the distribution ranges of 31 species of *Eucalyptus* gum trees that live at the northern tropical end of Australia and in New Guinea, in either moist or seasonally dry climates. They found that many species had distribution ranges that spanned less than 1°C. The amount of warming expected by the end of the century is likely to raise tropical temperatures by somewhere between 2° and 3° in this region, which apparently puts the climate throughout their ranges beyond the upper limits of their known temperature tolerances. Where these species will go when the temperature warms is an open question.

Along with the warming, there is also the likelihood that rainfall patterns and the moisture balance between rainfall and evaporation will also change across the tropics, although the amount of change depends very much on the particular climate model that is being used to make the prediction. Another study, by Miles and colleagues (2004), looked at 69 species of Amazon rainforest trees, and what might happen to them by the end of the 21st century given one widely respected climate model scenario (the Hadley Centre GCM). Given what is known about the present growth habits and climate requirements of the tree species in this sample, it looks like some 43% of them will not be able to survive anywhere within their present ranges, because of drought and high temperatures. Generally, climate models predict a lot of warming across the Amazon Basin during the next century. For example, the range of models reviewed by IPCC (2007) suggest anything between a 2°C and 6°C rise in temperature across central Amazonia during the southern-hemisphere winter half of the year. It is an open question as to what effect each of these various amounts of warming would have on the countless species within the forest.

Before we gloomily forecast wholesale disaster and mass extinction for the tropics, however, it is worth bearing in mind that part of the reason that so many tropical species end up beyond their present climatic limits by the end of the century is simply that there are not any areas of warmer temperatures that happen to occur in the lowland tropics right now (the last time significantly warmer temperatures occurred in the lowland moist tropics may be back millions of years ago). It is possible that the real tolerance ranges of these rainforest species extend beyond the temperatures we see at present, so that they could actually adjust quite nicely to the new warmer climate if they were presented with it. At present there is a lot of uncertainty about just how much warming the physiology of tropical forest in general can cope with. To really get a better idea we would have to study these tropical species in more detail, and we need more experimental studies of what happens to tropical forest plants if they are warmed. There is not much prospect of getting these sorts of answers in a hurry, though.

Perhaps a blunter, broader approach is needed. Rather than trying to model the tolerances of each species within the rainforest individually, we can try to observe the conditions that (say) tropical forest in general does well in, and then make forecasts about its overall survival in the future climate. Presumably if there is no tropical forest surviving in a particular place, most of the species within it will not be surviving there either.

If the climate shifts over the next few decades away from conditions that seem to favor tropical forest, and more towards those favoring open woodlands or grassland, we can certainly expect a lot of species of plants and animals to be lost. This sort of shift is forecast in at least some of the climate models, combined with models that try to represent how vegetation responds to the climate. For example, Cox *et al.* (2004) suggested that by about 2100 there will be much drier conditions across northeastern Amazonia, causing a large-scale die-back of rainforest. In their most pessimistic scenario, around 50% of the total Amazon rainforest could be lost in the die-back (although it is worth bearing in mind that the Hadley Centre climate model they use predicts more drying there than most other models). Part of the reason for this drying is a collapse of the "water-recycling" mechanism by which the Amazon Basin supplies itself with rain by evaporating moisture from the trees, which is then carried off to fall as rain again. And part of the reason that the recycling collapses is that the trees shut the tiny pores in their leaves that evaporate water, in a direct response to high CO_2 levels (an example of the "direct CO_2 effect", see below). Is there any sign of the forecast shifts in climate occurring so far? As yet there are no significant warming trends in the Amazon, at least away from the growing cities that tend to trap heat. There are some signs of droughts and heavy rainfall events both becoming more extreme, though it is difficult to put these down to any specific cause.

Even where forest does not disappear as the climate shifts, it is likely that the balance of species will change—with unpredictable effects (Morecroft *et al.*, 2001). Some species will be winners, others will be losers—and some or many of these might go extinct.

A very ambitious study has looked at a selection of 1,100 animal and plant species in regions scattered around the world, both in the tropics and elsewhere, and predicted what is likely to happen to them as the climate warms to the level it is expected to reach by 2050 (Thomas *et al.*, 2004). If one pessimistically assumes that species are not able to migrate to keep track with climate changes, this is predicted to lead to their ranges being squeezed down in many cases. Even if one assumes that they *can* migrate, climate change also squeezes the ranges of many of them. While very few species will immediately go extinct around 2050, Thomas *et al.* suggested that there are reasons for thinking that a lot of extinctions will eventually follow from this squeezing effect on ranges. MacArthur and Wilson's (1967) theory of island ecology (Chapter 4) suggests that if an island shrinks it size, it should begin to lose species because smaller populations are more at the mercy of chance fluctuations in numbers. The eventual number of species the island ends up with will be a result of a balance between species lost by bad luck and newly arrived or newly evolved species. Thomas *et al.* took equations that have been used to describe how the species richness in isolated areas such as islands decreases as the island size decreases (MacArthur and Wilson, 1967), and then applied them to shrunken-down ranges of species in 2050.

Following the line of species richness per unit area in relation to total island area, Thomas *et al.* projected a big increase in extinction rates as total species richness adjusts itself to the new circumstances of shrunken-down ranges. The amount of extinction they forecast is between about a fifth and a third of all the species in their sample, the exact number depending on the assumptions made. Note that these are not forecast to be immediate extinctions—many will take centuries or even thousands of years to occur as the reduced populations drop out one by one. But it is an alarming scenario: by extrapolation, as many as a third of the species on Earth may be threatened just by the amount of warming likely to occur by 2050. And climate is likely to keep on warming after that, as more greenhouse gases are pumped into the atmosphere. Bear in mind, however, that the theory of island ecology is very much a simplification (see Chapter 4 for a discussion of it), and that often species richness does not follow its predictions. Just as important, or more important, are such factors as topography, habitat variability, vegetation type, and detailed community structure. To me, it seems much too simplistic to pluck this equation out of its context (brushing aside all the reservations that ecologists have about it) and use it to make such grand predictions.

7.2.1 Clues from the past

Key to survival in a greenhouse world, then, may be the ability to migrate fast enough to keep track of climate, or at least not get left too far behind. But what are the chances of most species being able to make this shift, as global warming really kicks in over the coming decades? We have some clues from migrations of species which followed on from big warming events in the Quaternary (the last 2.5 million years)—the most recent one being the end of the Younger Dryas cold phase 11,500 years ago (Adams *et al.*, 1999). Many of these warming phases were probably about as large and as rapid as the "greenhouse effect" warming which is projected to occur in the near future.

If we take just the most recent climate changes at the end of the last glacial, the overall picture is in some ways reassuring. The vast majority of tree species survived the changes (Adams and Woodward, 1992). Of all the many hundreds of beetles known from the late Quaternary before the end of the last ice age, all survive in the present-day world (Coope, 1987). In fact many seem to have responded to the sudden warming that ended the glacial with ease, migrating to fill the new ranges almost instantly as the climate warmed at the end of the Younger Dryas cold phase 11,500 years ago. For example, temperate climate beetles managed to migrate hundreds of kilometers north from southern Europe—where they would have been surviving the last ice age—into England, in the space of just a couple of decades as the Younger Dryas ended (Coope, 1987; Adams and Woodward, 1992). Similarly, land snails (which one would have thought were slow-moving) appeared within 50 years of the sudden warming event (Adams and Woodward, 1992).

However, there were at least some losses associated with climate change in the late-Quaternary. One of these was a species of spruce tree, *Picea critchfeldii*, which had formed forests across the coastal plain of Mississippi and Louisiana during the last ice age up until about 14,000 years ago. Unusually for the spruces, which normally only live in fairly cool climates, this one was adapted to surviving in the mild temperate climates that prevailed through the Deep South at that time. During the ice age, *Picea critchfeldii* is one of the most abundant trees to turn up in the fossil record of the region, but then suddenly it vanishes. Right at the point when a sudden warming event ended the last ice age, *Picea critchfeldii* seems to have gone extinct.

Even though most species of animals and plants survived the last ice age without difficulty, there must be limits to how far the warming can go while still allowing most of them to survive. In addition to climate change, we also have in store the "double-whammy" effect (see Chapter 5 on human-induced extinctions). The fate of the large animals of Europe, Asia, and the Americas at the end of the last ice age seems to be a lesson in the ways in which multiple factors can work together to push a species to extinction. It was apparently the combination of climate change plus human hunting or habitat modification that was able to eliminate the mammoth, the woolly rhino, and many other species.

Past extinction that resulted from an interaction of climate change and human interference is a worrying sign for the future, because many animals and plants have already had their populations reduced and fragmented by humans hunting them or destroying their habitat.

7.2.2 Polar environments under global warming

Models of global warming generally make it clear that over the next century certain types of environment in the world's lowlands will get rarer. Tundra, at the colder end of the spectrum of the Earth's environments, is likely to be squeezed as conifer forests expand up to the northern coasts of Canada, Alaska, and Siberia. Species that are specialized to the tundra environment (e.g., musk ox) may be threatened by such changes.

The sea ice of the Arctic, and around Antarctica, is another environment likely to decrease under global warming. Already, northern sea ice in summer has undergone an unprecedented retreat in the last few decades. Forecasts based on models of climate and sea ice suggest that the whole Arctic Sea, including the North Pole, may be ice-free in summer before 2030.

If all the Arctic sea ice disappears each summer, this will have serious effects on the population of polar bears which feed mainly on and along the edges of sea ice. Already some reliable anecdotal reports suggest that the retreat of sea ice is causing difficulties for polar bears along Arctic coasts. It is hard to envisage how polar bear populations will survive if all the sea ice vanishes each summer. Although they are adapted to last through a starvation period in late summer each year after the sea ice has gone, polar bears are not adapted to cope with a very extensive meltback of all the ice within reach of land, lasting through the whole summer. Molecular evidence suggests that polar bears have existed as a distinct species for about 200,000 years, since they split off from their cousins the brown bears. If this is so they *did* already survive at least one other period that was warmer than present, 125,000 years ago during the last interglacial. However, it is not clear how extensively the Arctic's summer sea ice ever disappeared during the last interglacial, nor is it known how warm the present world is going to get with the present phase of greenhouse effect warming: our climate may well begin to exceed the last interglacial in warmth, within the next few decades. If the Earth's climates become warmer than they ever were during the past 200,000 years or even the last 2.5 million years, we are essentially into unknown territory as regards the polar bear and many other Arctic species.

Many of the same questions and worries apply to species at the other end of the world, off Antarctica. When sea ice retreats, how will penguins plus the abundant life under the sea ice be able to cope? While it is not hard to see that it could cause problems, it is difficult to model exactly what will happen. One possibility is that the warming of shallow-sea waters will allow an influx of warmer climate predators such as crabs which have been unable to adapt to the cold Antarctic waters. As a result of the lack of predators, many Antarctic species lack defenses such as hard shells: if predators arrive, these species will be put in jeopardy.

7.3 MOUNTAINS UNDER CLIMATE CHANGE

The effects of climate warming on the ecology of mountains are not merely hypothetical. We already see evidence of treelines—the upper limits of tree growth—moving upslope in many parts of the world. High on the slopes of at least some mountains, but below the treeline, there is also evidence for changes in species composition and a thickening of the forest in response to warming. For example, this is the case on Changbai Mountain in northern China, which was studied by my postdoctoral associate Yangjian Zhang (Zhang *et al.*, in press).

It is reasonable to suppose that climate warming could make extinctions happen on mountains. If the habitat of a localized, endemic species is eliminated completely as the mountain's climate warms, then of course it will go extinct. Mountain species in both the tropics and the higher latitudes each live within a particular band of temperature, which determines their altitudinal range. As things warm up, these species will need to migrate uphill too. On one hand it is not perhaps so hard for mountain species to migrate a few hundred meters vertically as it is for lowland species to move hundreds of kilometers horizontally. Indeed we already see signs that species are migrating uphill in response to warming in many parts of the world (e.g., treelines are rising in mountain ranges in Europe, Asia, North America, and Australia).

However, even if there is still some suitable climate available on a mountain and species are able to move up to reach it, that is not necessarily a long-term guarantee of survival. Mountains tend to be pointed towards the top, so the higher uphill a species has to move, the less area of potential habitat there will be. This reduction in habitat area could result in more extinctions of populations in the long term, due to stochastic processes (see above). But once again it is necessary to caution that there are many other things that affect the species–area relationship.

Already we can see that some populations of mountain plants may be suffering as a result of global warming. In Tasmania, several beautiful species of alpine flowers live isolated near the mountain summits. Although none of them is endemic to Tasmania (occurring on mountains on the Australian mainland and some also on New Zealand), they may provide a reminder of the precarious situation of many mountain species. These plants have only the very tops of the mountains on which to live, in small areas above the treeline. Already with a trend in climate warming, trees are spreading upslope, and it is likely that within a few more decades many of the mountains will be forested right up to their tops. With this, the habitats for these alpine plants will have been lost, and their local populations will surely go extinct.

So a big concern is what happens when the mountain is not tall enough to give any climate zone up above which is still cool enough for the species to migrate into. Tropical mountains in particular have locally evolved species that could be left without anywhere to live, as the climate warms beyond anything that has occurred in the past several million years. In the temperate latitudes, mountain species tend to be widespread, effective dispersers that occur over large ranges, so the danger of them being lost completely is not so great. Nevertheless, Thuiller *et al.* (2005) predicted that climate change will hit the mountain species of Europe particularly strongly, compared with lowland species.

7.4 CORAL REEFS

In another very different type of ocean environment, global warming could also have major effects. The world's coral reefs have often been called "the rainforests of the sea". On the scale of around a hectare they may indeed rival the tropical rainforests in species richness, although as one moves up to larger sample areas the rainforests actually prove to have many more species. Nevertheless, to a marine ecologist or even to a tourist snorkeling on vacation, the diversity of forms and colors in a shallow clear water reef is immediately obvious; far more so than in a rainforest with its rather uniform greenery. I myself have only ever seen coral reefs while snorkeling on vacation, but it was one of the most memorable experiences of my life. Corals, anemones, and other invertebrates grow with bizarre and varied forms, and in many different shades of orange, red, and yellow. Fish of many fantastic colors dart between the corals, or float in shoals unafraid of the diver. Psychedelically patterned nudibranch slugs and delicate glasslike shrimps pick at the sea floor for food.

The reefs look like a fragile environment, and indeed they are. They rest precariously in a narrow band of temperatures. For the maximum number of coral species to do well, the monthly average temperature of the water that washes over a reef must be around 28°C. Above 29°C, a difference of only one degree, many of the corals start to suffer. They lose their symbiotic algae (known as zooxanthellae) in a process known as "bleaching", which leaves the coral polyps white. If high temperatures persist, the corals will die—of starvation essentially, because the polyps rely on the algae to supply sugars for them. If the corals are lost, the whole intricate reef ecosystem collapses, and mats of green and red algae end up coating the dead coral skeletons, with most of the fishes and invertebrates gone. Over time, after the temperature has cooled down and if no more bleaching events occur, the reef species will tend to come back, but it can take many years to recover from a severe bleaching event.

Bleaching events like this have always occurred, but in recent years they seem to have become more common. For example, the Great Barrier Reef of Australia (overall the most species-rich area of reef anywhere in the world) has been hit by eight large bleaching events since 1979, the most recent in 2006. In several parts of the Indian Ocean, more than 90% of the coral reef area has been lost due to recent bleaching events (Berkelmans *et al.*, 2004).

Global warming due to greenhouse gases has been blamed for this, although it is difficult to definitively pin the blame on this one factor. Disease can contribute, and natural fluctuations and extremes in temperature have always occurred. Certainly, though, if the world warms as the models predict, the tropical seas will heat up too. Since only slightly warmer-than-normal temperatures are enough to set off bleaching in the warmest and most diverse coral reef regions, there can be no doubt that many reefs will suffer. The IPCC (2007) report on global climate change warned that even under a moderate warming scenario, by the end of this century the Great Barrier Reef—and many other of the world's reefs—will regularly be exposed to temperatures hot enough to bleach corals.

From the point of view of global species richness of coral reefs, an important question for the future is whether corals and other reef species will be able to reach areas of slightly cooler temperatures that could allow them to survive. Most of the world's shallow seas are too cool for real coral reefs to form, but as global temperatures warm, areas at the outer fringes of the tropics will become suitable for diverse reef communities. Yet these new areas of potential reef habitat may be hundreds of miles away from existing species-rich reefs. Corals and other reef animals are adapted to disperse via larvae that drift with the tides and currents, but they may have to make the jump quickly (over the next few decades perhaps) before many of the original areas they inhabited are lost.

There is some reason for optimism, though. Coral reefs have often been exposed to warming before, and their species richness has survived. There was apparently a warmer-than-present time around the world, including the tropics, during the last interglacial phase about 130,000 years ago, and no extinctions of corals resulted from that. More importantly, there have been many other times in the last million years when global temperatures warmed up suddenly, perhaps over just a few decades in a way that resembles what is likely to happen over the next century (Adams *et al.*, 1999). Likewise, these sudden warming phases did not result in coral extinctions. Such warming events usually had a different starting point: the much cooler-than-present temperatures of glacial phases, and then warmed up to conditions similar to the present. To what extent they offer a good analogy for the future—where things already start from the warm present-day world—is not entirely clear. But overall these past warming events do seem to offer an encouraging general analogy to the future.

Where there is warming, there is also sea level rise. The melting of ice on the land, plus the expansion of warmer ocean waters worldwide, is expected to produce a rise in sea level of at least 50 cm, and perhaps more than a meter, by the end of the century. This rise can be expected to drown the lower levels of many reefs, which depend on shallow water to get enough sunlight for the algae inside the corals to photosynthesize. Compensating for this, corals will be able to migrate to shallower areas that become drowned with the rise in sea level, but it will be an additional stress on coral reefs. Nevertheless, the lesson from the past million years or so is broadly reassuring: many times global sea levels have risen tens of meters or more as cold glacial phases ended and ice sheets melted back. At the Last Glacial Maximum some 20,000 years ago, sea levels may have been as much as 130 meters lower than they are now. Detailed geological records show that the rate of sea level rise at the end of the last glacial period was more than a meter per century—even three meters a century at some times. The sea level rises occurred in combination with short-lived rapid warming phases, much as we can expect for the next several decades. And these phases of sea level rise were sustained over thousands of years, which we might arguably expect if the worst greenhouse effect scenarios come true and a large part of the polar ice caps melts.

And yet, despite all this disruption, the surprising thing is that none of the more recent past warming phases (from the last million years or so at least) produced any extinctions of corals or other marine life that we know of, despite the excellent fossil record of the shallow seas which should enable us to spot them. It had been a different story earlier at the beginning of the Quaternary, around 3.0 million to 2.5 million years ago. At this time the onset of such large rapid climate changes apparently drove many corals (and many other shallow-sea invertebrates including bivalves) extinct (Chapter 3). This more recent lack of extinction may be because the species that we see in the present-day world are the survivors of so many earlier warming and cooling phases (probably hundreds of events, over the whole span of the Quaternary). They likely survived because they had certain traits that allowed them to move rapidly to colonize new territory each time the climate shifts. In this sense, the biota of our modern world has been selected and thus "pre-adapted" to survive the sorts of challenges we are now throwing at it.

I have talked about the "double-whammy" effect of climate change combined with other more localized pressures on large-animal populations on land pushing them over the edge into extinction. It could turn out to be the same for the species of coral reefs. I myself have seen the effect of development and over-exploitation on reefs in Malaysia. Existing reefs have often been devastated, and hemmed in by coastal development—and I do wonder how the reef species in some areas such as this will cope with this combination of pressures.

And as if all this were not enough—what with all the warming temperatures, human exploitation, and sea level rise—there is something altogether more insidious and ominous for marine life to contend with: the direct CO_2 effect on the growth of organisms that have carbonate skeletons (see Section 7.5).

7.5 GLOBAL WARMING IN THE LONGER TERM

How long the warming from the greenhouse effect will last is in itself a moot point. Over the next several thousand years, the extra CO_2 we have pumped into the atmosphere will tend to be soaked up into the oceans, reacting with calcium carbonate to form dissolved bicarbonate. But it is possible that other mechanisms set in motion by the initial warming will magnify it and sustain the greenhouse climate for a lot longer. These amplifying mechanisms are known as "feedbacks". For example, methane gas bubbling up from the ocean floor (from substances known as methane clathrates), in response to the effects of the warming, might still be keeping the atmosphere topped up with greenhouse gases for tens of thousands of years into the future.

There is evidence that something like this happened 55 million years ago during the late Paleocene, when alligators and palm trees reached nearly to the poles. Bubbling methane clathrates from the ocean floor may have poured greenhouse gases into the atmosphere, keeping the Earth in its hothouse state for 100,000 years (Chapter 3). In the end, though, we really cannot be sure what will happen as the Earth warms over the next century or two: it is essentially uncharted territory for us.

As the warming really sets in by the end of the century, we can expect increasing heat stress and changes in water balance to exert their toll on many of the world's areas of forest. Cox *et al.* (2004) suggested that much of the Amazon rainforest could be dying back before this stage, and increasing warming only seems likely to intensify the problems. What happens to the millions of tropical forest species will then be a balance between the rate at which they are eliminated from some areas where they lived before, and the rate at which they can reach new areas of potential habitat.

In some areas the climate will probably become more suitable for tropical rainforest: southern Louisiana, for example, may become potential tropical rainforest habitat before the end of the century. This does not necessarily mean that these newly available areas will actually be colonized by significant numbers of rainforest species, especially if these new habitats are far away from the nearest remaining areas of forest.

If the new greenhouse world lasts for a long time—hundreds of thousands of years or longer—evolution may enter into the picture. The rate of formation of new species in the tropics may accelerate as indicated by the long-term record of tropical diversity in South and Central America put together by Jaramillo *et al.* (2006). During the very warm phase 55 Myr to 40 Myr ago (see Chapter 3), rainforest tree species richness seems to have increased to reach a peak, and then declined later as the world cooled off. So in the longer term the increased greenhouse effect could actually *add* to the long-term species richness of the planet. However, it is necessary to bear in mind that we have only a few scattered records of rainforest diversity from this vast span of time, and what we see might not be representative of what actually happened. There are also many other aspects of the changes under long-term global warming that could complicate this. For example, there is the direct effect of very high CO_2 levels on ecosystems (see Section 7.6) which might in itself cause extinctions.

7.6 DIRECT CO_2 FERTILIZATION EFFECTS ON PLANTS

To live on land, plants must contend with the drying air. To avoid being killed by dehydration, they have to cover themselves with a layer of waxy varnish called a cuticle. At the same time, for plants to photosynthesize and grow they must take in carbon dioxide—but the cuticle almost entirely cuts off the flow of CO_2 into the leaf. Given that they must limit evaporation of water from their leaves, and yet allow CO_2 in, plants have come up with a very good compromise. They have microscopic pores called stomata in their leaves. These stomata are able to open and close as needed, when two sausage-shaped cells change their shape to cause the hole in between them to either open or close. When a plant is short of water or has done enough photosynthesizing for the time being, it keeps its stomata tightly shut to cut down on unnecessary evaporation. When it has both enough water and the need to photosynthesize, it opens its stomata allowing CO_2 to diffuse in, tolerating the loss of water that occurs at the same time. Natural selection has favored plants that do not keep their stomata open any longer than they really need to. Even if there is plenty of water in the soil, once a plant has got all the sugars from photosynthesis that it needs for now, it will usually close its stomata. This means that it does not draw upon the store of water in the soil around its roots, which it may need for another day.

If you do an experiment and add more CO_2 to the atmosphere around the plant, one immediate effect is that it increases its rate of photosynthesis. Because CO_2 going into the leaf means that photosynthesis is more efficient, the plant can make more sugars for itself. Another effect will be that the plant tends to keep its stomata closed more of the time. This is because it does not need to keep them open for long to get the CO_2 it needs; at higher concentrations the CO_2 pours into the leaf. So as soon as it has fixed enough CO_2 into sugars, the leaf clamps its stomata shut and cuts down water loss. So with more CO_2 around, plants are in a sense better fed but also better watered, since they don't need to evaporate so much water to get the CO_2 that they need. In effect, then, adding CO_2 to the air around a plant is a lot like watering it. And watering a plant even without raising CO_2 concentrations will tend to mean it does more photosynthesis, because it can keep its stomata open for longer before its water supply runs down. It turns out, then, that both CO_2 and water are flip sides of the same thing: add either and the plant responds by photosynthesizing more and growing faster. It will also be able to tolerate drought better, being able to do well on a more limited supply of water.

At present, we humans are carrying out a vast experiment on raised CO_2 effects, involving all of the world's plants. The effects that this will have—and the effects it may already be having—are very uncertain. Wild plant communities are extremely complex systems, and so they might behave in all sorts of unpredictable ways (see my other book, Adams, 2007). The best clues to what will happen over the coming decades, as CO_2 levels soar, are a few very limited experiments where plots of forest or other vegetation out in nature have had the air around them deliberately enriched in CO_2. These are known as free air CO_2 experiments, or FACEs for short. It would be encouraging if FACEs all showed one consistent result but they do not. In some cases the vegetation responds by increasing its growth for a year or two, and then stops responding. In others, after an initial burst of growth the response trails off but still remains detectable for many years while the experiment is run. In some cases, there is just no detectable effect of increased CO_2 on growth rate, right from the start of the experiment (Adams, 2007).

Observation of nature over the past centuries offers another approach to the problem, perhaps revealing what the buildup of CO_2 that has already happened has done to plants. This might either confirm or deny our expectations of how responsive vegetation will be in future as CO_2 builds up. Records of the growth rate of trees around the world—recorded in tree rings—have been looked at extensively to try to detect a speeding up of growth rate, which would be seen as wider rings each year. In the tropics where trees do not form rings, overall increase in girth of the trunk has been used as an estimator of growth rate.

In these long-term studies what do we find? Not what would have been expected, at least by modelers (see below). The temperate zones and the Arctic have failed to show any overwhelmingly convincing signs of faster growth in response to the extra CO_2 that has accumulated in the atmosphere over the last 200 years (Adams *et al.*, 2007). In many areas there are trends in growth of trees and other plants, but these can more readily be explained as a response to warming in that region, because they have fluctuated in the same way for centuries each time the climate became a bit warmer or cooler (Schweingruber *et al.*, 1993). In other cases, unintended fertilization of forest growth by other pollutants—apart from CO_2—looks like being the cause. For example, nitrogen and sulfur oxides from industry and cars can in many regions act to make up deficiencies of nitrogen and sulfur in the soils—actually promoting tree growth across Europe and parts of North America and Asia. To make the case that observations show a real direct CO_2 effect altering the growth of vegetation, we would need to find trends that are utterly anomalous, beyond what we would expect for climate warming or sulfur pollution, for example. Unfortunately so many different things are changing simultaneously in the higher latitudes, and it will always be hard to tease apart the threads to show convincingly that any particular part of the trend in vegetation growth is due to CO_2 fertilization alone. Direct CO_2 *must* be in there somewhere, altering the vegetation of the mid-latitudes and high latitudes, but we can have very little idea of how significant it is.

How about the tropics? This is where most of the world's biological diversity is, so it is certainly very interesting to know what CO_2 might be doing to the ecosystems there. Also the tropics have not been warming as fast as the higher latitudes, so maybe the relatively constant background of temperature over the last several decades will allow us to see the direct effect of rising CO_2 on plant growth?

An extensive network of small rainforest plots scattered across the Amazon Basin has been monitored since the mid-1970s (Philips *et al.*, 2005). One trend that has been found is a significant increase in the "turnover" of trees and bushes in the forest. More turnover involves more deaths, but also more young establishing, so it seems as if the woody plants are growing fast and reaching the end of their lives sooner. This is perhaps the sort of thing that would be expected if increased CO_2 was increasing the growth rate of plants—the life cycle of trees in the forest would be speeded up. What could this mean for the species richness of the forest? One possible explanation for the high species richness of tropical forests, and why some tropical forests are richer in species than others, is the delicate balance between rate of growth and outcompetition between strong and weak species (on one hand) and the rate of disturbance that cuts down competition (on the other hand). The faster growth of plants in the forest due to CO_2 is rather like throwing fertilizer on it, and it will increase the intensity of competition while disturbance stays constant. What happens in terms of diversity may depend which side of the "humpback" curve each piece of forest was originally on (Chapter 1)—something which we do not really know. Perhaps in some areas it will increase tree species richness, whereas in other areas it will decrease it? Another interesting aspect of the trend found in Amazon rainforest plots is that vines steadily became more abundant in the forest during the past decades. This may imply that trees and bushes are steadily being pushed out—leading to a loss of diversity.

The trend in Amazon forest plots is striking, but we cannot be entirely sure that it is due to increasing CO_2. Other changes across the region have occurred in the past decades too (e.g., an increase in cloudiness). Some areas also show an increase in temperature, and increased frequency of droughts, and either of these might have caused this trend. If the CO_2 effect is really important as Phillips *et al.* suggested, we probably should expect it to be operating everywhere. Putting a dent in this idea that the striking trend in turnover in Amazon forests is necessarily due to CO_2 is the finding that the opposite thing has been happening in parts of Southeast Asia and in Central America. Time series data of tree growth from the Pasoh reserve in Malaysia and from the Barro Colorado Island reserve in Panama, show that the growth rate of trees has been slowing down over the last several decades (Feeley *et al.*, 2007). It has been suggested that the trend is due to an increase in temperatures across Southeast Asia, causing the trees to waste more of the energy they fix in photosynthesis in respiration. The fact that other tropical forest regions can show an entirely different trend from the Amazon Basin suggests that whatever effect the recent CO_2 rise is having on tropical ecosystems, it is not strong enough to consistently overwhelm other background factors that can change. This tends to weaken the case that the recent trend in tree turnover in the Amazon is necessarily due to CO_2, rather than a combination of other factors.

The only really convincing sign of a direct CO_2 effect on the world's vegetation during the past two centuries has been a progressive decrease in the density of stomata on leaves. If you look at leaves of almost any species of plants preserved from the 1600s or 1700s, and compare them with plants of the same species growing today, it turns out that the present-day plants have fewer stomata on their leaf surfaces (the abundance of stomata is measured in terms of the ratio between the numbers of stomatal cells and the ordinary leaf surface cells: the ratio is called the "stomatal index"). My former boss at Cambridge, Ian Woodward, was the first to think of looking back in time in this way (Woodward, 1987), and his work has been backed up by many other studies since. Experiments growing plants at different concentrations of CO_2 show that as one increases the CO_2 concentration around the plant, the frequency of stomata on newly formed leaves decreases—so this ties in nicely with what is seen in nature, proving the point that this is a genuine direct CO_2 effect on plants. Why do the plants change the numbers of stomata in response to CO_2? Presumably because there is a "cost" associated with having stomata—they may use up energy, or leak a little bit of water vapor even when they are closed. So if stomata are not so necessary to let CO_2 in, from the plant's point of view it is best to do away with them as much as possible.

So much then for what we see in terms of present evidence for the increased CO_2 effects on plants. What about the future as CO_2 keeps on increasing? Mathematical models, mostly based on very short-term responses of plants to enrichment with CO_2 in a laboratory environment, have been used to try to predict the growth rate of all the world's vegetation for the coming centuries. They project something like a 20% to 30% increase in growth rate of global vegetation by 2050, relative to how things used to be when CO_2 was much lower 200 years ago (Adams, 2007). So far, the observations from CO_2 enrichment experiments, and the trends in vegetation growth over the last 200 years, do not seem to support what the models predict; it seems that the growth response of plants to CO_2 tends to be much weaker and shorter lived than expected. But this has not dented the confidence of many of the modelers, who are still adamant that they have most of the details right. This rather looks like just another case of "modeling mania"; a common problem in the environmental sciences!

What will be the effects of increasing CO_2 on species richness? Because we understand so little about direct CO_2 responses, there is plenty of room for speculation. One view is that the increased growth of the world's vegetation will actually help large numbers of species to survive and coexist. This is partly because more CO_2 may in a sense make the environment more "benign", less demanding from the point of view of many plants, particularly tropical rainforest plants. Rainforest plants need plenty of water, with the most species-rich areas of tropical forest tending to occur in the wettest climates. Since adding CO_2 is a lot like adding water to the plants, the increase in CO_2 over the coming decades should favor the growth of many different species of rainforest plants. I suppose that this is what would be predicted empirically from Axel Kleidon's GeDi model (see Chapter 2) to explain higher species richness in low latitudes. So with the perspective of Axel's model, perhaps adding CO_2 will actually aid the coexistence of larger numbers of species.

Tropical rainforest in general may tend to spread outwards into areas that would otherwise be too dry for it, helped by raised CO_2. This might counteract the effect of humans cutting down the forest, so even though the forest may be shrinking in one direction it might be expanding in another. Also if there are any unfavorable effects of greenhouse climate warming on rainforest (e.g., a drying of climate, or an increase in heat stress on rainforest plants) increasing CO_2 might help to make up for these effects. The increased efficiency of water use by rainforest under high CO_2 will mean that a given amount of rainfall goes further, as mentioned above. One major effect of heat stress on plants occurs through something called photorespiration, where oxygen becomes tangled up in the photosynthetic reaction so that the fixed photosynthetic products must be broken down and wasted. Raising CO_2 will tend to suppress this photorespiration, meaning that plants can grow better at high temperatures. This might help the rainforest trees to cope with another indirect effect of CO_2, climate warming through the greenhouse effect.

The idea that increasing CO_2 might actually be good for biological diversity has been taken up by certain lobby groups, many of them supported by the fossil fuel industry and all implacably opposed to limits on greenhouse gas output. One example is the Greening Earth Society which advocates improvement of the Earth's environment through adding more CO_2 to the atmosphere.

On the other hand, there are reasons to suspect that the direct effects of high levels of CO_2 could ultimately lead to a decrease in biological diversity, and even mass extinctions. As I mentioned briefly just above, local-scale patterns in species richness suggest that favoring more rapid growth of vegetation (e.g., through adding soil nutrients) can often cause a decrease in species richness in the plant community (see Chapter 2). The same pattern is often apparent when one just compares places which have different soils, and compares their species richness. Within the tropics or within the temperate zone, the most species-rich plant communities often occur on rather nutrient-poor soils. It is thought that this is because certain strongly competitive species in the plant community are able to take over and push out all the other weaker species. Occasional disturbance events which knock back all the competitors and even up the score are less effective when plants are growing fast. It is quite plausible that adding CO_2 to the world's plant communities will be just like throwing fertilizer onto a species-rich grassland or shrubland. The result of soaring CO_2 levels could possibly be a crash in plant species richness, occurring all around the world. This is little more than speculation at present: there are few experiments on raised CO_2 effects and no observations in nature that back up the idea. Nevertheless, it is a threat which is worth considering when trying to weigh up whether to control global CO_2 levels. We certainly have a lot to lose if it turns out to be true.

Another way that high CO_2 levels might affect species richness is through the interactions of plants with the insects that feed off them. Very often, plants growing experimentally under high CO_2 turn out to have less nitrogen and other nutrients in their leaves. Insect herbivores lead a precarious existence, feeding on plant tissues which are very low in the nitrogen-containing amino acids which they need to grow and maintain their tissues. It may not take much of a change in average nitrogen content of leaves around the world to make life far more difficult for many species of herbivorous insects. In experiments, plants that have been grown in high CO_2 often tend to suffer less from the insects that feed off them. As a proportion of their total leaf area, the high-CO_2 plants lose less to insect feeding, and the herbivorous insect populations on their leaves and branches do not become as abundant (Adams, 2007). Some (though not all) studies have suggested that insects feeding on CO_2-fertilized plants tend to grow more slowly and die more often, perhaps because of a shortage of amino acids in their food supply. It is possible, then, that at sufficiently high CO_2 levels many species of plant-feeding insects will go extinct because they can no longer sustain themselves on the food that is available. It is thought that the majority of the world's species are herbivorous insects: even a minor proportion of these species going extinct could mean hundreds of thousands of species being lost.

And there is another, more insidious, effect which might permeate back round to affect the plants that the insects had been feeding off. One of the possible reasons why so many species of trees are able to coexist in the tropical rainforests is that highly specialized insect herbivores (attacking only one or a few closely related species of plants) will always multiply up and eat back any one species which starts to become too abundant (see Chapter 2). This keeps strong competitors from ever becoming abundant enough to push out weaker species in the forest, allowing a hundred or more species of trees to coexist in a hectare of tropical forest, for instance. However, what would happen if these specialized herbivores start to become less abundant or even extinct, because of the direct CO_2 effect decreasing the nutritional value of their food supply? With less of the specialized herbivory that had previously held them in check, certain strongly competitive species of trees could then start to increase their populations and eventually push most of the other tree species to extinction. The result of all this then could conceivably (in the very worst case scenarios) be a spiral in which the species richness of the world's land ecosystems crashes down into a mass extinction.

Has the world's vegetation ever been through very high CO_2 levels before? If it has, we could use these times to get some clues as to what will actually happen to diversity over the coming centuries. Various lines of geological and fossil evidence suggest that during certain periods in the past 500 million years, CO_2 levels were much higher than they are now—and even a lot higher than they are expected to reach over the next 100 years or so. Not all of this evidence agrees on the exact timing of these high-CO_2 events, or the concentrations that were reached. According to one source—the stomatal indices of fossil leaves—it looks like CO_2 may have been substantially higher during a particularly interesting time around 50 Myr to 55 Myr ago in the late Paleocene and early Eocene (Rettalack, 2001) (I should point out that some other sources such as Berner and Kothvala, 2001 disagree with this conclusion on high CO_2 levels at the time). Already by that stage in Earth history, most of the modern families of flowering plant trees were already abundant, and modern families of insects had appeared—this is important because it was a world that was in many ways analogous to our own. It was also a much warmer world, perhaps hotter than the most extreme greenhouse effect scenarios of the next few thousand years: tropical and subtropical vegetation extended far beyond its present limits: there were even palm trees growing in the high Arctic at one point. Surely, If there were high CO_2 levels we might expect any effects on tropical forest species richness to show themselves—in a sense this was a "dummy run" for the future. We do not have many fossil floras from the rainforests of that time, but a good series of sites has been studied in careful detail for the American tropics (Jaramillo *et al.*, 2006). It turns out that instead of this being a time of extinction in tree floras, in the Eocene tropical tree diversity reached a high peak—and one which has never been equaled since (Chapter 3). This looks like reassuring news; for whatever other problems high CO_2 may cause, we have no evidence that the direct CO_2 effect resulted in a mass extinction in the tropical rainforest. If anything, high CO_2 seems to favor the generation and maintenance of very high species richness. But this reassurance depends on the assumption that that phase around 50 million years ago really did have higher CO_2 than at present—and that is something we cannot yet be entirely sure about.

7.7 THE "OTHER" DIRECT CO_2 EFFECT, ACIDIFICATION OF THE OCEANS

It is unlikely that increasing CO_2 levels will have much effect on the photosynthesis of marine plants. This is because there is already so much dissolved CO_2 available to them in the seawater, that any extra amount does not make much difference to photosynthesis. Another reason that more CO_2 will not increase the productivity of plants in the sea is that most of the oceans are very short of nutrients. When nutrients are heavily limiting, any potential increase in photosynthesis is not going to make much difference to the growth rate of plants, because they do not have the minerals that they need just as much as carbon in order to build new cells.

However, there is a more insidious effect of increased CO_2, one that could be far more damaging to marine life. Many marine organisms rely on calcium carbonate shells to provide support for their cells and tissues, and in protection. They range in size from tiny single-celled plankton, to large invertebrates such as sea urchins and crabs. Reef-building corals, which form so much the shallow-water species richness of the tropics, rely fundamentally on their carbonate shells for support and protection—the reefs themselves are built of the skeletons of dead coral skeletons. As the amount of CO_2 dissolved in the seawater increases, it is expected to make it more difficult for these marine organisms to grow and maintain their calcium carbonate skeletons and shells. This is because CO_2 dissolved in water is a weak acid that tends to etch away and dissolve calcium carbonate. The more CO_2 in the water, the more acidic it is, and the more trouble these marine life forms may be in.

There can be no doubt at all that each year the oceans are becoming a little more acidic. It is a simple prediction of physical chemistry: as CO_2 levels climb in the atmosphere, more CO_2 will be dissolving in the ocean water, and making the water more acidic. The increase in CO_2 in the atmosphere since the start of the industrial evolution should already have produced a small change in the pH of the surface layer of the oceans, from about pH 8.179 to pH 8.104 (a change of −0.075) between 1751 and 1994, for instance (Figure 7.4). Given the trend of increasing CO_2 in the atmosphere, it looks like the pH of the world's oceans will decrease by somewhere between 0.3 pH and 0.5 pH units by 2100 (Orr *et al.*, 2005).

What will this do to marine life? The potentially devastating effects of this eating away by CO_2 have been demonstrated in laboratory tank experiments (Royal Society, 2005). In many experiments, marine organisms from a variety of groups that use carbonate shells or skeletons (e.g., corals, bivalves, single-celled plants such as coccolithophores, and single-celled protozoans called foraminifera) showed reduced growth or erosion of their skeletons. The effects seem to get a lot worse

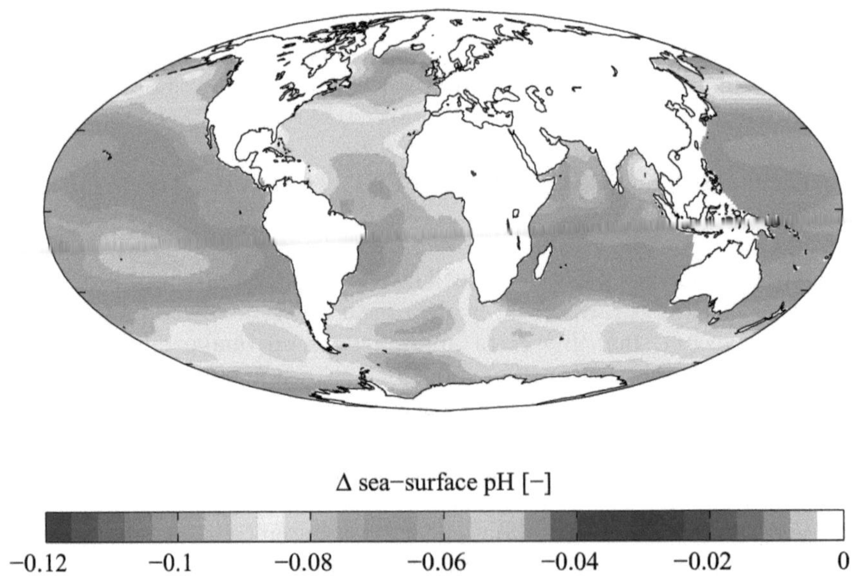

Figure 7.4. Estimated change in pH of ocean surface waters as a result of atmospheric CO_2 increase between the 1700s and the 1990s (source: "Plumbago", WikipediaCommons). *See also* Color section.

when marine organisms are exposed to a concentration of CO_2 dissolved in seawater that corresponds to an atmosphere of 1,000 parts per million CO_2 (a level which may be reached by sometime during the next century). This is where some organisms reach the point where they become completely unable to lay down calcium carbonate. Partial or complete loss of the skeleton is a disaster for most of these species, for their whole functioning depends upon having a carbonate skeleton as a part of their body. For example, single-celled plants known as coccolithophores (whose relatives laid down the chalk deposits in the late Cretaceous: Chapter 3) collapse in on themselves and die when they are unable to make the calcium carbonate plates that surround each cell. It is much the same with corals when they become unable the lay down a skeleton—the coral polyps become unable to feed, and quickly starve. Even if they were able to function and feed, many of these organisms would quickly be picked off by predators if they lacked their hard skeleton.

It is not hard to see that this has the makings of a biological catastrophe. Calcium carbonate–containing plants such as coccolithophores form part of the basis of many marine food chains. Many of the single-celled protozoan predators such as *Globigerina* that feed off these algae also have carbonate shells. There are thousands of species of bivalve mollusks, all secreting calcium carbonate shells, and thousands of species of hard corals. Many crabs, lobsters, and a host of other marine invertebrates (including bryozoans, brachiopods, starfish, and sea urchins) rely on skeletons that are made at least partly from calcium carbonate. If high CO_2 drives all of these species extinct, as it seemingly threatens to do, the effects will also cascade on down through food chains. It is impossible to estimate how many other species will disappear along with the ones that were directly affected, because they relied on these for food or shelter. The result could truly be a marine mass extinction, comparable with the worst in the geological record.

On the other hand, maybe things will not be as bad as these experiments suggest? It is important to bear in mind that laboratory experiments are not exactly the same as the real ocean, and organisms in less confined surroundings might be more resilient in terms of forming their carbonate skeletons. Certainly, not all species show a decrease in carbonate deposition with the sorts of CO_2 levels they are likely to be exposed to in the next 100 years—and some even show an increase in carbonate deposition!

In search of some perspective on what will really happen as a result of ocean acidification, it is worth looking back to the geological past when CO_2 levels may have been much higher than they are today. For example, in the Paleogene around 50 Myr to 55 Myr ago (according to Rettalack, 2001, for example), and back into the Mesozoic, CO_2 is thought to have been higher than it is now—often several times higher. Berner and Kothvala (2001) have come up with an independent set of estimates which do not show the same peak in CO_2 for the Paleogene, but agree that CO_2 was often very high during the previous several hundred million years. So there is a reasonable case to be made that CO_2 has been very high—several times higher than now—in the past.

What we see from these past times does not point to the disaster that laboratory experiments would suggest. The times of high atmospheric CO_2 were not times of mass extinction of carbonate-secreting organisms; in fact they often seem to diversify through these high CO_2 periods. The only times when such marine life forms went extinct *en masse* were the times when pretty much everything else was badly affected as well, such as at the end of the Cretaceous. Some have made a case that the end-Cretaceous boundary was a time of high CO_2, but so many other things were happening with the environment right then (Chapter 3) that it is hard to pinpoint any direct CO_2 effect in the ocean waters.

Why then is there a contradiction between many experiments in the laboratory (indicating strong effects of raised CO_2), and what happened in the fossil record (when marine life does not seem to have been very much affected)? As I have already mentioned above, a lab experiment is not necessarily much like the real world. If something in the simulated conditions is not quite realistic, that could invalidate the results of the whole experiment. Another possibility is that the geologists have got it wrong, and CO_2 levels were never really much higher than they are at present in the past several hundred million years. However, it is difficult to ignore various different lines of evidence which suggest that CO_2 really was a lot higher during at least some times in the past.

Another explanation why high CO_2 in the deep past may not have been associated with extinctions is that the organisms back in these times had evolved some sort of mechanism that allowed them to continue making calcium carbonate shells, even at high CO_2 levels. If the increases in CO_2 in the geological past occurred slowly enough, organisms might have been able to adapt gradually without extinctions occurring. But this would not offer much reassurance for the present-day world—a sudden rise in CO_2 on the time scale of a couple of centuries is certainly not long enough for the right genes to cope with it to arise by chance mutation and then spread.

So, as it looks at present, there is at least the possibility of a mass extinction of marine life over the next couple of centuries, resulting from the direct CO_2 effect. The case for this occurring is not overwhelming, but it is robust enough to be taken very seriously. Given how devastating it could be for biological diversity and the aspects of human welfare that depend upon marine ecosystems, there certainly ought to be a lot more research on this.

7.8 INTRODUCED SPECIES

In the past few thousand years, as humans have begun to travel and trade over vast distances, they have often brought living organisms along with them. Sometimes these animals and plants were taken as goods to trade, sometimes as pets, or useful crops and domestic animals for the farm. Often they came across as stowaways, hiding in the hold of a ship or in a crate of goods, or as seeds or molds among the roots or leaves of plants. When they reached their destination and escaped into the wild, they were breaking down the barriers of millions of years of separate history. Isolated by water or mountain ranges, land organisms in different parts of the world have often evolved along different evolutionary paths. There is an analogous situation with aquatic species, where land may instead form the barrier between organisms living on different coasts of a single landmass. For a species living in shallow marine environments, the open ocean itself may be an insurmountable barrier to other similar habitats on the far shore of another continent. Different species, genera, even completely different families and orders have appeared in different places. Often, ecological communities are assembled in strikingly similar ways, but with different lineages playing much the same role. One example is the native fauna of Australia, which evolved a whole range of marsupials (pouched mammals) that resembled particular mammals of the old world in appearance and lifestyle, despite each having evolved from unrelated groups of ancestors (below).

Throughout geological time, since life on Earth first evolved, organisms must occasionally have made it across barriers that had stood for millions of years, colonizing new lands and new ocean basins. Sometimes it was pure chance—a lucky dispersal event on a raft of drifting vegetation that was washed out to sea, for example. Sometimes it occurred when drifting continents collided, spilling their flora and fauna onto a new landmass. Invasions such as these must often have upset the ecological balance and driven some species extinct as a result of competition or predation by the new arrivals. But these were rare events, significant only on geological time scales. What humans have done is to speed up the pace of such events enormously, probably by several orders of magnitude. Over the past few thousand years, humans have brought about ecological disruption on a global scale, simply by bringing along thousands upon thousands of new species and letting them go in places where they do not come from. Already many native species of animals and plants have been lost as a result of introductions (Chapter 5), and we can see many cases where more extinctions are likely to follow.

It is not clear when the first extinction occurred as the result of an artificially introduced species, but one candidate is the loss of the thylacine and the Tasmanian native hen from mainland Australia as a result of the human introduction of the dingo, which is an Asian dog (Chapter 5). However, this was not a complete and true extinction, as both survived until recently on the nearby island of Tasmania—with only the Tasmanian native hen now still hanging on. Perhaps the first complete species extinctions that resulted from introduced species (and not from human hunting or habitat burning) occurred in the Pacific over the past 3,000 years as Polynesians spread between remote islands, carrying the Polynesian rat with them as a convenient source of protein. Escaped populations of these rats have been suggested as playing a key role in the loss of hundreds—perhaps even thousands—of species of ground-nesting birds that had evolved on isolated Pacific islands (Chapter 5).

In more recent history, a burst of extinctions likely occurred after goats were turned loose on the island of St. Helena in the mid-Atlantic, devastating the daisy trees that had evolved there without defenses against grazing mammals (Chapter 5). From around this point on, the pace of extinctions as a result of introduced species seems to pick up—or at least we have good records of them occurring. There is the example of the dodo on Mauritius, probably finished off by ships' rats. On Australia, some species of small rabbit-like marsupials seem to have gone extinct as a result of predation by introduced foxes, and competition by introduced rabbits. The Indian mongoose eliminated many of the native birds of Hawaii that had managed to survive into recent times. And in the western Pacific, the island of Guam lost most of its endemic birds to the brown tree snake after World War II (see Chapter 5).

I have mentioned these examples again because they illustrate how devastating introduced species can be. Around the world, there are many examples of extinctions that have already occurred due to introduced species, but it is possible that we have not yet seen the worst of it. In many places we see what looks like an extinction event unfolding, even if many of the species involved have not yet actually gone extinct. One of the worst instances is in Lake Victoria in the African Rift Valley, where already several tens of species of endemic cichlids may have gone extinct due to the Nile perch (Chapter 5).

7.8.1 Argentine ants and the fynbos

In South Africa, the species richness of the fynbos vegetation (with its thousands of endemic heathers, shrubby proteas, and bulb-forming restionids, see Chapter 4) is under threat from a tiny introduced ant. The Argentine ant (*Linepithema humile*) is pushing out various species of native ants, possibly condemning many of them to extinction too. Their disappearance may have an insidious effect on the whole ecosystem, due to the interdependence of native ants and fynbos shrubs. Around 20% of fynbos plants (including 50% of the proteas) have tiny pieces of fat, known as "oil bodies", attached to their seeds. The native ants gathered these seeds and took them into their nests several meters away, where they would prise off the oil body. The seed would then be discarded around the entrance to the nest, on top of finely tilled soil excavated from the ant burrows and all nitrogen-rich waste from the nest. The result was that the seed found a perfect seed bed in this very nutrient-poor environment. Now, it seems, the seeds of many fynbos species are laying unclaimed on the soil surface, because the Argentine ants do not gather them. Left out on the soil, the seeds generally die or are eaten by other creatures. It is quite possible that in the long term, this will affect the replacement of shrubs in this very nutrient-poor soil, and also the chances of each species reaching small patches of suitable habitat where it may survive without being outcompeted by its neighbors. Caroline Christian of UC Davis (Christian, 2001) studied the situation in the fynbos while it was being invaded by the Argentine ant, and found that some of the proteas whose native disperser ants have been pushed out do indeed seem to be getting rarer. The key difference occurs when these shrubs are unable to reestablish themselves from buried seeds after a fire which has knocked back the vegetation.

7.9 THE AMPHIBIAN DECLINE

All around the world in recent decades, a large proportion of the world's frogs have vanished. Many species have become much less common, and others have almost certainly become extinct. The "amphibian decline", as it is called, has been particularly strong in the tropical Americas, Australia, and New Zealand (Stuart *et al.*, 2004). However, it has been noted in all regions of the world where amphibians occur. Partly this decline may be due to the familiar patterns of habitat destruction and pollution around the lands of the world—draining of swamps, runoff of fertilizers, and contamination with pesticides. But there also appears to be something more insidious, something more unusual. It seems to sweep in as a wave, eliminating the frogs in a particular place within a few months at the most. Many of the areas from which the frogs have vanished were pristine habitat, apparently unaffected by human interference. By chance, the decline affected some areas that had been studied in detail just before the amphibian decline occurred there. For example, at one rainforest site in Central America, the researchers left a thriving and species-rich community of amphibians, and when they returned a few weeks later the forest was practically devoid of frogs. It is looking more and more like the main cause of many disappearances of frogs is a disease, caused by a type of chytrid fungus, *Batrachochytrium dendrobatidis*. The fungus has been isolated and studied, and found to cause the death of frogs that are exposed it. In many places where the amphibian decline has occurred, once the fungus appears, the frogs disappear shortly afterwards, sometimes being observed dying of it in the wild. It seems to kill them by infecting their skin and preventing them from getting oxygen in through it—since frogs rely on gas exchange through their moist skin. So far, 120 species are considered to have gone extinct as a result of the "amphibian decline", and many of these seem likely to be due to the fungus. Among the species lost is the brilliantly colored golden toad *Bufo periglenes* of Central America. Its color was so striking that Jay Savage, the first biologist to find it, really thought that it must have been dipped in Day-Glo paint by someone as a joke! It vanished some 30 years after it was first discovered. Also, there are the strikingly patterned harlequin frogs of the mountains of Central and South America—many of these are now thought to be extinct. The only reasonable explanation for such a widespread and simultaneous decline as a result of this fungal disease is that it is an introduced species, spread around the world by humans. Where it comes from is not clear, but presumably it is derived from the Old World where the signs of an amphibian decline are much less dramatic. The original host of the fungus seems to have been the African clawed frog (*Xenopus laevis*), which it does not harm, and it is possible that it was brought to other parts of the world by the export of clawed frogs, which are often used in school anatomy studies. It is ironic that the study of biology may have caused the extinction of so many species.

So far the amphibian decline has not reached all parts of the tropics, but there is a serious possibility that most of the amphibians of the New World and Australian tropics (e.g., the many beautiful species of poison arrow frogs of the Americas) will go extinct as a result of this. Recently (in the last few days as I prepare this book for the publisher), there has been a glimmer of hope in that one species of Australian frog and another from Central America, previously thought lost to the fungus, have been found surviving. They may have acquired resistance to the fungus, which would be a very encouraging sign.

7.10 TREE DISEASES

Trees in various parts of the world suffer analogous threats from introduced diseases and pests. The breaking down of geographical barriers by humans has led to certain tree diseases and insect pests spreading from one region of the world to another. If anything, these have the potential to be even more devastating to the world's biological diversity than the frog-killing fungus, for they can destroy not only one group of species but the entire habitat. If the frogs die, the forest is merely emptied of frogs (tragic though that is in its own sake). If most of the trees die, the forest itself no longer exists and all of the species within it lose their habitat.

This exchange of tree diseases has occurred most frequently between the three main temperate forest regions of the northern hemisphere. The scene for this disaster was set tens of millions of years ago, when the trees of the Arcto-Tertiary flora evolved and spread all across the northern hemisphere (Chapter 3). As the world cooled and became more arid, it became fragmented into three main temperate forest regions: eastern Asia, Europe, and eastern North America. Many of the same genera are found in all three places: for instance, the oaks (*Quercus*), the ashes (*Fraxinus*), and the maples (*Acer*), to name just a few of the scores of genera that these regions have in common. After these three regions were broken off from one another 20 or so million years ago, evolution took different paths. Tree species took on different appearances and diversified. There was also another less conspicuous sort of evolution going on: between trees, and the pests and diseases that attacked them. In each region there arose certain fungi, bacteria, viruses, or insects that specialized in attacking each tree species. Over millions of years these reached a rough sort of balance with their host trees: the host trees were to some extent susceptible to these little enemies, but not usually killed because they had evolved mechanisms to defend themselves. The trouble started when humans began to transport goods and materials—including young trees—between these three regions. The roots and leaves of imported tree saplings could carry the diseases which had attacked them in their home range, and the insects and pathogens could also hide in packing material, food, and all manner of other stuff transported between the continents.

The first big disaster involving a tree disease brought between continents was the loss of the American chestnut (*Castanea dentata*). In Asia, a fungus known as chestnut blight (*Endothia parasitica*) infects the Chinese chestnut (*Castanea mollisima*). The Chinese chestnut is not especially harmed by this fungus, and it often carries it on its leaves—which is probably how the chestnut blight reached America, on the leaves of imported saplings of Chinese chestnut. The disease proves to be lethal to American chestnuts, killing each tree over a series of years. First hopping across from Chinese to American chestnuts at the Bronx Zoo in New York in the early 1900s, the chestnut blight spread out from the city and into the forests of the eastern United States. The American chestnut had been one of the most abundant and widespread trees in the eastern forests, and one of the most important economically. Yet at the present point, it is essentially extinct within its native range. A few young chestnut shoots continue to grow as suckers from old stumps here and there, but as soon as they reach any appreciable size the fungus (which lives on in their roots) breaks out again and attacks and kills them. Eventually even these few survivors will be shaded out of the forest, and that may well be the end of the American chestnut in the wild (Burnham *et al.*, 1986).

Another fungal disease from Asia, elm disease (actually three species of the fungus *Ophiostoma*), has been steadily reducing the populations of the various species of native elms (*Ulmus*) in the eastern U.S.A. Over the past century or so, more than 90% of the American elms have died and the disease continues to take its toll, picking off the survivors one by one. It is not certain how far the process will go, but the extinction of the various American species of elms is a possibility.

In the past 15 years or so, ominous new threats have broken out. Sudden oak death is a disease caused by a fungus (*Phytophthora ramosum*), which may have been brought from Europe on cultivated rhododendrons. It has been steadily killing evergreen oaks throughout California and shows no sign of slowing down. It is feared that the same fungus could eventually make the leap across to the eastern U.S.A., where some 35 other species of oaks fill the forests. Experiments indicate that the eastern oaks are mostly just as susceptible to the fungus as their western cousins, so if it were to reach the east it would presumably devastate the forests throughout the region, perhaps eliminating entire species of oaks.

One species whose existence is very clearly threatened by an introduced disease is the native *Torreya* in the southern United States. It has suffered badly from an introduced fungus that has devastated its populations. Always having had a restricted distribution, it is now down to a few score individuals in its native range (although fortunately populations planted in other areas are doing better).

In the past several years an insect called the emerald ash borer (*Agrilus planipennis*) has arrived in North America from China, where it feeds off the various native ash (*Fraxinus*) species. The emerald ash borer seems to have reached the U.S.A. in goods or packing material from China at the port of Chicago. From there it has spread out into the eastern North American forests, killing every ash tree in its path. The extinction of the six species of ash trees native to the region is a distinct possibility.

Another wood-borer, the Asian long-horned beetle (*Anoplophora glabripennis*), has reached the eastern U.S.A. several times by stowing away in materials from China. It kills American maples (*Acer*), buckeyes (*Aesculus*), and poplars (*Populus*), filling them with holes until they collapse. Maples—along with the oaks—are the dominant trees of the eastern North American forests, and there is the distinct possibility that we will lose them all, as well as the various other genera that this beetle attacks. After breaking out and then being contained by ruthless felling of street maple populations in New York, the long-horned beetle has now reappeared across the other side of the Hudson River in New Jersey, and also as a very large outbreak in Worcester, Massachusetts. My colleague Gareth Russell has been working on a model to help predict its spread and guide the authorities to areas in New Jersey where they should keep a close watch for it. So the future of maples in North America is partly being determined by events a few miles from where I sit writing in Newark, New Jersey: and also perhaps by the man who works in the office next door to my own.

And so it goes on. The great hemlocks (*Tsuga*) of the eastern U.S.A. are being steadily eliminated by an aphid from Japan, the woolly adelgid. An introduced canker has broken out on beech (*Fagus*). European sycamores (*Platanus*) are being destroyed by a fungus that has come over from American sycamores. The elm in Europe is already rare, as a result of the effects of elm disease. One wonders where it will all stop. The likelihood is that the diseases will just keep on coming, one after another. Global trade is increasing, and there is no reason to think that we have exhausted the supply of pests and diseases which could devastate the trees of other continents.

As far as we can tell, the flora of the tropics is rather less susceptible to introduced pests and diseases. At least there are no recorded cases of tropical trees being wiped out in their native range by an introduced species. Perhaps with the constant attack by so many different kinds of enemies in the tropical forests, the defense mechanisms of tropical trees are better primed to cope with new enemies? There may also be defense in the sheer diversity of tropical trees. Problems with introduced pests do tend to occur when closely related host plants are living on different continents. The astonishing diversity of tropical trees, and their long separation since the Paleogene some 50 or more million years ago, may mean that they each tend to be too different from the plants on other continents to be attacked. The curse which the northern temperate trees have is that because of their more recent common history, they mostly have close relatives on other continents that carry diseases which can infect them. On the other hand, if a species of tropical tree was being attacked and died out, would anyone really notice that it had vanished? A species of tree in the tropical rainforest is only one of many thousands, mostly living at very low population densities. Unless a botanist specifically went looking for it, it would probably never be missed, and even then its failure to turn up in collections would probably be put down to bad luck in the search—not the fact that it was now extinct.

7.11 HABITAT CLEARANCE

Throughout the tropics, forest is presently being cleared, for reasons which vary from place to place. In many areas the loss of forest is due to small farmers expanding or establishing their farms. In other cases, large corporations are clearing land—planting valuable export crops such as coffee, cocoa, or oil palm (Figure 7.5). Around 6 million hectares of rainforest are being lost each year, according to a 2002 estimate by Frederick Achard and colleagues.

Such wholesale loss of the forest is easy to spot from a satellite image or aerial photograph, but there is another more subtle form of human damage to forests that is harder to quantify, but nevertheless may be almost as important in terms of its effects on biological diversity. This is timber extraction, a major industry in many tropical countries (Figure 7.6). To satisfy the world's demand for both high-quality and low-quality hardwoods, logging companies move in to areas of old-growth forest, clearing tracks for the logging trucks and caterpillars to pass along. Big trees are felled and dragged out of the forest, leaving large gaps in the canopy and a tangled mess where

Figure 7.5. Tropical forest recently cleared for pasture—Sabah, Borneo (source: Author).

Figure 7.6. Tropical logs awaiting processing into composite board, in a factory yard. These ones are from plantation-grown trees, but in many other areas the trees are harvested from old-growth forest—Sabah, Borneo (source: Author).

they fell and were brought out. It is hard to say for sure what this does to the long-term species richness of a forest, but there can be little doubt that it reduces it. Another more destructive form of clearance is wholesale clear-cutting of the forest for pulp or chipboard. Everything woody from the forest is simply chewed up for further processing. After this sort of exploitation, there may not be much left that could resemble the original old-growth forest. According to the satellite-based estimates of Achard *et al.* (2002), as well as the forest that is completely removed, a further two-and-half million hectares of forest is visibly degraded each year by these sorts of processes.

While trees may grow back after intensive logging, they are typically not the original species of the old forest, but different types suited to colonizing open ground, to growing in bright sunlight and in soils enriched by the rotting roots and leaves of the trees that were there before: such genera as *Macaranga* in Asia, and *Cecropia* in the tropical Americas. These "weed trees" were always abundant anyway throughout each region, so conserving them is not a priority, and the forest that they form lacks the species richness and complexity of the old-growth forest. For example, along with the "weed trees" go "weed birds" and "weed amphibians": species that do well in the recently regrown forest, and which in any case are widespread and common across the region. The specialized birds or frogs of the old-growth forest, with their intricate web of connections with the ecosystem, are no longer present. It will be many decades, even centuries, before a regrowth forest once again approaches the species composition and richness of the original old-growth forest—and even that depends on there being surviving pockets of primary forest for those species to recolonize from.

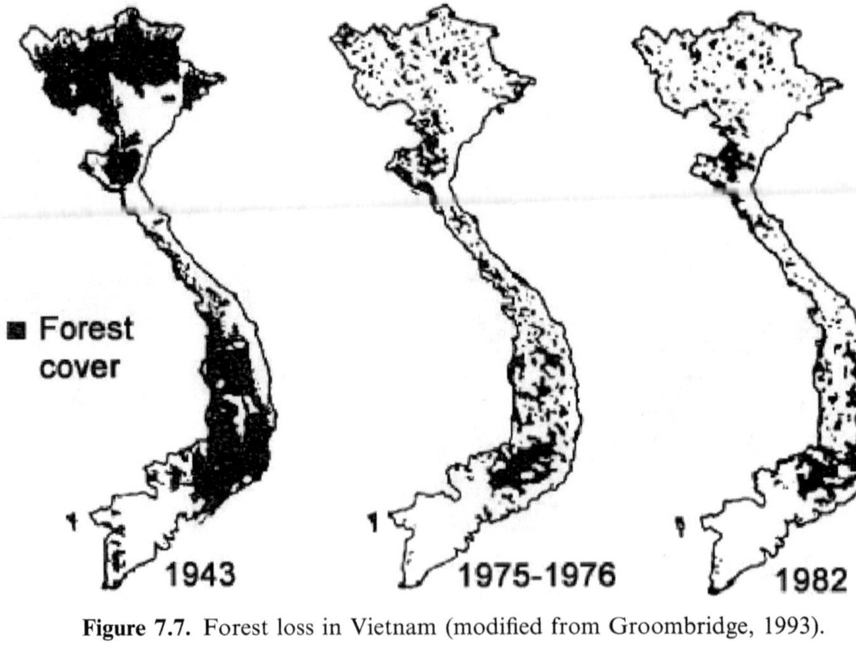

Figure 7.7. Forest loss in Vietnam (modified from Groombridge, 1993).

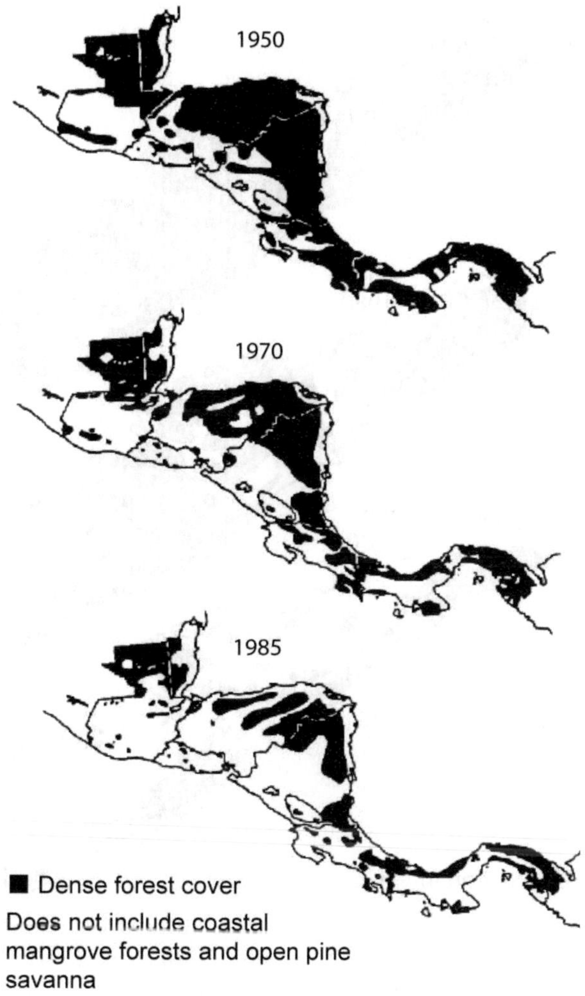

Figure 7.8. Forest loss in Central America (modified from Groombridge, 1993).

To understand just how much of the world's tropical forest cover has been lost over the past few decades, take a look at the maps shown above and below (Figures 7.7–7.9). In many of these areas the forest loss has now been halted or even reversed, by careful replanting of forests with a mix of native species. More worrying right now are particular places where the loss is continuing unchecked and may even be accelerating. In Madagascar (Figure 7.9), only scattered fragments of the original forest now remain, diminishing each year as farmers clear more land to feed their families. The Atlantic rainforest of southeastern Brazil has almost gone, with only 1% of the old-growth forest remaining. It stands to reason that many species must still just be hanging on in those remaining fragments, but that as these pieces of forest too are mopped up, the rate of extinctions will increase enormously. Even if the destruction finished right now, many species left isolated in the remaining forest

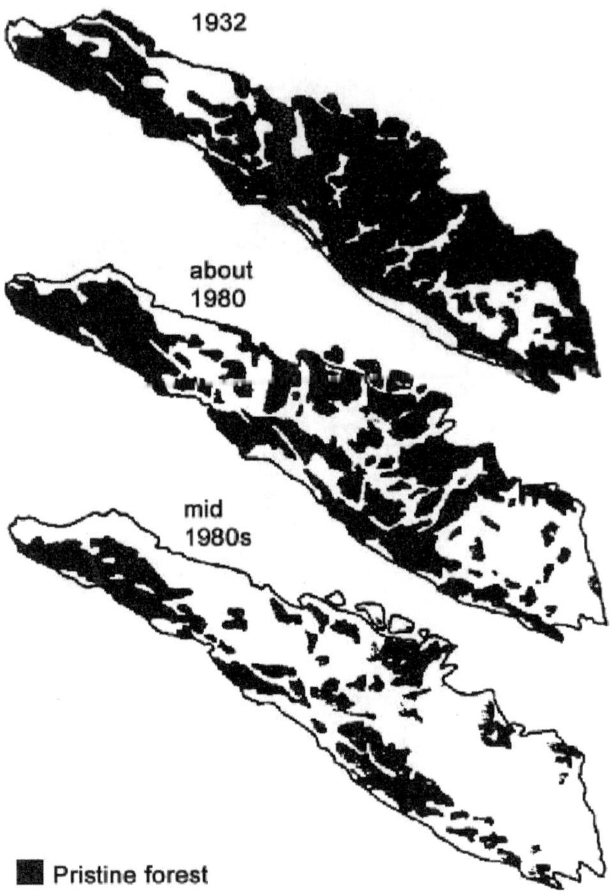

Figure 7.9. Loss of old-growth original forest in Sumatra, Indonesia (modified from Groombridge, 1993).

fragments may be doomed to extinction in the longer term as the result of random fluctuation in populations.

Brazil still has a huge amount of forest left in its Amazon sector, but this has been steadily eaten away over the past several decades. At first, in the 1970s and 1980s, large government programs deliberately promoted the clearance of forest and its conversion to farmland or ranch land. More recently, with a rising sense of conservation awareness in Brazil—plus the economic failure of most of the land conversion programs—the government has emphasized protection of the forest. There was a shock recently when it was found that forest clearance in the Amazon is now back up to levels that last occurred during the 1990s, due to the continued migration of landless settlers and extensive illegal logging in the forest. The Brazilian government has sent troops in to enforce the laws, though it remains to be seen what effect this will have. Despite the efforts to protect the forest, last year (2007) seems to have been worse than ever based on the latest satellite data.

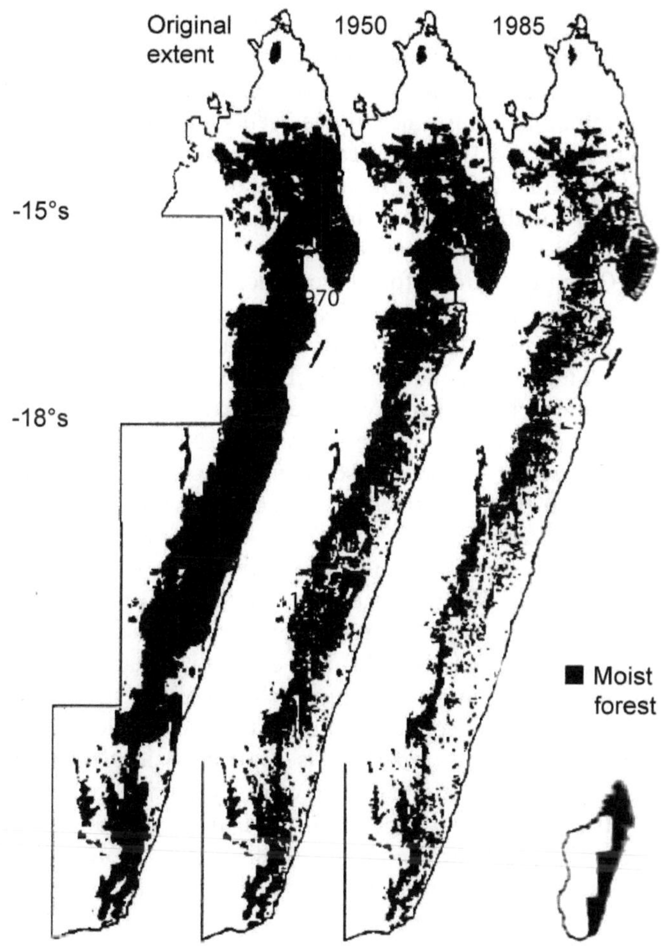

Figure 7.10. Forest loss in Madagascar (modified from Groombridge, 1993).

Most of the clearance of old-growth forest in the mid-latitudes and northern latitudes has now stopped. This is partly because there is not very much of the original forest left to log after thousands of years of exploitation, and partly due to the efforts of conservation agencies in buying up land or lobbying governments to pass laws to protect the few remaining areas. Nevertheless, logging of virgin forest is continuing in Russia, where large areas in Siberia are being swept away by clear-cutting to supply high-quality dense woods from the slow-growing trees. However, in terms of extinction of species this is not very much of a threat. These boreal forests are poor in species, and those species which do occur there tend to be very widespread. Hence, it is unlikely that this forest destruction in the northern latitudes—reprehensible though we may feel it is—will in itself be responsible for the loss of any species.

There are broader effects of forest destruction, beyond the loss of the forest species themselves. CO_2 is released from the soil and the rotting wood left behind, and CO_2 of course is the source of much concern for world's species diversity at present (see above). There are also the influences that habitat clearance itself can have on the broad-scale climate of a region, feeding back on the few areas of intact habitat that remain. For instance, it has been suggested that habitat clearance in the tropical lowlands of Central America contributed to the extinction of various frog species in the still-intact forests of nearby mountains—because the reduction in evaporation of water from the lowlands meant less cloud and drier conditions in the mountains (see the description of this effect in Adams, 2007).

Given the rate of forest clearance in the tropics, and given just how species-rich the tropical forests are, we can be sure that species are being lost every year. In addition to this, there are, for example, the ongoing losses of frogs due to chytrid fungi, and the decimation of cichlid fishes in Lake Victoria.

But exactly how many species go extinct each year? At the outset, it is important to admit that we cannot be sure of this. No red light blinks on each time a species is lost. Instead it happens unseen and unnoticed: often these will be species that have never even been described by science. However, it is generally accepted that at least several thousand species a year are being lost (Chapter 5).

Depressing though these losses are, we have every reason to think that this is only the beginning. There is the effect of the final mopping up of the last fragments of habitat in some areas where they are already badly fragmented (e.g., in the rainforests of Madagascar). This will surely extinguish many species that were up to this point just hanging on in the remaining islands of forest. And as both the greenhouse effect and direct CO_2 effect begin to bite, it is reasonable to expect an increase in the rate of extinction. In addition there is the future loss of species due to introductions—including introductions that have not yet occurred.

Is all of this inevitable, or is there anything that can be done to save what we still have? Chapter 8 will consider some of the opportunities that are available to us, before so much more is lost.

8

Holding on to what is left

Despite all that we humans have done to the world, a wonderful legacy of biological richness remains. Almost everyone agrees that it would be a tragedy to lose more of this, and yet, as I explained in Chapter 7, it is quite obvious that the threats are intensifying.

In some parts of the world, many species seem to be clinging on in the last fragments of old-growth forest, in mainly deforested landscapes (e.g., in the Madagascar rainforest and the Atlantic coast rainforest of Brazil; Chapter 7). If the forest loss continues, we can expect that many species which had just about been surviving will finally go extinct. Some countries (e.g., China) are industrializing rapidly. Despite the economic growth, there is also a cost in terms of pollution of air and water which is affecting not only humans but also wild species. Because of the sum total of fuel burning and deforestation of all the countries in the world, greenhouse gases are accumulating in the atmosphere and changing the global climate, and even the composition of the oceans (Chapter 7).

If we want to hold on to what we have, despite these threats, how should we best go about it? There are various opinions, none of them mutually exclusive. However, they do tend to involve some difficult choices. Most of them are expensive and many will surely make life inconvenient, for somebody at least. Conserving species against extinction involves not just tackling scientific questions but also dealing with formidable economic, political, and cultural challenges.

8.1 CONSERVING HABITAT

One of the most widely used means of ensuring the survival of species is to put aside protected areas, preserving them against habitat change and exploitation by humans. This is a commonsense method with a long history. In Europe it has to some extent been used since medieval times, seven or eight hundred years ago. At that time, the all-powerful monarchs declared large swaths of land as hunting preserves. These were intended to maintain abundant populations of game animals for the sport of aristocrats: few people imagined in those days that any of God's creatures could go completely extinct as a result of human actions. These medieval hunting preserves often survive today as parks (e.g., the famed Sherwood Forest in England, which is reputed to have been the home of Robin Hood). Habitat preserves were also set up for strategic purposes. Since warships needed huge numbers of large trees in their construction, the naval power of a country rested partly with its forests. Large areas of trees were planted and preserved against cutting—such as the Landes of south-western France, planted with oaks by Napoleon to ensure a future supply of wood for building warships. Thus most of these earlier preserved areas had purely practical purposes: ensuring the continued supply of something that was needed either for sport, for making money, or for warfare. There was no sense in conserving natural areas as a habitat for threatened animals and plants that might exist within them. All this reflected the old view of nature as something existing purely to be exploited, or even something threatening which should be tamed and made to serve some practical purpose, whenever possible.

Yet there are a few early examples of attempts to try to preserve a particular rare species: not for hunting, but because it was valued for its own sake. In medieval Poland, kings passed laws to try to conserve the last truly wild population of cattle—the aurochs (*Bos taurus primigenius*)—in the Jaktorowski Forest. Royal hunting of the animals stopped, and poaching was made punishable by death. This was a brave attempt, destined ultimately to fail, but as far as we know it was one of the first of its kind—a set of laws and a designated nature reserve, set up for nature's sake. The European bison, or wisent (*Bison bonasus*), was luckier and is still with us. In the mid-1500s the last remaining herds of the European bison were protected by King Sisigmund of Poland against poaching, again under penalty of death. By the 1600s a large reserve had been set up for them in Bialowieza Forest in Poland which allowed to population to grow to nearly 2,000, although despite this protection they were decimated in the instability and lawlessness of a succession of European wars, and at one point reduced to only a handful of individuals—from which the population has now successfully been bred up.

It would be several centuries more—the 19th century—before the next nature reserves that we know of: Yosemite and Yellowstone National Parks in the U.S.A. Even then, these were first established to preserve the scenic beauty and majesty of nature, rather than to prevent anything from going extinct. The next modern nature reserves to be set up with the express aim of saving species from extinction were established in the early 20th century in Europe and North America. An early example was a reserve established in 1907 for the American bison (*Bison bison*) on Elk Island National Park in Canada, using surviving herds that had been kept in semi-captivity on ranches. At that time, this species—like its European cousin—hovered precariously on the edge of extinction, and the Elk Island reserve played a major role in saving it.

This general model for preservation of species was followed many times over the next few decades of the mid–20th century. Some of the early attempts seem clumsy in retrospect, because modern-day ecological knowledge that would have helped had not yet developed or become widespread. I recall older acquaintances from my undergraduate days, who had been part of the first wave of modern conservation in the 1940s and 1950s in southern England, telling me how nature reserves were set up to preserve populations of rare orchids growing in the limestone chalk grasslands. In these earliest attempts, patches of turf containing these orchids were simply fenced off against sheep grazing, the assumption being that nature would look after itself. Against expectations, many grasses and other common species in the grassland now grew tall and smothered out the orchids that were supposed to be protected. Instead of benefiting from being released from grazing, the orchids suffered more competition. Previously, these orchids might have suffered grazing, but at least they were preserved from the worst effects of competition—for the grazing was keeping other species of faster growing plants in check. This illustrates what ecologists now call the "intermediate disturbance mechanism" for species coexistence (see Chapter 1). Too little disturbance is actually a bad thing for most species in a plant community. Nowadays, any nature reserve that includes grassland is always carefully managed to ensure that either grazing or fire is a regular occurrence.

In the 21st century, far more of the Earth's land surface has some degree of official protection against exploitation that might destroy species and their habitat. In some cases the main aim of setting up a reserve has been to protect just one particular species, often a cultural icon and an object of tourism. Examples include the reserves for the giant panda (*Ailuropoda melanoleua*) in China, and the tiger in several Southeast Asian countries. In many other cases, though, the reserve is set up for a whole assemblage of hundreds or even many thousands of species that will likely be under threat if habitat destruction continues.

On paper, reserves always look impressive, but the reality can be very different. They were often first declared in a sweeping gesture, as much for the sake of international approval as any genuine concern for nature. If a reserve does not have backing in terms of enforcement, by legislators or by rangers on the ground, it can be next to useless. For example, in many of the world's poorer countries there is not much government money available to train and maintain rangers who will prevent illegal poaching, or timber cutting or farming within a reserve. Sometimes government officials themselves seem to have sold logging concessions to their friends, within areas that were supposed to be protected. James and Green (1999) estimated that the world's existing major protected areas are underfunded by about $1.5 billion a year, money that is needed to help them hang on to their species richness effectively.

Even if a reserve is well protected at one point in time, this does not necessarily mean that it will stay protected forever. An economic slump can lead to funds for protecting it being cut back sharply. A civil war can mean a complete breakdown of authority and widespread poaching by troops or by local people, in the absence of any law enforcement. This situation seems to have occurred in recent years in central Africa, where civil wars have raged for several years. The mountain gorilla population in parts of the Congo and Rwanda has been devastated as a result, although recent reports suggest that the situation may be improving as government control over the reserve areas has been reasserted. On the time scale of the survival of species over centuries and millennia, all it takes is a temporary breakdown of authority once every few centuries, for a species to be lost from its last refuges. These are the sorts of events that in Europe led to the aurochs being lost, and the wisent almost to be lost, despite the efforts of the aristocracy to protect them. Most of the decline in populations of both species occurred during times of war.

At present, species are protected in reserves at several levels of national or international importance. There are local or provincial nature reserves, drawn up and administered on a fairly local level. In the U.S.A., for example, there are hundreds upon hundreds of state parks and state wildlife refuges. In the U.K., there are many small nature reserves owned and administered by private trusts, associations which survive off charitable donations from the concerned public. Working for an agency like this (the Worcestershire Wildlife Trust in England) in my early 20s I first cut my teeth as an ecologist, was paid almost nothing, but did very satisfying and worthwhile work. And then on a broader scale, there are national nature reserves or parks, a source of pride and prestige and set up and funded by central governments.

The setting up of global networks of reserves began in the 1970s with the Ramsar Convention (named after a key meeting in Ramsar in Iran, on the Caspian Sea coast) to protect wetlands which were being drained and polluted at an alarming rate. Later, a global network of biosphere reserves was set up to preserve a broader range of the Earth's biological diversity, focusing especially on "hotspots" (Chapter 4). In addition, overlapping with the biosphere reserves and Ramsar sites, but also taking in some nature reserves of their own, are world heritage sites (Groombridge 1993). In such protected areas, hunting and collecting are banned, and so is development which might threaten the species within. This top tier of international reserves—the "crown jewels" for conservationists—are government-run in each country, but they are jointly administered with the international bodies that helped to set them up.

8.2 INTERNATIONAL BIOSPHERE RESERVES AND WORLD HERITAGE SITES

International biosphere reserves and world heritage sites are usually areas that have something particularly special about them. Both of these networks of reserves are administered by the United Nations educational and cultural agency, UNESCO. As nature reserves, they have mostly been designated based on what was known of levels of species richness, or the presence of some particularly desirable and threatened species (e.g., the giant panda).

These reserves tend to be set up in biodiversity hotspots (Chapter 4), and examples include Lake Baikal and the Cape Fold Mountains. However, they are in some ways a mixed bag—some world heritage sites have nothing to do with nature (e.g., they are beautiful and historic cities). Also some biosphere reserves do not have exceptionally high levels of diversity compared with the rest of the biome in which they are located, but instead they offer a fairly accessible protected area that is representative of the region in general. They also offer the reassurance of being at least one protected area that is typical of the region, that could be maintained if all other similar habitats in the region were destroyed. As well as helping to preserve diversity, biosphere reserves and heritage sites also act as a magnet for international collaboration, for scientists coming to study the ecosystem. An example of a biosphere reserve that has more this sort of role—as a back-up and a center of research—is the Pasoh Biosphere Reserve in peninsular Malaysia: a place where I have done a fair amount of fieldwork and study.

Many of the international biosphere reserves are set up with a sort of internal structure: a series of concentric zones. At the core of the reserve is an area that is designated as pristine, and to which access is controlled and limited, usually to scientists working on research permits. Surrounding this core is a buffer zone of habitat that is as much as possible maintained in a natural state, except with laxer conditions. This serves as an extra area to help maintain populations of animals and plants, as a protection again the "edge effects" such as dry winds blowing into forests, and as a barrier against poachers or illegal loggers who might otherwise penetrate directly into the core area of the reserve.

The general philosophy of biosphere reserves and world heritage sites has been criticized on the grounds that they actually only take in a small proportion of the world's total biodiversity, and distract attention from the need to conserve what is left outside the reserves. As such they might give governments a sense of satisfaction and the aura of respectability, when they are actually doing very little to look after most of the diversity within their country's borders. Nevertheless, I am convinced that by acting as a focus for attention and action, they have done a lot of good towards protecting the world's biological diversity.

8.3 UNCERTAINTIES ABOUT SPECIES RICHNESS: A PROBLEM FOR CONSERVATION

As I explained in Chapter 6, there is a lot of uncertainty about how many species actually exist at present out in nature. When we say we want to preserve the diversity of the rainforests, we do not even know within an order of magnitude how many species we are trying to save! When we try to prioritize our conservation efforts into hotspots, there is the nagging doubt that we might have missed the most species-rich places because not enough collecting work has been done there.

On a far more local scale, uncertainties about species richness can also make it more difficult to choose the location of nature reserves wisely, to include populations with the largest number of species. The only way to find out the level of species richness in a potential reserve is to go out and sample it—but how much sampling is enough? If the sampling is at too small a scale or too cursory, we might miss the true levels of diversity that are present in that particular area of habitat.

As more and more samples are taken from a particular locality—more sweeps of a collecting net, or successive or larger areas of forest examined—we can expect more and more species to be added to our list. Generally, it is good idea to keep on sampling until the rising graph of numbers of species identified begins to level off (this plateau in the graph is known as the asymptote) (see Chapter 6). Then at least we can roughly say how many species there are of the type of organism we are sampling for, in the area of habitat we are studying. In the most species-rich habitats (e.g., in some tropical rainforest areas) it might well be necessary to give up the sampling before any asymptote is reached, because extra species keep on accumulating almost endlessly—far exceeding the time and the budget available to identify them all (Groombridge, 1993).

If one was to sample the plant species richness of a square meter of tropical rainforest, it would probably be quite low (easily exceeded by many temperate grasslands). Only as we increased the sample size would it become apparent that species richness keeps on rising steeply with successively larger samples, far exceeding anything found in a temperate grassland. Because of this effect of scale, cursory sampling of only a small patch of forest can likewise give misleading results. Secondary rainforest (recovering from damage) may seem to have a greater species richness than primary (old-growth) rainforest because its stems are smaller and packed together—which gives more trees overall to show up as different species (Gotelli and Colwell, 2001). Only as the sample area increases will we find that the primary forest is in fact far more species-rich than the secondary forest. Another way around this problem of the size of individuals and density of stems is to count the rate at which species accumulate per number of stems counted, not the size of the area sampled. This would clearly show from the start that the primary forest is far richer in species.

When comparing different areas as potential nature reserves, it is important to use closely standardized methods to compare their species richness. So, for example, for plant diversity we must compare results from sample plots of equal size, or (better) for samples of similar numbers of individuals (Gotelli and Colwell, 2001). Often the most helpful way to do the comparison is to plot graphs on the same axes showing how more species accumulate with successively larger sample sizes.

8.4 MINIMUM VIABLE POPULATION SIZE

Simply having a few individuals of a particular species and protecting them in a reserve is no way to guarantee that they will survive. If only a small population of—say—five or ten individuals is present to start with, then despite a perfect environment it could well go extinct soon after the reserve has been established. One of the main potential problems is stochasticity (Box 8.1), but the risk of the population being wiped out in sudden disasters is also important. The priority must be to try to get the population growing as quickly as possible to get it above these hazardously low levels. The level at which we can start to feel comfortable that the population stands a good chance of surviving is called the "minimum viable population size".

A population size of 30 or 40 seems like a good safe size for a population to most people. Yet the calculations of the effects of stochasticity suggest that if the population is unable to grow its numbers any further (e.g., due to lack of space on the reserve), its extinction will be more or less inevitable at some stage over the coming decades or centuries. While it is a good idea to get the population well above this minimum level, there are often practical constraints. In some reserves, for large animals especially, there is just not enough land available for a large population even if they do reproduce well enough to keep their numbers growing.

Box 8.1 Stochasticity

Part of the reason that small populations can be expected to die out frequently is the effect of chance, bouncing their numbers around. Of course the population only has to reach zero once, and that will be the end of it.

A very small population of say half a dozen individuals—whether tigers, orchids, or whatever—will be extremely susceptible to going down to zero, due to the effects of what is basically nothing more than bad luck. For example, imagine a tiger population of just six individuals in a nature reserve. Of three females of breeding age, one may happen to miscarry repeatedly. Of the two females which do give birth to live cubs, by chance for several years most of them die and the only ones that do survive are males, with no females that they could eventually mate with. It might then happen that one of the adult females gets sick and dies. If this run of misfortunes continues, the population might quickly end up reduced in size and also in being all males; a definite recipe for extinction. A run of chance events like this can knock down a small population, and the smaller it gets the more susceptible it is to further chance events—continuing a spiral down towards extinction. This is quite aside from the harmful effects of inbreeding in small populations.

Not all the bad luck that hits populations can be termed stochastic in this sense. Purely "stochastic" changes in numbers come from *within* a population, but in reality small populations are also likely to be wiped out completely by unpredictable and unfortunate events that come from the broader environment. For example, a forest fire or a flood might sweep through and kill off the whole of a small population. These external bad-luck events also act against small populations more severely than large ones. If the population had been larger, by chance a few individuals might have survived and been able to re-establish a population.

Bad luck can hit a small population from various directions, sometimes coming from within the population and sometimes from its general environment (see Box 8.1). One example of the way in which a run of bad luck can knock a population down is the story of the heath hen (*Tympanuchus cupido cupido*), which we considered in Chapter 3. Even starting from several hundred individuals well-protected on an island, this subspecies was driven to extinction by a series of unfortunate events including a fire and several hard winters, ending up in its final days as nearly all males. And this was an example of a population which actually had the advantage of starting out quite large, at around 2,000 individuals. Populations which start out much smaller can be expected to be at far greater risk.

These sorts of chance events affect all populations, and there is no doubt that a large population of—say—a thousand individuals is much better placed to survive the effects of a run of misfortune over a few years than half a dozen or 20 or 50 individuals. But bad luck need not happen very often to be sure of extinguishing a population: if it only happens once in several thousand years, the population is still gone forever. All of the good times do not matter, if bad times can wipe out a population so easily.

Another problem that enters into the conservation of small populations is the effect of inbreeding. All diploid species (species with two sets of chromosomes and thus at least two copies of each gene) carry many hidden copies of "bad" non-functional genes—or at least they will if they normally had high population sizes in the past. These "bad" copies are usually masked by the "good" copy on the other chromosome. The bad copies can exist almost like parasites, only exposed and harming the organism when by rare chance offspring inherit two copies of the recessive gene. If a population is shrunken right down to less than 30 or so individuals, eventually relatives will end up mating with one another and the chances of offspring ending up with two copies of a particular "bad" gene are magnified enormously. The effects of high death rates or sterility among offspring caused by inheriting these recessive genes could quickly wipe out a very small population.

The effects of inbreeding have been well known by humans for thousands of years; of brother–sister matings, about half of these children are in some way disabled. Successive generations of inbreeding among farm animals are known to reduce viability enormously; in one experiment with successive brother–sister matings with ducks, the offspring simply did not survive after several generations of crossing. Cheetahs (*Acinonyx jubatus*) seem to have survived episodes of severe population reduction and inbreeding at some time in the last few tens of thousands of years, and they still suffer the effects in terms of continued high mortality rates of their cubs, and low fertility of males. Many thoroughbred varieties of dogs and cats have accumulated harmful genetic traits as a result of inbreeding from a few founders. As many examples of highly inbred crop plants show, a genetically uniform population is also very susceptible to being wiped out by epidemic disease. But if a population can make it through the bottleneck of severe inbreeding and then increase its numbers, over time it may recover its genetic health as harmful recessive genes are lost, and as new sets of mutations come along to diversify it (giving a greater range of raw material for natural selection). Not all populations are going to make it through this stage, though.

The risks of stochasticity, natural disasters, and inbreeding have to be considered in the design of nature reserves. There are also, incidentally, much the same problems involved keeping small populations in captivity in a zoo or botanic garden, for they suffer many of the same risks despite their cosseted environment. If at all possible, a reserve (or zoo population) must be made big enough to give a population a good chance of long-term survival.

But how big does a population have to be for it to be assured survival? In a sense there is no absolute answer, because risk never goes away entirely and anything involving risk must involve subjective decisions about what is an "acceptable" level. However, ecologists can at least attempt to provide some quantitative back-up, on which these subjective decisions can be made in a more informed way. Given a few basic details such as the time required to reach sexual maturity, the number of offspring produced per unit time, etc., it should be possible to come up with some approximate measures of the survival chances of populations of different sizes, for any given species. For example, for a fairly typical mid-sized mammal species (e.g., about the size of a goat), a population that can never get above 30 individuals (because the low carrying capacity of the reserve limits it at that number) is likely to drift down towards extinction within a few decades (Groombridge, 1993; Groombridge and Jenkins, 2002). This is simply as a result of stochasticity—chance variations—coming from within the population. And that is not even counting the possibility of natural disasters and inbreeding that could also wipe out a small population.

So, based on calculations and a few observations, for our "medium-sized" mammal, a population of around 200 individuals is likely to survive for several centuries—a good start for conserving a species. As the uppermost population size that can be contained on the reserve approaches 1,000 individuals, it becomes likely that it will instead survive for tens of thousands of years (though the heath hen was not so lucky!). Even though it may drift downwards for a while, a large population like this will probably recover and bounce back. This recovery can be aided by the increased survival of young that occurs when population falls below its carrying capacity, so that there is less of a scramble for food and nesting space. A survival time of thousands of years is more than enough to ensure survival of a population on the sort of time scales that humans can easily visualize; essentially it seems like "forever" to most people. But thinking on the scale of the normal lifetimes of species—millions of years—this is not a particularly long-term form of security.

Although I have tried to put out some average numbers as a general illustration, the minimum viable population size can be expected to vary a lot from one species to another, because species are each so different from one another. Calculating the number for minimum viable population size (or MVP as it is called) has become something of a craft in itself. In calculating it, first you need to get clear how long you want the population to be viable for, before it becomes likely to drift far enough in its numbers to go extinct. One can never be absolutely sure of this survival time, because the buffeting of population size is random, so it is only possible to talk in terms of likely (say, 95% assured) survival time. If individuals in that species don't have a very long natural lifespan, then that surely means that with deaths happening all the time anyway, the population is less likely to survive for a given number of decades or centuries. Because of this, their minimum viable population size will need to be higher. So something like a mouse will need to have a minimum viable population size larger than a rhinoceros which lives a long time. On the other hand the mouse could recover its numbers quicker after a decline, and that affects its population's long-term chances too. These sorts of arguments turn up again later in the chapter in the context of the likelihood of extinction in small reserves.

Even if you needed fewer rhinos than mice for a viable population, it is clear that it would be a lot more trouble to accommodate the rhinos than the mice. Large herbivores like rhinos need a huge amount of space, and large carnivores will need even more space to be able to find enough food for them to eat. So for this reason, when designing reserves it is often more useful to talk about the "minimum area required" (or MAR) to sustain a viable population for whatever period one is interested in. And this is often where the money crunch hits: How can you afford to buy and protect that much land?

A related index—the "genetic minimum viable population" or genetic MVP—is meant to tackle not the stochastic effect of population numbers bouncing around, but the effect of stochastic changes in gene frequencies, and the likelihood of genetic variation being lost. Genetic variation surely affects the long-term survival prospects of a population (e.g., its vulnerability to epidemic diseases or its likelihood of being able to evolve to cope with climate changes). Franklin (1980) suggested that a genetic MVP of around 50 is needed to guard against serious inbreeding effects on the time scale of decades. Even if we aim for a genetic MVP of 50 or above, genetic variation will still tend to be lost over time and Franklin (1980) suggested that around 500 individuals are needed to compensate for loss of variation, by providing a big enough source of new mutations. However, bear in mind that the genetic MVP is going to vary a lot from one species to another. Comparing same-sized populations, fast-breeding creatures like mice will lose genetic diversity much faster than slow-breeders like elephants. This is because each new generation that replaces an old one is a chance for the variation to be lost.

Having some general idea of the risks of populations going extinct allows us to make a more informed judgment about what to conserve, and how best to do it. If only a tiny reserve area can be bought with the funds available, it might be worth diverting the funds elsewhere into another project that stands a better chance of success. However, lack of long-term security is not necessarily a reason to abandon any attempt to set up reserves. Even though a small population of just 30 or 100 individuals may not stand much chance of surviving in the longer term, it may still be well worth preserving as a "stop gap" until a larger reserve can be established. Even populations that were well below the MVP (both population and genetic MVP) have also recovered to make the leap back to health and abundance. Examples include the northern elephant seal (*Mirouga angustirostris*) and captive populations of the golden hamster (*Mesocricetus auratus*) (Groombridge, 1993; Groombridge and Jenkins, 1992). So even if the surviving population of a species is very small, that is no reason to give up on it.

In addition, various clever tricks can be used to help increase a small population's chances of remaining viable. For example, one way to try to prevent loss of genetic diversity in small captive populations is to establish studbooks, similar to those used by horse-breeders and dog-breeders. These can be used to ensure that every mating that takes place is between individuals who are as far apart genetically as possible, helping to prevent inbreeding and loss of gene diversity.

8.5 METAPOPULATIONS: A COMPLICATION TO MINIMUM VIABLE POPULATION SIZES

A population is not always what it seems. Sometimes it behaves as a small part of a larger interlinked network of populations, known as the "metapopulation". "Meta" comes from a Greek word meaning "beyond", and it is indeed a further level of population beyond the scale that we would normally notice. For example, a set of butterfly populations of the checkerspot (*Euphydryas editha bayensis*) at Jasper Ridge Preserve in California turned out to be linked into a larger "metapopulation" (Ehrlich and Murphy, 2005). Other parts of the metapopulation supplied individuals and genes to each local population, and if a local population died out these allowed it to re-establish itself. One of the three main local populations had a tendency to go extinct each time there was a big drought, but when this happened the population would be re-established from another neighboring population. Thus the smaller local populations were effectively supporting one another against extinction.

From a conservation viewpoint, the importance of these wider networks of metapopulations is that they help to maintain species against extinction. It is really an example of the "rescue effect" that ecologists who look at islands have often talked about—where individuals washing up on the shore help prevent an island population from going extinct. Even just a few individuals now and then filtering across from neighboring populations can make a big difference to the prospects for a population, if they happen to arrive at times when the population is very low. They can prevent the local population from dying out, or if it does die out then they can re-establish a population in the same place. To take advantage of the sustaining effects of metapopulations, conservationists need to think on a broader scale. Making an extra effort to conserve other populations in the surrounding areas may pay off in terms of the longer term survival of the population that one was originally trying to conserve.

Detailed study has shown that some local populations—known as "sink populations"—cannot actually keep up their numbers without the continuous addition of individuals they get from other nearby populations, which are known as "source populations" (Pulliam, 1988). For whatever reason, members of the sink population cannot breed well in that local habitat, or their death rate is higher there. One example of a species that has been observed to have "source" and "sink" populations is the white-crowned sparrow (*Zonotrichia albicollis*) in high mountains in the U.S.A. (King and Mewaldt, 1987). Obviously from the point of view of conservationists, it would be a big mistake to concentrate resources into preserving a sink population, while letting a nearby source population go extinct!

8.5.1 The shapes and sizes of reserves

Given what we think we know about the biogeography of islands (Chapter 4), it might be possible to design nature reserves in particular ways that help to reduce the probability of extinction of the populations within them. Conservationists generally agree that it would be better to set up one large agglomerated nature reserve, rather than several disconnected small ones. If there are several separate small reserves each containing—say—30 individuals, the populations in each little reserve must run the gauntlet of stochastic events, inbreeding, and natural disasters. If they cannot exchange individuals (i.e., they are not linked into any larger metapopulation), these small isolated populations are individually far more likely to go extinct. Each one may blink out of existence separately until eventually there are none left. If they are all lumped together into one large population of a couple of hundred, the population is far more likely to survive after a few centuries. Another very practical reason for having one large reserve is the very fact that it costs less to set up and patrol a single large reserve than several small ones (Groombridge, 1993).

Despite this argument, there is still something to be said for having spatially scattered reserves rather than just one compact, unified reserve. Natural disasters such as droughts, volcanic eruptions, or fires, plus introduced diseases or outbreaks of civil war are less likely to wipe out everything scattered across a broader area.

For example, Simberloff and Abele (1982) mentioned the example of the Seychelle Islands off eastern Africa—which in effect resemble several unofficial small, scattered nature reserves from a nature conservationist's point of view. Despite centuries of severe fires, introduced predators and overgrazing, these islands have lost only 2 of their 14 endemic bird species. Simberloff and Abele suggested that this is probably because the specific types of bad luck that hit populations were different from one island to another, and occurred at different times—so there was always somewhere for each bird species to survive. The demise of the North American heath hen probably has much to do with the fact that the whole population was in one place (on Martha's Vineyard), where disease and fire could easily spread throughout.

Buying up several small areas may also take in a greater amount of species richness overall, compared with setting up one large reserve. For example, among coral reefs scattered around through the sea, several small reef crests may have more species richness overall than one large crest of the same combined area (Simberloff and Abele, 1982). This might be because of subtle differences in the environment from one reef to another, giving different niche opportunities for species on each of them. It might also just relate to the effect of chance colonizations—on some reefs one species was lucky enough to establish itself, while on a different reef another species is there in its place. Similarly, several small scattered forest fragments in New Jersey have been found to have more species of birds overall than one large forest of the same area (Simberloff and Abele, 1982)—something that might be expected as a result of differing forest composition and structure on different soil types scattered across the broader landscape.

Weighing up the benefit from not having all one's eggs in one basket (so to speak) in one large reserve has to depend on how large the populations can be in each smaller scattered reserve. It must also depend on the overall likelihood of something dramatic such as a volcano, fire, or war destroying any one area. A larger reserve that is all in one place might be wiped out by such an event, whereas if there were smaller scattered reserves at least some of them would have survived. It is quite possible that each of the small scattered reserves can still maintain large enough populations of the most desirable species to avoid worrying about them going extinct through stochasticity and inbreeding anytime in the near future (Simberloff and Abele, 1982)—and in this case why not conserve populations that are spaced out and unlikely to all be destroyed in a single unlucky event? From a purely financial perspective, one point in favor of buying up small reserves is that smaller land parcels of forest or agricultural land tend to be cheaper per unit area than large ones, because of the greater trouble involved in moving equipment between them. For example, forest fragments under 32 ha in England became progressively cheaper to buy (per unit area) as they became smaller (Simberloff and Abele, 1982).

This dilemma over whether to go for one large reserve or several small ones has often been mulled over in the literature on the theory behind nature conservation. It has become known as SLOSS, which stands for "single large or several small?" (Simberloff and Abele, 1982).

In fact, it may be possible to have something of the best of both worlds—spatially scattered reserves that do not lose as many species because they remain connected by narrow strands of habitat, known as "corridors". The idea is that animals and plants from one relatively small patch can always filter through the corridors to other patches, sustaining each population against loss of numbers and genetic diversity, and so preventing local extinctions. This way, populations within the reserve can be repeatedly "rescued" by one another, ensuring the long-term survival of populations throughout. This is not quite like establishing one large united population, but it is better than nothing. These "wildlife corridors" have become a central idea in nature reserve planning around the world in the last 20 years or so, and are used now in all sorts of different reserves from the very local to the internationally important.

While it makes intuitive sense that corridors could help prevent population extinctions, it is always a good idea to put anything to the test, to check if things really do work as we expect they will. The most rigorous test of the corridor idea so far was a study set up by Damschen *et al.* (2006). They preserved swaths of native open pine woodland vegetation (grassland with just a scattering of trees) within a uniform area of dense forest in South Carolina, U.S.A.—leaving different-shaped "islands" of the open habitat surrounded by dense forest (Figure 8.1). In some cases the forest islands were out on their own, unconnected to anything else. In other cases, the researchers left the islands interconnected with thin corridors of remaining forest habitat. Then they let the slow process of adjustment and extinction of populations play out, closely monitoring all the species of plants within the islands. What they found was exactly as the corridor idea predicts—significantly less extinction of plant populations in the interconnected islands, compared with the completely isolated ones. Corridors really did seem to be rescuing populations against extinction.

Something to bear in mind is that the theory of nature reserve design depends largely upon the equilibrium model of island biogeography of MacArthur and Wilson. Because it has been handed down to them indirectly, few conservationists are aware of the uncertainties and doubts that surround this theory (see Chapter 4). While the basic idea—that small populations are more likely to go extinct, and isolated patches of habitat are less likely to be reached by colonizers—is really indisputable, there are many other factors that determine whether an island (or a

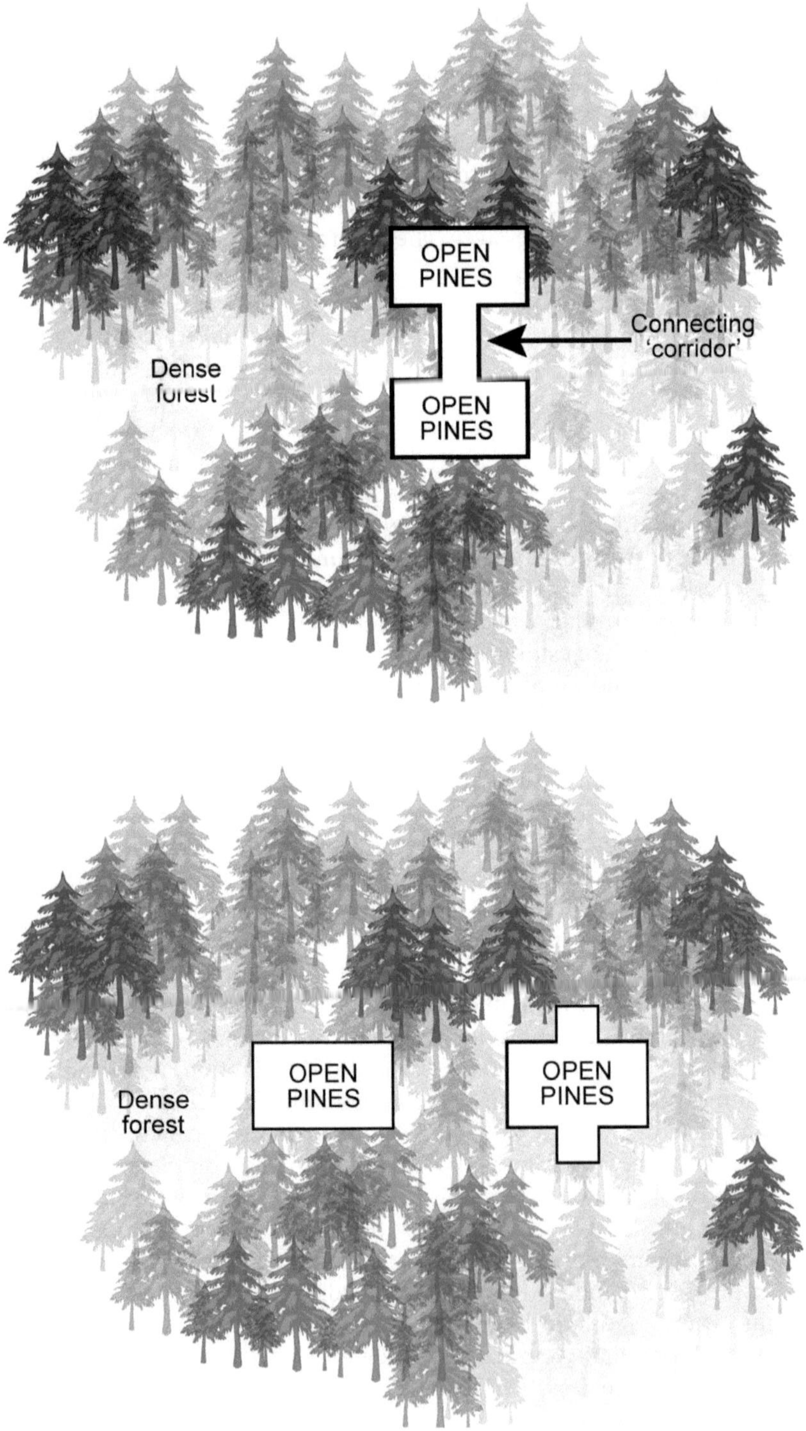

Figure 8.1. An experiment by Damschen and colleagues compared isolated blocks of habitat (bottom) with the effects of connecting blocks to other adjacent areas by thin strips known as corridors (top). The connected blocks (top) lost fewer plant species over time than the isolated ones (bottom), demonstrating the beneficial effect of corridors in conservation.

nature reserve) holds onto species. For example, differences in the detailed structure and age of vegetation are likely to be as important as reserve size in determining what can persist there. In each habitat, there is likely to be a relatively resilient "core" of species that are going to be present in the habitat almost no matter what happens—often because they are such good colonizers that can always reach there. These then are relatively immune to the chance effects of extinction and colonization. Scheffer *et al.* (2006) found that small isolated pond habitats were actually more species-rich in a range of types of organisms, compared with large ponds—which is exactly the opposite of what island biogeographical theory predicts! This is apparently because the small ponds were less likely to have fish which exerted a disruptive, suppressive effect on the species richness of other groups. While this is a rather unexpected result, complications to island biogeographic theory are often just a matter of commonsense to anyone who has studied nature and its workings. Even though nature conservationists are by inclination commonsense people, they are sometimes over-awed by a theory that comes equipped with graphs and mathematical equations. Sometimes, it is necessary to be reminded of the sort of things that you knew all along!

Given what we are learning nowadays about the feedbacks between forests and climate, there may be another dimension to consider in terms of the sizes of reserves that are needed. It seems that in at least some areas of the tropics, such as the eastern and southern Amazon Basin, getting enough rainfall to make rainforest depends upon being surrounded by hundreds of kilometers more forest (Adams, 2007). This is because of a recycling mechanism involving the trees that juggles water between the forest and the clouds without allowing it to escape down rivers to the sea. This may mean that to sustain any sort of viable reserve will require conservation of forest more or less intact over very large areas: anything less and the forest will die and its constituent species will be lost. This need for extensive intact forest areas to maintain climate is not thought to apply to all areas of tropical rainforest. For example, in central Africa, areas of forest tend to be more able to "stand alone" in terms of climate, not requiring the support of large areas of intact forest round about them.

But in any part of the tropics, at the smallest, most highly fragmented scale, very small patches or individual tropical forest trees that are left standing isolated in fields do not tend to do well. This is because the dry air and raised temperatures over areas of bare ground or fields tends to be more than these rainforest plants can cope with. In Malaysia, for example, I have often seen big trees—sole survivors of the old-growth rainforest that was there before—dying as they stand surrounded by newly planted oil palms. The altering, impoverishing effect of dry hot air blowing into the edge of a rainforest patch can extend tens of meters into the interior: this is known as the "edge effect". So if the remnant patch of forest is only a few tens of meters in width anyway, it is essentially not much use as a rainforest habitat.

8.6 HOW MUCH DO NATURE RESERVES LOSE? RELAXATION EXTINCTION

Another of the various things that MacArthur and Wilson's theory of islands predicted is that after a piece of habitat first becomes isolated (e.g., when a small patch of forest is left standing after all the forest around it has been removed) it will gradually begin to lose species. This is what MacArthur and Wilson predicted for real islands in the sea, anyway, and it should occur because the population sizes of the species that are left isolated there are smaller, so those populations are more likely to go extinct through stochasticity. The species richness of the isolated patch of habitat will decline, until it is roughly balanced by the rate at which species are colonizing or recolonizing from the nearest areas of habitat scattered around in the landscape.

The important message from this is that the species richness we have in a nature reserve when we start out may not be the species richness we will end up with in a few decades' time. Say, for example, there has been a recent burst of deforestation, and only one small area of forest has been saved. We should not expect this patch to remain as species-rich as it was immediately after the deforestation had happened. This process of loss of species one by one until the diversity reaches a balance is known as "relaxation". It is the enemy of conservationists because it destroys what they are trying to hold onto. Clearly, it is something they need to be aware of and guard against if possible. For one thing, naively buying up a very small reserve that is still brimming with species (because it is newly isolated) is not a sure way to conserve species. It might be better to buy up a larger reserve that has fewer species but which has been isolated for a long time, so we know that it is not going to be losing species rapidly.

Are there any examples that allow us to know whether "relaxation extinction" really occurs out in nature after an area of habitat is newly isolated or made smaller in size? The classic demonstration was Simberloff's experiment on mangrove islands where he actually resorted to chopping the islands up into smaller remnants (Chapter 4). Just as expected, the remnants lost insect species over time until they reached a new stable, and lower, level of diversity. Diamond (1972) also considered he had found evidence of this process of progressive impoverishment of species diversity over time on islands that had been cut off by sea level rise at different times in the past. The authors each explained these changes purely in terms of the theory of island biogeography, with smaller populations going extinct more often due to stochastic effects.

Another study that might also show relaxation extinction has been the progressive loss of species of plants from Singapore's rainforest. It is a remarkable thing that in the midst of the small island of Singapore, surrounded by several million people, there is a nature reserve that contains several square kilometers of tropical rainforest. This reserve is a remnant of the forest that originally covered the whole island of Singapore, now pushed back to an area on the central uplands. Although its survival is a remarkable feat of coexistence of humans and nature (and shows what can be done if one really tries to conserve natural habitats), the remaining forest has apparently lost at least some species because of its isolation. Old species lists from the 1800s, when the reserve area was first isolated, show various rainforest plant species that were there but have since vanished (Turner *et al.*, 2002). It will be interesting (though perhaps depressing) to see how many more are lost over time.

When isolated fragments of forest were left behind by deforestation in the Usanbara Mountains of Tanzania, the number of understory bird species that they had ended up with decades later showed the general sort of relationship that would be expected from MacArthur and Wilson's theory. Compared with the original, unbroken forest cover, smaller fragments of just a few hectares or less than a hectare had lost many more bird species than large ones that were tens or hundreds of hectares in size (Newmark, 1991). Relatively isolated fragments had also lost more species than those of similar size that were close to other forest areas. A similar pattern was found for the dung beetles in remnant forest patches in the Amazon.

The idea that relaxation extinction occurs as a reserve "adjusts" to its new size and isolation has been used to make some fairly dire warnings. For example, Soule *et al.* (1979) predicted that the famous Serengeti reserve in East Africa, subject of countless wildlife documentaries, would lose 50% of its large mammals (15 herbivore species) within 250 years of being cut off from the broader landscape of savanna and woodlands that had existed across the region. However, Newmark (1996) focused on both the needs of each species and the range of habitats present, and suggested that actually only one of these species was likely to be lost. There is a general mood amongst ecologists, in fact, that blindly treating species as numbers in a calculation (as MacArthur and Wilson's theory would suggest) will just lead to inaccurate conclusions. It is necessary to bear in mind the characteristics of each species, its needs, and its interactions with others.

Island biogeography has been used to make broader, more generalized estimates of how much will be lost as the tropical rainforests are fragmented. If bigger areas of forest contain more species per unit area, we can expect a lot of extinction among forms that survive in more isolated patches. One widely quoted generalization is that reducing the area by 90%, one will lose half of the species that were present throughout—not even counting the narrow endemics whose ranges will be completely lost (Reid and Miller, 1989). But the parameters that such sweeping predictions are based on are rather shaky, depending on many different assumptions about the patchiness of species distributions, their actual resilience against dying out within patches once they are isolated, and the precise pattern of deforestation.

In general, there is plenty of evidence that isolated patches of habitat tend to lose diversity over time. The only question is whether the processes (chance extinction vs. chance colonization) that are invoked by MacArthur and Wilson's theory actually determine this decline in diversity, and whether the theory has any useful predictive value. Abele and Connor (1982) concluded on various grounds that many supposed examples of the predicted "relaxation extinction" (e.g., Diamond's, 1972 study) were likely due to other complex and more local habitat factors. So far, there has not been enough time and there have not been enough examples to know how much use MacArthur and Wilson's theory is for guiding conservationists about the numbers of extinctions that they can expect from a newly isolated reserve. But the general logic of the idea that "the larger and less isolated the reserve, the better" is pretty compelling, for many reasons in addition to the ones involved in the original island biogeography theory. The overall message that is important for conservation is that a newly isolated habitat loses species over time, whatever it is that causes this.

As well as reserve size and isolation, there are other factors which determine how likely it is that extinctions will take place in the reserve. A lot of it depends on the characteristics of the individual species that one is trying to conserve there. It is necessary to focus on these features of each species to predict its chances of survival (Groombridge, 1993). This is especially crucial if the reserve was created for just one or a few key species—one wants to know how likely they are to survive.

One factor that can be expected to affect the chances of extinction of a particular species is how rare it normally was in the intact habitat before it was fragmented. Some species never have high population densities at the best of times, and when they are confined to a reserve this means that their populations are especially small—which surely increases the likelihood of extinction. For example, some of the birds in Newmark's (1991) study on forest fragments in Africa naturally had lower population densities even in unbroken forest areas, so they tended to have especially small populations within the fragments. Animals high up in the food chain, mostly active predators, always have low population densities so they are likely to be the most vulnerable to extinction in an isolated, small reserve. This means that the glamorous top-predator species such as tigers or cheetahs are also likely to be among the hardest to conserve.

Species that are good dispersers, able to make the leap across from one fragment of habitat to another, are surely less likely to go extinct. These new arrivals may be able to "rescue" the numbers of a local population, before it can die out from that particular fragment.

If a species can tolerate the interface environment between (say) forest and farmland or forest and grassland, then it is much less likely to go extinct in the highly fragmented reserves that are nearly all "edge".

Species that have relatively stable populations are likely to be less vulnerable than populations that go through large fluctuations. This is because they are less likely to decline below a critical threshold where the recovery of their population becomes unlikely because of inbreeding or chance extinction. One thing that may favor stability in populations is the potential longevity of the adults: species that do not continuously need to replace themselves with vulnerable young are surely less susceptible to fluctuations in the environment (just as they are less susceptible to stochastic fluctuations in population, above). So long-lived species like tortoises might be better placed to survive than short-lived rabbits. On the other hand, though, if a species can increase its numbers rapidly after a disaster, that is also likely to give it a better chance of avoiding extinction in the reserve. So in this sense, fast breeders like rabbits might be better placed to survive than tortoises which would recover their numbers slowly.

All of this is really just commonsense, without much data to back it up. One interesting study, however, did look in detail and gather data on species lost from isolated fragments of habitat. Davies *et al.* (2000) focused on the land mammals of the Queensland rainforests in northern Australia and noted the likelihood of extinction of local populations among 17 species of land mammals in forest fragments, in relation to such characteristics as their body size, longevity, level in the food chain, and reproductive rate. They found that in fact the only good predictor of survival of populations was how abundant they usually were in the disturbed or fragmented rainforest. So this is rather like MacArthur and Wilson's theory would have predicted: population size determines how likely it is that extinction will eventually occur! Except in this case, the species are not all equally likely to go extinct as MacArthur and Wilson's theory assumed. Instead the probability of a particular species dying out is weighted by its own normal level of abundance, which depends on its niche and what its ecological tolerances are.

8.7 ACTIVE MANAGEMENT OF RESERVES

As conservationists in the mid–20th century sometimes found out to their cost (above), it is often not good enough to simply fence off an area of habitat, and then let it look after itself.

Nature is full of disruptions, and often these disruptions help to maintain species richness. Fires, grazing, floods—each of these can provide specific niches, or knock back competitors to allow species to hold on within the community. Fencing off an area of grassland that is rich in wild flowers will merely allow common species of fast-growing plants to push out most of the rarer species (Chapter 1). For this reason, many nature reserves in scrubland or grassland areas are managed to provide intermediate levels of disturbance that will allow the greatest number of species to thrive. In England, sheep grazing on chalk meadows is used to maintain species richness: the density of stocking of sheep has to be carefully calculated based on experience, to allow just the right amount of grazing disturbance for the growth rate of the vegetation. In the American prairies, the natural background of lightning-caused fires that used to sweep across whole landscapes no longer occurs in small isolated nature reserves. That natural frequency of burning has to be recreated by setting fire to the grass when it is dry. In addition to burning, or instead of it, bison may be used to provide the grazing disturbance that the prairies also experienced in their natural state.

In past centuries, humans going about their daily business in a traditional rural landscape often created species-rich habitats purely incidentally. Now that these traditional customs have died out and rural economies have collapsed, it is rather ironic that a great deal of time and effort is poured into maintaining what no-one particularly cared about in the days when people were just trying to survive off the land.

The traditional chalk grasslands of southern England were almost entirely a human-made product of grazing with an introduced species (the sheep, which came from the Middle East) in cleared landscapes that would naturally be forested. Many of the wild flowers that inhabited the meadows were likely not even native to Britain: they had filtered across from mainland Europe mixed with crop seeds or on clothing, to occupy the open habitat over the past several thousand years. Many of them originally come from the natural steppe grasslands of southeastern Europe (e.g., in Bulgaria and Ukraine).

In the surviving forest fragments of Britain, there was a traditional custom of cutting the trees down to ground level and letting them regrow as long straight poles which would be harvested every few decades. Known as coppicing, the sudden periodic opening of the woodland canopy allowed an abundance and richness of wild flowers to thrive (e.g., the bluebell which forms beautiful blue carpets in the English and Welsh woodlands in spring). Nowadays, there is little economic need for coppicing and the poles that it provides, and so woodland nature reserves are coppiced by volunteers. I remember myself, from when I was a conservation volunteer in my 20s, the hard work of hacking or sawing away at coppice trunks. I used to think how strange it was that now we were just "playing" at what our ancestors had needed to do as part of their hard scrabble existence. I probably spent too much time thinking about such things when I should have been paying attention to what I was doing, and after several near accidents I wonder how in all my clumsiness I managed to avoid losing any fingers or get my skull broken by a falling coppice trunk (some people just seem better suited to writing books than doing practical things, and I suppose I am one of them).

Often it seems that conservation is really about turning nature into gardens. Ecosystems that once looked after themselves without humans must be intensively managed because the important disturbance factors and predators are no longer present in these small fragments, which are surrounded by a landscape that is dominated by humans. Even in places where the habitat was always a human creation, like the chalk meadows or coppiced woodlands of England, what once happened readily as a by-product of everyday life now takes a concerted effort by armies of enthusiasts.

Sometimes, things that humans have added to the ecosystem make it inevitable that we will always need to work hard to prevent everything going extinct. Introduced species are one of the greatest threats to the world's biological diversity (Chapter 7). The situation is at its worst on some small oceanic islands, where invasions of introduced plants have almost entirely pushed out the native flora. In these cases, the battle to try to preserve species in their native habitat is particularly arduous. In the case of Mauritius, in the Indian Ocean, the only way to save the endemic flora is to make weeded plots a few hectares at a time, where the invaders are continually grubbed out by hand. On larger landmasses, invasive species also often require intensive work. In Wales and in southeastern England, the introduced *Rhododendron ponticum* takes over woodlands on acid soils, pushing out many of the native plants. Volunteers spend their vacations "rhododendron-bashing" on nature reserves, trying to keep it under control and give the natives a chance. Recently, the U.S. government had to earmark $100 million for grubbing out an introduced Australian tree, *Melaleuca*, from the Florida Everglades National Park. It was clear that if nothing was done, the Everglades would eventually have become one continuous forest of *Melaleuca*.

As societies have developed and priorities have changed beyond mere survival and the gain of wealth, there has been a growing realization that too much of nature has been lost. In many instances over the last few decades, habitat that had been destroyed by human alteration of the landscape has been deliberately restored to something like its original state—potentially providing a refuge for species that had been under threat. After losing much of their area to deliberate drainage schemes in the 1950s, the Florida Everglades are now being expanded as a result of farm land being bought up by the State of Florida and the U.S. National Parks Service. Drainage canals dug with great effort by the Army Corps of Engineers some 50 years ago are being deliberately stopped up to allow this land to flood again.

Some areas of tall-grass prairie in the mid-western U.S.A. have been reclaimed from farmland by stopping cultivation and sowing with mixtures of seeds consisting of the species of the original prairie (combined with careful management regimes of burning and fire). In natural and semi-natural grassland systems such as these, an important part of the process of rehabilitating the land is to reduce the nutrient content of the soil. Agricultural fields and modern productive grazing lands are intensively fertilized with agricultural fertilizers—nitrates, phosphates, and suchlike. Even after these lands have been taken out of intensive agriculture, the soils can retain very high levels of nutrients in the soil for many decades. Such high fertility is, as we know, the enemy of species richness for it leads to a few strongly competitive species squeezing all the others out (Chapter 1). The only answer is to deliberately rid the land of its legacy of fertilizer by removing a hay crop each year. Without the recycling of the nutrients back into the soil, it eventually becomes depleted and suitable for high plant species richness.

Even tropical rainforest can be deliberately restored, as in the case of an area of forest that I am familiar with at the Forest Research Institute of Malaysia (FRIM), on the outskirts of Kuala Lumpur, the capital of Malaysia. The FRIM site was species-rich old-growth forested before World War II, but many of the timber trees were logged out under Japanese occupation, leaving the forest badly depleted in terms of species richness. Following the end of the war, the site was deliberately replanted with a mosaic of different native tree species. These have now grown up into a diverse canopy, which has been supplemented by the natural dispersal of many additional tree species into the forest over time. In places within this particular reserve, the forest at FRIM now approaches the plant species diversity of primary rainforest, an encouraging sign for the potential for restoring parts of this the most diverse ecosystem on Earth.

8.8 MAINTAINING RESERVES IN A CHANGEABLE CLIMATE

Humans naturally tend to think that whatever is around them, here and now, will always be so. It is easy to forget that we live in a world in which the climate can shift—and until recently this was not something that conservationists gave much thought to.

The geological record shows that the Earth's climate has changed countless times over the last two and a half million years, and it also appears to be warming rapidly at present due to the increasing greenhouse effect. Against this background, no population that is protected in a reserve can be expected to be able to exist in one place "in perpetuity", because the ecological ground under its feet is always shifting. Rather, as in the "Red Queen" model of evolution (Chapter 3), most species may have to keep physically moving themselves around to avoid extinction. As one suitable zone of climate disappears, each species must find another area that has now become more hospitable for it. Nature reserves present a problem in this changeable world, because they have definite boundaries and tend to be surrounded by unfavorable habitat—which is often what led to the setting up of the reserve in the first place. Ideally, then, a reserve should be designed bearing in mind the possibilities for migration routes of suitable habitat out to other areas, if and when the climate changes. This is, of course, easier said than done.

Species that live on mountains in the tropics may be especially in need of the right nature reserve designs. They often seem confined to quite a narrow altitudinal zone that corresponds to their right temperature band. If they cannot move uphill as warming occurs, they will be cut off from their natural climate zone and die out. Hence the need for connections between reserves that extend not just horizontally, across the lowlands, but also vertically across a range of altitudes (Hannah and Lovejoy, 2005).

Another factor that should be considered in relation to the Red Queen model is whether species that are constrained in reserves will have enough genetic variability and enough new mutations to evolve, in order to cope with the long-term challenges that they will face. Both environmental change and biotic factors such as disease are going to be affecting every species on the time scale of tens of thousands of years, just as they always have done in the past. This is the reason, presumably, that hardly any species have been able to abandon sexual reproduction and then live long enough to be found by us. If only a small population of a particular species survives in a nature reserve, there is going to be less chance that is contains the new mutations and the inherent variability that could enable it to survive. However, there are usually much more pressing problems to consider than the long-term evolutionary survival of a species. Just getting it through the next few decades will often be difficult enough.

8.9 AIDING PLANT MIGRATION—PLANTING

If climate is going to warm as much as the computer models say it will, many mid-latitude and high-latitude plant and animal species will be left far away from the edges of their new potential ranges by the end of this century. Animal species tend to be remarkably mobile, and many could probably keep pace with the shifting climate zones as the planet heats up. There are many case studies of invasions by introduced species which moved tens of kilometers per year, which is certainly fast enough to keep pace with climate. For example, the collared dove (*Zenaida macroura*) expanded by as much 50 km a year when it invaded North America in the last century (Adams and Woodward, 1992). Plants, however, are more likely to be left far behind—especially the long-lived species such as trees which are slow to mature and bear seed. At the same time, the warmer climate limits of ranges will tend to contract towards higher latitudes as climate becomes too warm for the species to survive or compete with other species better suited to the new climate. If the warmer end of the species' range keeps moving away from the equator faster than the species can migrate to take advantage of new zones of favorable climate, it may ultimately go extinct.

Conserving species against extinction may not then simply be a matter of fencing off areas of habitat and preserving them there. It will be necessary to get species to new areas of suitable climate where they can become established, and survive in the long term. In the end, this may come down to deliberately transplanting seedlings or young plants into their new range, directly into natural habitats. To do this on any large scale would take either armies of volunteers or considerable government funding. There is no precedent in nature conservation for such an effort, and the choice of exactly what to plant where would probably result in a lot of disagreements. For example, should we be careful to match our seed stock between areas of similar soils in the old range and the new range? If there are several distinct varieties or subspecies, which ones should we use? For plants there might also be problems with photoperiod adjustment for the seasonal rhythms, which might result in the poor adjustment of leaf break and leaf fall in many of the southerly species, even if the mean climate is the same as it was in the south where daylengths in early spring and winter are different. There is evidence that populations of some trees (e.g., elms—*Ulmus*–in Europe) are genetically attuned to the precise daylength that marks the coming of spring at each latitude. We simply do not know how this would affect the survival of populations planted at a new latitude.

We also have to consider whether species should be planted further north now, in anticipation of the full impact of climate change, or only planted after the climate zones have actually moved. One might not expect that a species could survive if planted beyond its natural climate range, but in fact many woody plants can grow quite well beyond this. For example, here in New Jersey I often see southern evergreen magnolias (*Magnolia grandiflora*) thriving in gardens several hundred kilometers north of their natural range in the eastern United States. These planted trees and shrubs can survive, and grow, but they will only exist as poorly reproducing or poorly competitive individuals until the climate around them warms up. When it does warm, they may be able to come into their own and take their full place in the forest ecosystems. The growth to maturity and potential rate of increase in population of many trees and shrubs is very slow, and they could get a much needed headstart if we do the planting in advance now, setting them off towards playing a role in the new plant communities that will be suited to a warmer world. Already, the hardiness zones that gardeners use as a guide for planting are moving north rapidly: even just between 1990 and 2006, the hardiness zone in most places in the United States shifted to the next warmest category.

Yet if we plant species from warmer climates out in the native woods or grasslands right now, and the climate fails to warm up as much as expected, we will have ruined the character of many of our native ecosystems. These are essentially alien species, even though they do come from within the same general geographical region. By inter-planting new species beyond their natural range we would in effect be reducing our natural habitats to a vast garden. This does sound disturbing to me—it goes against the grain of my thinking as an ecologist, which is to want things to remain as natural as possible. On the other hand, considering it some more, most of these habitats have already been extensively managed for centuries by humans, in their various ways. There is reason to suspect that even as far back as the end of the last ice age 11,000 years ago, when forests established themselves once again after the cold and aridity of the glacial, humans dominated the migration of certain important tree species. For example, it has been suggested that rapid northwards migration of beech (*Fagus*) and hazel (*Corylus*) in Europe at the end of the last glacial was due to humans transporting the edible nuts on their migrations, and accidentally dropping them beyond the advancing treeline. It may be that the way the world is changing inevitably will lead us towards "gardening" the ecosystems that we care about. This may have to involve not only bringing in populations from warmer climates to allow for the greenhouse effect, but also releasing genetically modified and hybridized trees to replace those that have been lost to introduced pests and diseases (below).

8.10 TAXONOMY AS THE ARBITER OF FATE

Through most of the 20th century, taxonomy (the classification of types of organisms) seemed to be an especially esoteric branch of biology. What could be further removed from the most interesting and relevant questions in science than endless examination and debate of subtle differences between species? Taxonomy began to whither, starved of money, attention, and in particular new graduate students—who are the lifeblood of any scientific profession. The pool of taxonomists in universities, zoos, and botanic gardens steadily aged, and were not being replaced as they retired or died.

Suddenly, some time during the 1990s, taxonomy gained a new and urgent importance. It became apparent that much of the world's biological diversity was under threat, and that one cannot know what is most threatened unless one is clear about its classification. Also, a conservation program is more likely to fail if we cannot be sure about the identification of the organisms we are trying to preserve.

An example of how important careful taxonomy can be to conservation is the case of the Ridley turtles. Originally there was thought to be just one endangered species known as Ridley's turtle, but molecular evidence showed that in fact it consisted of two cryptic species, now named as Kemp's Ridley (*Lepidochelys kempii*) and the olive Ridley (*L. olivacea*)—each of these species being in even greater danger of extinction. Similarly, the northern sportive lemur (*Lepilemur septentrionalis*) turns out to have within it a separate cryptic species, distinguishable from its mtDNA and chromosome configuration, with only 150 or so individuals and non-protected status. The tuatara (*Sphenodon*), a lizard-like reptile found only in New Zealand, was shown to consist of two separate species in a genetic study in 1989—one of these far rarer and more endangered than the other.

Without the careful taxonomic study that led to this reclassification, the rarer species of sportive lemur, turtle, or tuatara could have been left to go extinct under the assumption that it was just another slightly different-looking population, and we would be none the wiser that we had really lost something quite unique. Such cases illustrate just how important it is to have taxonomists to help us interpret the world's biological diversity. Unfortunately, the numbers of graduate students training in taxonomy are now nowhere near enough to match the rate of retirements—just at a time when taxonomists are needed more than ever to help us understand just what is under threat, and what we should give priority to.

In various cases, reclassification has gone in the other direction by finding that what were previously thought to be distinct species are actually just subspecies or varieties of something more widespread. The idea of a subspecies going extinct does not have quite the same gravitas about it, compared with a species being lost. So, in a taxonomic sleight of hand, what was once a precious species can sometimes be relegated to a "mere" subspecies of something more common. If this means that it will be less valued now that it is just a subspecies, that could ultimately well be the end of it. This fear was voiced by conservationists some decades ago when the tiger was reclassified, from several closely related species into a group of subspecies of a single species, *Panthera tigris*. However, there is no evidence since then to suggest a loss of interest by authorities or the public in conserving each of these subspecies (which have each remained a source of national or regional pride)—although it is difficult to know how much keener the public might have become by now if each type of tiger was still regarded as a separate species.

8.11 BOTANIC GARDENS AND ZOOS

Zoos were first established for their entertainment value (e.g., aristocrats would assemble a menagerie of exotic animals to amaze visitors). It was only in the 1800s that such collections of animals became the basis of serious study, of taxonomy, and of animal behavior. It was still later, in the 20th century, that another potential purpose of zoos was realized: as a temporary repository for rare species, where they could be reared in captivity and used to restock wild populations. Early attempts to save species through zoos were not very successful, because not enough was known about the requirements of each species. Animals are often reluctant to breed or to rear their young in the "wrong" environment, and many of them are very particular about this. They tend to die young too if they are stressed by their environment, or receiving the wrong sort of food. The epitome of a choosy eater and reluctant breeder is the giant panda, *Ailuropoda melanoleuca*, which eats mainly bamboo and also for many years resisted all attempts to make it breed in captivity (although the art of panda breeding now seems to have been perfected after many years of careful research and observation in China).

Nowadays, zoo-keeping has become an extensive and sophisticated field of knowledge, and zoos are equipped to play a role in the survival of species. At the outset, however, it is clear that they can never hope to save more than a tiny fraction of the world's species away from their own natural habitat. To preserve a population of any one species in captivity requires a great deal of skill and attention, and money. Hence the role of zoos in conservation can only be limited to a few relatively glamorous species, mostly birds and mammals. Most of the threatened species of mammals that are held by zoos still have very small populations—far fewer than are still present in the wild. Only 9 species of threatened mammals have captive populations of more than 500 (Groombridge and Jenkins, 2002). There are at least some cases, however, of creatures that would certainly be extinct now if it were not for the populations that survive in zoos. Examples of apparent success stories in the preservation of species include the Socorro dove (*Zenaida graysoni*), with about 100 individuals surviving only in captivity.

It is a moot point though, how much good the efforts of zoos really do in conservation. In order to really help an endangered species survive in the longer term, the zoo-bred animals must eventually be released back into the wild, to supplement the numbers and the genetic diversity of a population there—or replace an extinct population. Re-release is not always easy to achieve, because animals raised in zoos will tend to lack the subtle survival skills that would have been taught by their mothers out in the wild. When a zoo-reared panda was released recently back into the wild in China, it died within a year despite the efforts of its zoo-keepers to train it. Similar experiences have occurred with captive-reared cheetahs, and other species. Out of 80 reintroduction projects reviewed in Groombridge (1993), only 44% were judged a "success".

As well as lack of hunting and foraging skills, the tameness of released animals can be lethal to them when humans regard them as a threat or something worth hunting. And also, some species can probably never be released again because something has fundamentally changed about the environment they used to live in—often the presence of an invasive species or disease. For example, several species of colorful harlequin frogs from the tropical Americas survive only in captivity, as far as we know, because of the introduced chytrid fungus that has apparently wiped out all the wild populations. If these captive-bred frogs were released back into the wild, they would presumably only be killed by the fungus—so there is essentially no hope of ever re-establishing them in the wild, unless they are in some way bred or genetically engineered to resist it.

There have, however, been some success stories involving the release of captive-bred individuals into the wild. One example is the preservation of the Californian condor (*Gymnogyps californianus*), which was down to only 22 individuals by the 1980s as a result of hunting, pollution, and habitat loss. The remaining small population was captured and raised in both a private sanctuary and in San Diego Zoo. They bred, and in order to maximize the rate of population increase, some of the chicks were hand-reared by humans. The condors were released into the wild after an extensive training program, which included teaching them to avoid collisions with power lines (which had been a common cause of death for condors). The release of the Californian condors was preceded by a trial release of Andean condors (*Vultur gryphus*), to see whether released condors could indeed probably survive in the wild. As a further ingenious twist, only females of the Andean condor were used so that they would not establish a population, and these were recaptured after they had managed to survive for two years, before the release of the Californian condors. The Californian condor population in the wild is now more than 120 birds (plus another 150 in captivity), and it has expanded its range into northern California and northern Mexico. So far the whole program is estimated to have cost around U.S.$35 million, illustrating how costly it can be to save a species through intervention with zoo-rearing.

Another example of successful reintroduction seems to be the Arabian oryx (*Oryx leucoryx*), a large desert-living antelope. Once present throughout the Arabian peninsula, its numbers were whittled down by hunting until the last wild individual was killed in 1972. Fortunately, significant numbers survived in captivity scattered around the Middle East, plus in a small herd of nine oryx in Phoenix Zoo, Arizona which had been established in the 1960s. With the patronage of the Sultan of Oman, a project to breed the oryx in captivity at Phoenix Zoo and re-establish it in wild was started in 1974. The genetic diversity of the Phoenix herd was supplemented bringing in individuals from elsewhere, and as the herd grew it was dispersed out to zoos across the Middle East. By 1980 a small population was readied for release, acclimatizing it to the environment in Oman in a special pre-release enclosure which, though fenced, resembled the open landscape of the country. Eventually in 1982, 10 individuals were released from the enclosure and did well, followed by several more releases over the next few years. The herd in Oman reached a peak of 450 in 1997, but this declined to only 65 by 2007—apparently related to illegal live capture of individuals for collections, oil drilling in the area, and natural migrations of the oryx across into Saudi Arabia. This illustrates the vagaries that can affect a re-released population, even after it is established. However, using the expanded captive herds and gene pool that had provided the basis of the Oman reintroduction—plus the experiences gained from re-releasing the oryx—the United Arab Emirates and Saudi Arabia have now established much larger populations of hundreds of individuals in the wild. At present, then, the prospects for the Arabian oryx as a wild species are looking very hopeful.

The successful examples of the Californian condor and Arabian oryx can be put down to careful planning and persistence, and also follow-up programs which monitor the populations. Also important in the case of both of these species has been co-ordination with landowners and herders, to avoid conflicts which could lead to the re-established populations being hunted down.

These last two examples show the important role that zoos and private collections can play in preserving animals against extinction, and then getting them back out into the wild. Nevertheless, having to fall back on this is not a very satisfactory way of handling problems of impending extinction. It would be far better to avoid these problems in the first place by establishing reserves and passing laws, and emphasizing public education to help preserve the surviving populations in the wild. It has been calculated the cost of keeping African elephants in captivity is 50 times that required for keeping equivalent populations in Zambian national parks (Groombridge, 1993). Generally speaking, given the cost and the difficulty of getting captive populations re-established in the wild, zoos and other such sanctuaries can only act as a last resort, helping to preserve a rapidly dwindling population and gene pool after everything else has failed.

8.12 BOTANIC GARDENS

Botanic gardens are descended from both the herb gardens of medieval monasteries and universities, and from the private gardens of the wealthy. The herb gardens of medieval universities such as Oxford were an important part of medical training, used for familiarizing students with the most important plants used in healing. Formal planted gardens of flowers and trees as places for leisure have existed for thousands of years, but as contact with distant lands increased, wealthy people took to planting more and more new species of flowers, trees, and shrubs for novelty and show.

Out of these beginnings, from teaching gardens and places of recreation, the first botanical gardens open to the public originated in the mid-1500s, associated with universities in northern Italy. The idea of botanical gardens as extensive, eclectic collections of plants—some useful, some merely interesting—spread throughout Europe. As voyages of trade and exploration increased, especially from the 1700s onwards, botanical gardens in Europe began to send collectors on the ships to bring back specimens. Many were preserved parts of plants for taxonomic classification (see Chapter 6), but some were live specimens of seeds, bulbs, and even whole rooted plants. Nowadays, almost every major city in the world has its own botanic garden, open to the public. Some of these are very extensive, the largest in the world at present being the Queensland Botanical Gardens in Australia. A much larger botanic garden (some 50 km wide) is presently under construction in Shaanxi Province in China, scheduled to open in 2008 and concentrating on plants from across China. A particularly interesting feature of this new botanical garden is that it will traverse some 2,500 meters in altitude at the front slope of a mountain range—allowing plants from a wide range of climates to be grown there.

As time has gone on, botanic gardens have played an increasingly important role in both research and conservation. By cultivating specimens of certain rare plants, in at least some cases they have been able to aid the survival of species on the brink of extinction. An example is the conservation of the strange trees from the daisy family—known locally as "cabbage trees"—of the island of St. Helena, which lies isolated in the middle of the Atlantic Ocean (Chapter 1). When humans arrived on the island they brought goats, which devastated the native flora. Many species of the cabbage tree group (or "daisy tree" group as I like to call them: Chapters 4 and 5) went extinct, while others were reduced to just a few individuals. Cultivation of seeds and cuttings from some of the last few individuals of each species has established populations that can be replanted to supplement the surviving population. A total of 14 species of cabbage trees and other endemic species have been replanted on St. Helena. One example is the tree *Trochetiopsis melanoxylon* on St. Helena, not actually a cabbage tree but a member of the cocoa family Sterculiaceae. This species had been reduced to just two individuals, from which propagation by cuttings and seeds on the island yielded several thousand more, now widely planted across the island (Groombridge, 1993; Groombridge and Jenkins, 2002).

On Reunion Island in the Indian Ocean, another member of the Sterculiaceae, the near-extinct endemic shrub *Ruizia cordata*, was propagated at Brest Botanic Garden in France, and then re-established on the protected cliffs of a ravine on Reunion. On Easter Island in the Pacific, the endemic tree *Sophora toromiro* (a yellow-flowered member of the pea family, Leguminosae) went extinct in the wild under the onslaught of rats that humans had introduced. Fortunately it was already in cultivation in various botanic gardens. The seeds and cuttings of these scattered survivors have allowed the tree to be re-established on Easter Island.

On many oceanic islands, the endemic plants are in such a dire situation that this sort of approach seems to be the best hope of saving them. Mauritius (in the Indian Ocean close to Reunion) is one such example—it has 112 threatened species with either less than 20 individuals, or found in only 1 or 2 localities.

There have been at least 210 projects in 22 countries aimed at re-establishing or bolstering wild populations of plants using artificial propagation. However, because there has been relatively little follow-up observation to see how the plants fared, it is not clear how successful this approach actually is. In one follow-up study in California, it was found that out of 15 attempts to re-establish plant species, 10 failed (Groombridge, 1993). It would clearly be beneficial to do some clear, cold assessment of the facts, to figure out whether the success rate of re-establishing endangered plants justifies the expense—money that might be better spent on protecting the wild habitat of these plants. For all we know, the lack of genetic diversity in plant species that have recently been reduced to a handful of individuals could be impairing their ability to establish and make a comeback—something which should be carefully allowed for in trying to propagate them before planting, to maximize genetic diversity.

Generally speaking, plants seem less particular than animals about the conditions they require to breed, so releasing them back into the wild is less of a chancy business. But as in the case of zoos, the potential of botanic gardens for directly saving the world's biodiversity is extremely limited. Though plants are less expensive to look after than animals, any one botanic garden can only cultivate a very small proportion of the world's endangered plants. Also, the populations must by necessity be small, and their situation is always precarious. With just one short-lived unfortunate event (e.g., a climate extreme or temporary neglect in watering the plants), a population could be wiped out. Potentially a larger contribution of botanic gardens to conservation comes from the taxonomic collections and expertise housed at botanic gardens. Millions of dried plant specimens have been stored at botanic gardens worldwide. In order to conserve biological diversity effectively, it is necessary to know what is where, and this is how botanic gardens may ultimately play their most useful role.

8.13 SEED AND EMBRYO BANKS

What with all the uncertainties about the fate of populations and whole habitats in the face of deforestation and climate change, many conservationists are interested in having a further last resort beyond nature reserves and collections. This back-up plan involves storing material from plants or animals, freezing it, and hopefully keeping it safe from all the world's problems, until a future time when there is enough safe habitat for it to be released back into the wild.

Stores of dried and frozen seeds of plants, and the sperm and eggs of animals, have been set up to try to conserve particular threatened varieties of crops and domesticated animals. However, there are also efforts underway to use it as a last-ditch attempt—or at least an insurance policy—to save whole species.

At Wakehurst in England, one of the world's most extensive wild gene banks has been set up for seeds of plants from around the world. The Millennium Project, as it is called, is funded by the British Government through the proceeds of a national lottery. At present more than 24,000 species are preserved there, including both rare and common species and among them the entire flora of the United Kingdom. Seeds obtained from wild plants are first dried, and then flash-frozen in liquid nitrogen and afterwards stored at −20°C. The seeds of most species of plants, dried to 5% to 8% water content and stored close to freezing, will last for between 5 and 25 years. Stored at about −20°C, they seem to remain viable for around 100 years. Although it has been going for only around a decade, the Millennium seedbank has already played a key role in saving at least one species. The Gibraltar campion (*Silene tomentosa*), a delicate white flower, only occurred on the Rock of Gibraltar in southern Spain, but recently died out in the wild. Its population was re-established using plants grown from seeds that had earlier been collected and stored in the Millennium seedbank.

Many botanic gardens have supplemented their role in the conservation of species by setting up their own seedbanks. By 1990, 528 botanic gardens around the world already had facilities for seed storage. Many of these are geographically focused—for example, the "Artemis" seedbank at the botanic garden of the Universidad Politécnica de Madrid, Spain. This has 1,000 of the 1,300 endemic Spanish taxa (species and subspecies) in its collections.

Frozen seeds do not remain viable forever. It seems that over time some sort of damage accumulates in their cells, despite the low temperatures. As an ultimate long-term limit of viability, natural radioactivity within the seeds and cosmic rays from space will tend to break up their DNA and prevent cells from functioning and dividing. How fast this will happen is uncertain. Seeds of lotus from Egyptian tombs have been germinated successfully after several thousand years, and a radiocarbon-dated 2,000-year-old date palm seed from Herod the Great's Palace in Israel was also grown into a healthy tree. However, examples of longevity such as this seem to be exceptional among the world's plant species. In order to check how well the viability of a preserved stash of seeds is holding up, small samples of the seeds of each species are periodically germinated to check for any decline in germination rates. If viability seems to be declining rapidly, the remainder of the stock is germinated and then as best as possible grown on to maturity. This is much easier to do with herbaceous plants than with trees, which normally take years to get to reproductive age. The seeds of the re-sown plants are then harvested and added back into storage. There is some danger here that compared with the original wild population, each sowing and harvesting event could select for certain variants that are better suited to being grown under artificial conditions. We could unintentionally be domesticating the plants that we are trying to save, leaving them unable to survive back in the wild.

Some other drawbacks of this method of seed banking include the possibility of accidental loss of the seeds (e.g., if power is cut off and the seeds thaw out). The main seedbanks have back-up power supplies, but of course if the electricity were cut off for long enough or if there were to be enough of a breakdown of civil order (e.g., as in a war), they could be lost.

In addition, there are many species which cannot so far be stored in the typical seedbank, dried, and kept at low temperatures. These are plants whose seeds normally germinate within a few days of falling from the tree, and have no adaptations to survive drying or cold storage. Many plants from the moist equatorial tropics come into this category, known as "recalcitrant" seeds for all the problems they cause. Some temperate-zone plants, too, have seeds which must germinate almost immediately: Willows (*Salix*) are one example. It is estimated that about 20% of the world's plant species (thus comprising about 50,000 species) have recalcitrant seeds (Groombridge, 1993). This rather limits the potential for seedbanks to preserve the world's botanical diversity.

For those seeds which *do* preserve well at low temperatures, one way to get around the problems of keeping them frozen and safe might be something like the Svalbard International Seedbank. Located on the Arctic Island of Svalbard (at 74°N), this seedbank is buried deep in a series of vaults at the end of a 120 m tunnel excavated into the side of a mountain, protected by thick blast-proof doors. In rocks that maintain a steady temperature of −3°C year-round, and chilled further to −30°C using locally mined coal to power the refrigeration units, the Svalbard Seedbank seems likely to survive most of the catastrophes that could befall the human race. However, the Svalbard Seedbank is not in itself a project that will help to preserve the Earth's species diversity. At present it exists solely to preserve cultivated varieties of crop plants, not wild plants.

8.14 *IN VITRO* STORAGE OF PLANT GENETIC MATERIAL

The term *in vitro* literally means "in glass", as in the glassware of a test tube or beaker. In science it has come to refer to any experiment that occurs in a liquid or gel in a small container in a laboratory. This environment has also begun to be used to try to preserve samples of living plants, not as seeds but as small fragments of tissue preserved in a growth medium. The growing tips of plants (known as meristems) can survive months or years in the right medium, kept close to freezing or at cool temperatures—and then when they are warmed up form young plants with roots and shoots that can be planted into soil. If they are not grown out into fully-fledged plants, eventually the tissue samples age and die. Before they age too much, they must be grown on in the culture medium until they produce shoots, from which new samples of growing tips can be cut out and preserved in culture for more months or years. This method provides some hope of preserving the tens of thousands of plant species whose seeds cannot survive in storage. So far it has mainly been used to preserve cultivated plant varieties, but some 1,500 species of wild plants are currently being kept in storage in this way. A problem with *in vitro* storage is that it is much more labor-intensive than seed storage, and the stored tissues are more susceptible to perishing if a cooling or heating system breaks down, for example. There is also the worry that subtle genetic errors could arise and be passed on over many rounds of growing on and culturing, to the point where the plant that comes out at the end is quite different from what was started with, and is also unable to survive in the wild.

However, the *in vitro* method could prove to be a very useful stopgap for many species whose last wild populations are being wiped out.

What about other groups of organisms? What are the prospects for preserving them safe from the world outside? So far there seems to have been little progress with the idea of sperm or embryo banks for endangered animals—the only ones in existence are for threatened domesticated breeds. Remarkably, however, there are some 345 known repositories of microbial diversity, where fungi and bacteria are kept in pure culture or spore form.

8.15 RESCUE THROUGH GENETIC ENGINEERING AND BREEDING, AGAINST INTRODUCED PESTS AND DISEASES

One of the greatest threats to species richness in the temperate forests is caused by introduced pests and diseases, which have been brought across from other lands. Examples of ecological disasters caused by introduced species include the near-extinction of the chestnut (*Castanea dentata*) in eastern North America, and a similar near-extinction of elms (*Ulmus*) in Europe. Both were wiped out by different fungi that were accidentally introduced from Asia (see Chapter 7). There are now many more introduced pests and diseases that threaten to do the same thing to maples, buckeyes, hemlocks, ash, oaks, and others (Adams *et al.*, 2002).

It is not really clear how much of a threat these various introduced enemies pose to the long-term future of the tree species that they affect. There is evidence from the fossil pollen record that in the past several thousand years, new strains of pre-existing tree diseases have sometimes broken out naturally, and wiped out almost all the populations of their own host tree species. For example, this is what most likely happened to the hemlock (*Tsuga canadensis*) 6,000 years ago in North America (Fuller, 1998), and may have contributed to a decline of the elm (*Ulmus* sp.) in Europe about 5,000 years ago (Parker *et al.*, 2002). Each of these trees suddenly went from very abundant to rare in the space of a few decades, or at most a couple of centuries. The various "black oak" species in eastern North America show the genetic signs of having been through a severe population bottleneck sometime in the last several hundred thousand years, perhaps caused by disease (Jeanne Romero, pers. commun.). Yet in each case, they came back to be abundant—and in the case of elm and hemlock it took only a couple of thousand years.

We cannot be sure that the tree species affected by the new wave of introduced pests and diseases will be so lucky. As far as one can see, there is little prospect of the American chestnut ever making a comeback by itself. Even though the roots of the killed trees produce young shoots (known to foresters as "suckers") that persist in the forest understory in some places, the fungus lives on in the roots and attacks the growing young trees once they reach about 3 or 4 meters height. It seems unlikely that the chestnut will be able to survive like this indefinitely, especially since the trees never become large enough to reproduce by seed. There are just a handful of the wild chestnut trees in eastern North America that have somehow escaped the blight, but it is not clear whether these really will prove resistant in the long term. Whether these isolated individuals can expand up to establish viable populations is also uncertain—they could already have lost most of the genetic diversity that they would need to survive in the long term. All considered, it is quite possible that we are facing the complete extinction of the American chestnut in the wild.

The various European species of elms are faring better, though not very much better. While a widely planted strain of cultivated elm known as ''English'' elm seems to show no prospect at all of gaining resistance, some other Europan elms such as the wych elm (*Ulmus glabra*) are only able to persist in the extreme north of their former range, where summers are too cool for the beetle (*Scotylus* sp.)—which spreads the spores of disease. But if climate warms further, this remaining range is likely to become occupied by the beetles and these remaining trees will die.

There is a strong chance that several other temperate species or genera of trees will be lost over the coming decades (see Chapter 7). The extinction of a large proportion of the temperate tree flora of North America is actually a serious possibility.

The only sure way to save and bring back these species seems to be some form of artificial breeding. One possible way forward which has already been attempted is crossing with other resistant species, often from the region of origin of the pathogen or pest. For example, in the case of the American chestnut, hybridization with the Chinese chestnut (*Castanea mollissima*) has yielded trees with a high degree of resistance to chestnut blight. The problem is, these hybrids look different and are plainly not the original American chestnut. The American Chestnut Foundation based in Vermont, U.S.A. has arrived at an ingenious plan for reconstituting the American chestnut from these hybrids, using repeated crossing back with the American chestnut. To start with, the American–Chinese hybrid is crossed with surviving American chestnuts (protected from the blight using fungicide) to give young trees that are now three-quarters American chestnut. These trees are then exposed to the fungus, and those which survive (because they happen to carry the resistance genes against the fungus) are grown on to reproductive age. They are then crossed back with American chestnut again, to give a tree which is now seven-eighths American chestnut, and so on; with each generation the genetic contribution from the Chinese trees is halved and yet only trees carrying the resistance genes are selected, by their survival of exposure to the fungus.

Eventually with enough generations of back-crossing the Chinese component of the genome will be negligible, except for the blight resistance genes. This process has now gone through several generations, growing on a new generation for crossing and selection every 5 to 10 years. Even after an almost complete American chestnut has been reconstructed bearing the Asian resistance genes, it will take time to multiply up the population and then plant it widely throughout the eastern forests. The Chestnut Foundation is taking seriously the idea that different genetic strains of America chestnut trees (''provenances'') might be adapted to local climate and soil circumstances, and is deliberately hybridizing different strains of the chestnut to plant back into these source regions. Overall, the prospects are good that the American chestnut can be saved, albeit as a slightly modified form of the original wild tree. However, even in the best-case scenario, it will be centuries before American chestnut trees can once again tower in forests throughout eastern North America.

In Europe the hybridization method was used in a program, based in the Netherlands lasting several decades, to introduce genes to native elms (devastated by elm disease) from various species of Chinese elms. Although the hybrids are resistant to elm disease, they do not have the shape of the original European elms. This then is another clear case for using multiple generations of back-crossing to dilute the number of imported foreign genes in the trees.

One other option besides the classic method of hybridization has become available in the past few years. This is to use "genetic engineering" (more properly termed biotechnology) to directly transplant genes from resistant species into the susceptible trees. While the capabilities of genetic engineering are certainly impressive, it is first necessary to ask whether the same ends cannot be reached through traditional hybridization techniques. This could be cheaper and require less effort. However, in some circumstances, genetic engineering might prove to be the only option for saving a particular species of tree. There are cases where the nearest relative that has resistance cannot hybridize with the susceptible species of tree. Taking genes out directly and transplanting them might then be the only option. An additional potential advantage of genetically engineered material is that it can be transferred very quickly, without the necessity of waiting for several generations of back-crossing to dilute the genes from the foreign species.

On the other hand, the disadvantages of genetic engineering include the expense of setting up a research project to identify and then transplant the genes. In many plants, disease resistance is not simply down to a handful of genes, but many genes scattered through the genome. In general, in plants the disease resistance mechanisms are poorly understood and it would be hard to identify genes which play a key role. In at least some plants, a major part of the DNA in the cell nucleus consists of repeated copies of genes involved in resistance producing, for example, lectins—proteins with linked polysaccharides—which may govern cell surface interactions with invading microbes (Adams *et al.*, 2002). Many of these repeated copies of lectin genes are subtly different from one another and some seem to be nothing more than inactive "junk" copies. It is difficult to imagine how one could ever begin to transplant such a huge and diverse part of the genome from one species into another, without disrupting other functions too. Even if resistance is achieved, when the plants are grown from the tissue culture they can sometimes look very different from the parent tree, because of mutations in the cells that occur in the tissue culture. This is known as "somaclonal" variation.

So far, genetic engineering for disease resistance followed by widespread planting has only been achieved in one tree, the cultivated papaya (*Carica papaya*), against ringspot virus. This has led to a resurgence in papaya production from Hawaii, which had previously seen its papaya crop devastated.

At present, there are considerable challenges to using genetic engineering to try to save trees from pests and diseases. However, as understanding of plant resistance mechanisms develops, it may become easier to identify the right genes for transplanting. Once it becomes a viable option for many species, the decision whether to use genetic engineering is likely to generate a lot of controversy. The environmental movement is inherently suspicious of genetic engineering (understandably perhaps, given the history of disasters caused by the release of introduced organisms, including the tree diseases themselves), and yet this may actually provide the only viable option for saving certain species. Ultimately, it may be necessary to weigh up the concerns about using such methods against the prospect of losing a species—a difficult decision for any conservationist or environmentalist to make.

8.16 BACK FROM THE DEAD: CAN WE REGAIN ANIMALS THAT HAVE ALREADY GONE EXTINCT?

For animals, there have been various attempts to "reconstruct" species or subspecies that have been lost in their original wild form, from hybrids with other species, or with highly bred domesticated descendants. So far these have only involved extinct subspecies of forms which still exist, rather than fully-fledged extinct species (although the taxonomic classification is often moot). The first attempt to do this aimed to bring back the extinct wild ox of western Eurasia, known as the aurochs (*Bos taurus* subsp. *primigenius*). This animal was the ancestor of the domestic cow, sometimes viewed as a separate species from the modern cow (which is usually classified as a separate subspecies *Bos taurus* subsp. *taurus*), because of the profound differences in its anatomy, physiology, and behavior. Once widespread through the European forests, the aurochs suffered the effects of hunting and habitat loss, and despite efforts to conserve it (see above), became extinct in its original form around the 1620s. A project to "bring back" the aurochs was begun by the Heck brothers in Germany in the 1920s. Their idea was that the characteristics of the aurochs remained scattered through various modern-day cattle breeds, and could be brought together by selective breeding. From old descriptions of the aurochs and from its bones, the Heck brothers knew more or less what they were aiming for: an animal which was larger than modern breeds (up to 180 cm at the shoulder in the bulls), with forward-pointing horns. Females and young had reddish-brown coats, while the males had a dark color, and a distinctive pale stripe along the spine. In an effort that took about a dozen years, the Heck brothers each reconstructed what they felt were reasonable approximations of the aurochs, each with their own separate breeding program. Only the cattle of one of the brothers, Heinz, survive today, and these are generally said to bear a poor resemblance to the "true" aurochs. While larger than other cattle breeds, these are still smaller than the aurochs, and do not bear the forward-pointing horns or white stripe. They do, however, have a darker coloration in the males, with females and young being reddish-brown. It has been said that, if anything, the Heck brothers made a poor choice of breeds and should have used Spanish fighting cattle, which more closely resemble the aurochs in their horns and coloration. Generally nowadays the Heck brothers' basic idea—that it is possible to bring a species back by breeding from its domesticated descendents—is not very well thought of. The Heck cattle, while bearing some passing resemblances to the aurochs, are clearly not aurochs.

Another attempt to reconstruct an extinct form focused on the quagga (*Equus burchelli quagga*), a subspecies of the plains zebra (*Equus burchelli*). Unlike the plains zebra, which normally has stripes all over its body, the quagga had stripes only on its front quarters and neck. This was not unique to the quagga and occurs to some extent in modern plains zebra populations. A breeding program in South Africa selected from within the natural variation of the plains zebra, selecting the most "quagga-like" individuals. After several generations of crossing and selection from among these, animals bearing the basic coloration patterns of the quagga have been arrived at. However, again this cannot really be said to be *the* quagga, just an animal that looks a lot like it.

Instead of trying to pull together genes and characteristics from descendents and close relatives of an extinct species or subspecies, would it be possible to bring it back from the DNA surviving in preserved remains? It has been shown that DNA can survive, at least in a fragmentary form, for hundreds of thousands of years if the remains dried out quickly after death or remained at a low temperature. For example, parts of the DNA of our extinct relatives the Neanderthals have been sequenced from well-preserved bones and teeth buried in cold caves in Europe, dating back some 40,000 years. The oldest DNA fragments of any life form that have been recovered and published in a respectable journal are from bacteria in 250-million-year-old salt deposits, although this study has been heavily criticized as based on unreliable techniques that could have led to contamination. In Michael Crichton's novel *Jurassic Park*, dinosaurs were reconstructed from the blood in the guts of mosquitoes preserved in amber. From all that we know, nowhere near enough DNA survives from fossils that age to reconstruct a whole living organism, although there have been published claims of the extraction of certain genes from dinosaur remains. However, for far more recently extinct organisms, a lot more DNA does survive and it might eventually be possible to reconstruct them using this DNA. The greatest hope is for museum specimens of species which have died out in the last few hundred years, and which have been preserved carefully under dry and cool conditions. Already the DNA of museum specimens has been used to show, for example, that the dodo was basically a type of pigeon, and that the thylacine was most closely related to the Tasmanian devil among the living marsupials.

A possible candidate for resurrection from DNA is the subspecies of zebra known as the quagga (see above). Several skins of the quagga still exist in museum collections. Another candidate is the thylacine: a pup preserved in formalin still exists and could yield a lot of DNA. However, even if most of the animal's DNA is present, it is likely to be in a highly fragmented form. Even if it is not fragmented, the drying or pickling will have distorted the intricate arrangement of DNA in the nucleus, plus the protein coating known as chromatin that controls the expression of DNA. At present, there is simply no practical way to get all the DNA in the nucleus of a dead, dried cell out and then getting it to work by transplanting it into a new egg cell that can then host the developing embryo. It would be unwise to say that this can never be achieved, for many things that once seemed equally far-fetched have already been done in the last couple of decades. All we can say is that for now, these extinct animals must remain extinct.

8.17 WARNING LABELS: ALERTING GOVERNMENTS AND THE PUBLIC OF A SPECIES IN TROUBLE

How best to tell the world that a particular species is in danger of extinction? If hardly anyone knows of the trouble that the species is in, then its conservation will not be given the priority that it deserves. One way to get the message across is to publish lists of the species that are known to be especially threatened, and rank them into different categories according to just how threatened they are.

The first attempt at doing something like this was started by the redoubtable conservationist Sir Peter Scott in the 1960s, leading to the *Red Data Book*. This book concentrated on listing the land vertebrates that seemed to be in danger of extinction, and placed each in a category from most threatened and in imminent danger, to least threatened but still a cause for concern. The idea was taken up by the U.N.-linked International Union of Conserving Nations (IUCN) and a new edition was put out each year, listing the world's most threatened vertebrates with updated and adjusted threat categories. The IUCN also began publishing separate red data books for invertebrates and plants.

Presently, the IUCN has switched away from publishing books to instead producing less costly and less cumbersome red data lists of threatened species, readily available on the Internet. The lists are updated and put out every two years, and the work of researching and compiling them (bearing in mind all the latest information of how each species is doing) is done through the World Conservation Monitoring Center (WCMC) in Cambridge, England, and through some other groups such as the International Council for Bird Preservation. The raw information for these lists is sent in by many hundreds of scientists, amateur wildlife enthusiasts, and conservationists from around the world.

On a red data list, a species is assigned its threat category based on several different types of information. These include the abundance of that species, how localized its distribution is, the intensity of the threats, and whether its population biology suggests it is likely to be able to recover its numbers readily.

In the IUCN Red Data List, the worst categorization a species can have is (not surprisingly, really) "Extinct". This is when it has not been located in the wild for at least 50 years, which should have been plenty of time for someone to find it again if it was still around. The next worst is "Endangered", which is when survival is unlikely if the current situation continues. "Vulnerable" means that the species is likely to become endangered in the near future, if the current situation continues. "Rare" is when the population is fairly small and at risk. "Indeterminate" means that the species could be in any of the above categories, but not enough is really known about it to be able to say. The most non-committal category of all is "Insufficiently known", which basically states that it might or might not be in the top threat categories, but there is only a fairly vague suspicion about this.

Confusingly, there are also national red data lists which deal only with the status of a species within the borders of a particular country. Looking on a country scale, a species might be badly threatened but actually in no danger on a broader scale that takes in its range elsewhere. For example, several species of orchids such as the lady's slipper (*Cypripedium calceolus*) are threatened nationally in the U.K., but remain much more common elsewhere in Europe.

8.18 LAWS

Formal rules that are concerned with conserving animals and plants have probably existed as long as there have been human societies. When Europeans first discovered them, hunter-gatherer societies such as the Australian aborigines had their own sets of unwritten laws, and many of these apparently related to the management of wild animals and plants that supplied food and other important resources. For example, some aboriginal tribes had taboos against men gathering certain resources such as ants' eggs, which only women were supposed to gather. There were "closed seasons" for certain prey, during which it was forbidden to hunt them. To most hunter-gatherer people, the underlying reasons for these rules being in place were not clear—it was simply ordained by the gods that such and such a thing was forbidden. Yet in effect many of these tribal laws are like resource management rules designed to ensure that that populations of prey did not crash down or become extinct. Restricting the proportion of the human group that could harvest a resource might prevent over-exploitation. Not hunting during a breeding season would allow prey populations to recover. The origin of these laws must have been many bitter experiences over tens of thousands of years, which empirically demonstrated what would happen if a living resource was exploited in the wrong way. It is even possible that the over-exploitation and extinction of the Quaternary large mammal fauna (Chapter 5) left its mark in folk memories of the more recent past.

In the early recorded history of civilizations there were more laws and superstitions to protect certain natural resources. The cedar forests along the rivers in ancient Sumeria, in Iraq, were said to be protected by guardian spirits that would destroy anyone who tried to cut them down. There were other sacred groves and forests in many parts of the ancient world (e.g., in Britain on the island of Anglesey).

In medieval times, kings would often decree that a certain animal was off-limits to hunting. While the motives were mainly selfish (to ensure that more was left for themselves to hunt) in many cases this probably prevented or delayed the local extinction of an animal, such as wild boar or black bear. One example I have already mentioned is the aurochs, which was protected by kings in northern Europe, to ensure a continued supply of the animals for hunting. But when the aurochs population declined, hunting by the nobility actually ceased by choice.

As interest in nature for its own sake (not just as a resource) has grown, so have the number of laws passed just to preserve species. In the early days these laws tended to come too late. The heath hen was the first taxonomic grouping (actually a subspecies, not a species) to be protected against hunting by law in the U.S.A., in the mid-1800s, although this did not ultimately prevent its demise. The passenger pigeon in the U.S. and thylacine in Australia were both protected by law when they were almost extinct, but it was also too late to save them. The first plant to be protected by law was the edelweiss (*Leontopodium alpinum*), the collecting of which was banned in 1911 in the Swiss canton of Zug. During the course of the 20th century, more and more species were protected. Many are still only protected by law within one country, but others are protected by international agreements, which are enforced through national-level laws.

Just how threatened a species should be to deserve the protection of law is a moot point. Some species perceived as rare may turn out (with better data) to be quite common—one possible example being the great crested newt (*Triturus cristatus*) in England, which inhabits murky ponds. Though it has legal protection, there is evidence that it is not so often seen because of the unpleasant habitat it lives in, rather than any real rarity. When I worked in nature conservation in England we seemed to find the great crested newt in every stinking, junk-filled pond we waded into. It just seemed that no-one else had ever wanted to go in and look for them in such places. Another and very encouraging example is the lowland gorilla, which has recently been discovered to have much larger populations in central Africa than had previously been thought. In this case, a population of perhaps 125,000 has been living hidden in the vast swamplands of the Congo Basin.

Over the past few decades, as knowledge and concern about the loss of species has spread, the number of laws intended to protect them has multiplied. These laws can differ quite a lot in their content and style from one country to another. Take the example of legislation designed to protect the wild flora. In some places, there is a near-blanket ban on collecting any wild plant, while in others (e.g., parts of Switzerland, and in Sweden) a certain limited quantity of most species can be picked—often just the flowers, and then only a certain number of flowers are allowed per person. In some countries (e.g., the U.K.), the right to collect wild plants depends on having the landowner's permission. In most countries, there are also laws passed to protect certain species which are seen as being under particular threat—with the severity of the penalty for uprooting or damaging them depending on how rare they are.

However, just because a species is rare and threatened in one country does not necessarily mean that is in danger of complete extinction. Some species that are protected on a national scale are simply at the edge of their geographical distribution range, and are actually quite abundant in adjoining countries. For example, as I mentioned above, some species of orchids which are rare and strictly protected in Britain are far more common in other parts of Europe. In such cases, it is simply seen as desirable to preserve as much of a country's natural diversity as possible, for the people in that country to see and enjoy—whether or not the species is threatened in the rest of its range.

The incentive for taking plants or animals from the wild is often cash, and many countries have outlawed aspects of the trade in wild species. Sometimes the trade involves parts or products of species that have already been killed (e.g., the ivory of elephants or the skins of big cats). In other cases, the trade ban focuses on live individuals. For example, in Costa Rica there is a ban on the export of wild orchids due to the problem of mass-collecting of orchids for sale abroad. In the U.S.A., the Venus flytrap (*Dionea muscipula*) is still often collected illegally from its remaining wild populations in the Carolinas.

While animals and plants that are collected live, sold locally, or exported may be appreciated and well looked after by those who ultimately buy them, they are unlikely to be able to reproduce and establish long-term viable populations in captivity or cultivation. For example, the Carolina parakeet (Chapter 5) was often captured and kept in captivity, where it was said to breed easily. Nevertheless, it went extinct anyway—both in the wild and in captivity. As regards private collectors and breeders, there are just too many accidents, whims, and changing fashions to provide a species any long-term security. In general, each individual taken from the wild is effectively dead, as far as the future of the species is concerned.

To preserve species more effectively against complete extinction, various international conventions have been signed, and matching co-ordinated sets of laws have been passed across many different nations. There are now international bodies that operate through the United Nations and other agencies to try to get countries working together in preserving their species richness. One important mover in all of this is the IUCN.

The efforts of international bodies have resulted in a series of conventions—standardized sets of laws. Among other activities, they have helped set up legally protected reserve areas that are intended to preserve as many species as possible (see above).

As well as setting aside broad areas of habitat to allow species to survive, it is often necessary to pass laws that target specific threats which come from hunting or collecting of species, or unintentional destruction of them during development. Some of these co-ordinated laws have been aimed at controlling the trade in live individuals of threatened species (e.g., parrots or orchids).

Ultimately, the more we know about what is going on with the species in our world, the better equipped we are to save those in danger of going extinct, and head off serious problems before they start. This is why organizations that collect and disseminate accurate information—such as the WCMC—are so important.

There is another fundamentally different type of law that is also important in preserving the Earth's biological diversity. In this case it is not about what can be taken out, but what can be brought it. Introduced species (Chapter 7) are one of the greatest threats to species richness worldwide. Most countries now have strict laws prohibiting the import of live species for these reasons, and also bans or restrictions on import of fresh fruit, vegetables, or topsoil. The introduced species of animals and plants may themselves run rampant, or the diseases and pests they carry may devastate wild species in the new range. It is up to each of us to respect these laws, and not bring any materials between countries that could unleash such ecological disasters.

8.19 WHERE WE STAND NOW

Things have changed. It staggers anyone to hear nowadays that back in 1938 the Chicago Mill and Lumber Company willfully, knowingly felled the Singer Tract, the last extensive area of forest that the ivory-billed woodpecker was known to nest in

(Chapter 5). Perhaps it shows how far humanity has come that such actions by any corporation against the last of a valued, iconic species would be inconceivable.

The old medieval thinking—that species in the Creation cannot go extinct and that nature exists only to be exploited and tamed—has been replaced by the view that it is in fact fragile, and that wilderness is also something desirable in itself. Increasingly, the view that species should be protected has become the mainstream. It is interesting that although most people will never visit a tropical rainforest, and many would not even want to visit one for all the heat and bugs, they are nevertheless saddened by the idea that species are going extinct. There is a widespread feeling that even if we never see them, we should try to preserve species—just because they are there. Ultimately, the decision as to whether most species are worth conserving is a subjective one. It is hard to make any convincing case that most of the millions of species in the world are useful to us in any way, even though some of them might turn out to be in the long run.

The change in attitudes seems to be partly the result of becoming an affluent, educated society. Once people have enough food to fill their stomach, and material goods to fill their house with, their minds can look beyond such things and become concerned with nature. This change has already gone a long way in North America and Europe. It will be interesting to see if it now occurs in other countries, such as China, India, and Indonesia which are now industrializing so rapidly. While each has taken great strides in the conservation of its forests, and in the preservation of certain species, they are all still suffering severe environmental degradation in at least some natural habitats.

And despite all their concern, the richer nations of the world are still a large part of what is probably the greatest threat of all to the world's species richness—global climate change, and CO_2 enrichment. Europe, North America, and Japan produce most of the world's greenhouse gases—although China and India are now catching up with them fast (and, in fact, China may already now have surpassed the others, according to recent news reports). Even with all the goodwill in the world, it is still not certain that *Homo sapiens sapiens* has the determination to do what it will take to preserve most other species.

References

Abele L.G. and Connor A. (1982) Refuge design and island biogeographic theory: Effects of fragmentation. *American Naturalist*, **120**, 41–50.

Achard F., Eva H., Stibig H-J., Mayaux P., Gallego J., Richards T., and Malingrau J-P. (2002) Determination of deforestation rates of the world's humid tropical forests. *Science*, **297**, 999–1002.

Adams J.M. (1988) The geographical ecology of tree species richness: A study of the northern temperate zone. Unpublished Masters thesis, University of Wales, U.K.

Adams J.M. (1989) Productivity and tree species diversity. *Plants Today*, **2**, 85–91.

Adams J.M. (1997) *Global Land Environments since the Last Interglacial.* Oak Ridge National Laboratory, Oak Ridge, TN. Available at *http://www.esd.ornl.gov/ern/qen/nerc.html*

Adams J.M. (2007) *Vegetation–Climate Interactions: How Vegetation Makes the Global Environment.* Springer/Praxis, Heidelberg, Germany/Chichester, U.K.

Adams J.M. and Faure H. (1997) Palaeovegetation maps of the Earth during the Last Glacial Maximum, and the early and mid Holocene: An aid to archaeological research. *Journal of Archaeological Science*, **24**, 623–647.

Adams J.M. and Faure H. (1998) A new estimate of changing carbon storage on land since the last glacial maximum, based on global land ecosystem reconstruction. *Global and Planetary Change*, **16**, 3–24.

Adams J.M. and Woodward F.I. (1989) Patterns in tree species richness as a test of the glacial extinction hypothesis. *Nature*, **339**, 699–701.

Adams J.M. and Woodward F.I. (1992) The past as a key to the future: The use of palaeoenvironmental understanding to predict the effects of man on the biosphere. *Advances in Ecological Research*, **22**, 257–314.

Adams J.M., Maslin M., and Thomas E. (1999) Sudden climate transitions during the Quaternary. *Progress in Physical Geography*, **23**, 1–36.

Adams J.M., Foote G.R., and Otte M. (2001) Could humans have existed undetected in North America before the LGM? An inter-regional comparison. *Current Anthropology*, **42**, 563–565.

Adams J.M., Piovesan G., Strauss S., and Brown S. (2002) The case for genetic engineering of native and landscape forest trees against introduced pests and diseases. *Conservation Biology*, **16**, 874–879.

Adams J.M, Green W., and Zhang Y. (2008) Leaf margins and temperature in the North American flora: Recalibrating the paleoclimatic thermometer. *Global and Planetary Change*, **60**, 523–534.

Al-Mufti M.M., Sydes C.L., Furness S.B., and Grime J.P. (1977) A quantitative analysis of shoot phenology and dominance in herbaceous vegetation. *Journal of Ecology*, **65**, 759–791.

Alroy *et al.* (2008) Phanerozoic trends in the global diversity of marine invertebrates. *Science*, **321**, 97–100.

Alvarez L.W., Alvarez W., Asaro F., and Michel H.V. (1980) Extraterrestrial cause for the Cretaceous–Tertiary extinction. *Science*, **208**, 1095–1108

Anhuf D., Ledru M-P., Behling H., Cruz Jr., F.W., Cordeiro R.C., Van der Hammen T., Karmann I., Marengo J.A., De Oliveira P.E., Pessenda L. *et al.* (2006). Paleo-environmental change in Amazonian and African Rainforest during the LGM. *Palaeogeography, Palaeoclimatology, Palaeoecology*, **239**(3/4), 510–527.

Arinobu T., Ishiwatari R., Kaiho K., and Lamolda M.A. (1999) Spike of pyrosynthetic polycyclic aromatic hydrocarbons associated with an abrupt decrease in delta ^{13}C of a terrestrial biomarker at the Cretaceous–Tertiary boundary at Caravaca, Spain. *Geology*, **27**, 723–726.

Bambach R.K., Knoll A.H, and Sepkoski J.J. (2002) Anatomical and ecological constraints on Phanerozoic animal diversity in the marine realm. *Proceedings National Academy Sciences U.S.A.*, **99**, 6854–6859.

Barnosky A.D., Koch P.L., Feranec R.S., Wing S.L., and Shabel A.B. (2004) Assessing the causes of late Pleistocene extinctions on the continents. *Science*, **306**, 70–75.

Barbour C.D. and Brown J.H. (1974) Fish species diversity in lakes. *American Naturalist*, **108**, 963–991.

Basset Y.S. (2008) Diversity and abundance of insect herbivores foraging on seedlings in a rainforest in Guyana. *Ecological Entomology*, **24**, 245—259.

Becker L., Poreda R.J., Hunt A.G., Bunch T.E., and Rampino M. (2001) Impact event at the Permian–Triassic boundary: Evidence from extraterrestrial noble gases in fullerenes. *Science*, **291**, 1530–1533.

Beerling D.J. (2000) Increased terrestrial carbon storage across the Palaeocene–Eocene boundary. *Palaeogeography, Palaeoclimatology, Palaeoecology*, **161**, 395–405.

Beerling D.J., Lomax B.H., Royer D.L., Upchurch G.R., and Kump L.R. (2002) An atmospheric pCO_2 reconstruction across the Cretaceous–Tertiary boundary from leaf megafossils. *Proceedings National Academy Sciences U.S.A.*, **99**, 7836–7840.

Benton M.J. and Twitchett R.J. (2003) How to kill (almost) all life: the end-Permian extinction event. *Trends in Ecology & Evolution*, **18**, 358–365.

Berger J., Swensson, and Persson I. (2001) Recolonizing carnivores and naive prey: Conservation lessons from Pleistocene extinctions. *Science*, **291**, 1036–1039.

Berkelmans R., De'ath R., Kininmonth S., and Skirving W. (2004) A comparison of the 1998 and 2002 coral bleaching events on the Great Barrier Reef: Spatial correlation, patterns, and predictions. *Coral Reefs*, **23**, 74–83.

Berner R.A. and Beerling D.J. (2007) Volcanic degassing necessary to produce a $CaCO_3$ undersaturated ocean at the Triassic–Jurassic boundary. *Palaeogeography, Palaeoclimatology, Palaeoecology*, **244**, 368–373.

Berner R.A. and Kothvala Z. (2001) Geocarb III: A revised model of atmospheric CO_2 over Phanerozoic time. *American Journal of Science*, **301**, 182–204.

Betancourt J.L. (2000) The Amazon reveals its secrets—partly. *Science*, **290**, 2274–2275.

Bickford D., Lohman D., Sodhi N.S., Ng P.K.L., Meier R., Winker K, Ingram K.K., and Das I.I. (2007) Cryptic species as a window on diversity and conservation. *Trends in Ecology & Evolution*, **22**, 148–155

Bond W. (1983) On alpha diversity and the richness of the Cape flora: A study in southern Cape fynbos. *Ecological Studies*, **31**, 567–578.

Brook B.W. and Bowman D.M.J.S. (2002) Explaining the Pleistocene megafaunal extinctions: Models, chronologies, and assumptions. *PNAS*, **99**, 14624–14627.

Brook B.W. and Bowman D.M.J.S. (2004) The uncertain blitzkrieg of Pleistocene megafauna. *Journal of Biogeography*, **31**, 517–523.

Brook B.W., Bowman D.M.J.S., Burney D.A., Flannery T.F., Gagan M.K., Gillespie R., Johnson C.N., Kershaw P., Magee J.W., Martin P.S. *et al.* (2007) Would the Australian megafauna have become extinct if humans had never colonised the continent? Comments on "A review of the evidence for a human role in the extinction of Australian megafauna and an alternative explanation by S. Wroe and J. Field". *Quaternary Science Reviews*, **26**, 560–564.

Bromham L. and Cardillo M. (2003) Testing the link between the latitudinal gradient in species richness and rates of molecular evolution. *Journal of Evolutionary Biology*, **16**, 200—207.

Bromham L., Rambaut A., Fortey R., Cooper A., and Penny D. (1998) Testing the Cambrian explosion hypothesis by using a molecular dating technique. *Proceedings National Academy Sciences U.S.A.*, **95**, 12386–12389.

Brown J. and Kodric-Brown M.G. (1977) Turnover rates in insular biogeography: Effect of immigration on extinction. *Ecology*, **58**, 445–449.

Brown P., Sutikna T., Morwood M.J., Soejono R.P., Jatmiko Wayhu S.E., and Rokus A. (2004) A new small-bodied hominin from the Late Pleistocene of Flores, Indonesia. *Nature*, **431**, 1055.

Burnham C.R., Rutter P.A., and French D.W. (1986) Breeding blight-resistant chestnuts. *Plant Breeding Reviews*, **4**, 347–397.

Bush M.B., Gosling W.D., and Colinvaux P.A. (2005) Climate change in the lowlands of the Amazon Basin. In M.B. Bush and J.R. Flenley (Eds.), *Tropical Rainforest Responses to Climate Change*. Springer/Praxis, Heidelberg, Germany/Chichester, U.K.

Canfield D.E., Poulton D., and Narbonne G.M. (2007) Late-Neoproterozoic deep-ocean oxygenation and the rise of animal life. *Science*, **315**, 92–95.

Cardillo M. (1999) Latitude and rates of diversification in birds and butterflies. *Proceedings of the Royal Society of London B*, **266**: 1221–1225.

Cardillo M., Orme C.D.L., and Owens I.P.F. (2005) Testing for latitudinal bias in diversification rates: An example using New World birds. *Ecology*, **86**, 2278–2287.

Chen J.Y., Bottjer D.J., Oliveri P., Dornbos S.Q., Gao F., Ruffins S., Chi H., Li C.W., and Davidson E.H. (2004) Small bilaterian fossils from 40 to 55 million years before the Cambrian. *Science*, **305**, 218–222.

Christian C.E. (2001) Consequences of a biological invasion reveal the importance of mutualism for plant communities. *Nature*, **413**, 635–639.

Colinvaux P.A. and De Oliviera P.E. (2001) Amazon plant diversity and climate through the Cenozoic. *Palaeogeography, Palaeoclimatology, Palaeoecology*, **166**, 51–63.

Colinvaux P., Bush M., Liu K-b, De Oliviera P., Steinitz-Kannan M., Reidinger M., and Miller M. (1989) Amazon without refugia: Vegetation and climate change of the Amazon

Basin through a glacial cycle. Paper given at *International Symposium on Global Changes in South America during the Quaternary, Sao Paulo, Brazil.*

Colwell R.K. and Hurtt G.C. (1994) Nonbiological gradients in species richness and a spurious Rapoport effect. *American Naturalist*, **144**, 570–595.

Colwell R.K. and Lees D.C. 2000. The mid-domain effect: geometric constraints on the geography of species richness. Trends in Ecology & Evolution 15: 70-76.

Connell J.H. (1970a) A predator–prey system in the marine intertidal region, I. *Balanus glandula* and several predatory species of this. *Ecological Monographs*, **40**, 49–78.

Connell J.H. (1970b) On the role of natural enemies in preventing competitive exclusion in some marine animals and in rain forest trees. In P J. den Boer and G.R. Gradwell (Eds.), *Dynamics of Populations: Proceedings of the Advanced Study Institute on Dynamics of Numbers in Populations, Oosterbeek, 1970*, pp. 298–312. Centre for Agricultural Publication Document, Wageningen, Netherlands.

Connell J.H. (1972). Community Interactions on Marine Rocky Intertidal Shores. Annual Review of Ecology and Systematics. 3: 169-192

Connell J.H. (1974) Diversity in tropical rain forests and coral reefs. *Science*, **199**(4335), 1302–1310.

Coope G.R. (1987) The response of late Quaternary insect communities to sudden climatic changes. In J.H.R. Gee and P.S. Giller (Eds.), *The Organization of Communities.* British Ecological Society/Blackwell, Oxford, U.K.

Coyne, J.A. and H.A. Orr (1989) Patterns of speciation in *Drosophila. Evolution*, **43**, 362–381.

Cox P.M., Betts R.A., Jones C.D., Spall S.A., and Totterdell I.J. (2000) Acceleration of global warming due to carbon-cycle feedbacks in a coupled climate model. *Nature*, **408**, 184–187.

Cox P.M., Betts R.A., Collins M., Harris P.P., Huntingford C., and Jones P. (2004) Amazonian forest dieback under climate–carbon cycle projections for the 21st century. *Theoretical and Applied Climatology*, **78**, 1438–1441.

Crane P.R. and Lidgard S. (1989) Angiosperm diversification and paleolatitudinal gradients in Cretaceous floristic diversity. *Science*, **246**, 675–678.

Cronk, Q.C.B. (1989) The past and present vegetation of St Helena. *Journal of Biogeography*, **16**, 47–64.

Culver S.J. and Buzas M.A. (2000) Global latitudinal species diversity gradient in deep-sea benthic foraminifera. *Deep Sea Research, Part I: Oceanographic Research Papers*, **47**, 259–275

Currie, D.J. and Paquin V. (1987) Large scale biogeographical patterns and species richness of trees. *Nature*, **329**, 326–327.

Currie D.J., Mittelbach G.G., Schemske D.W., Cornell H.V., Allen A.P., Brown J.M., Bush M.B., Harrison S.P., Hurlbert A.H., Knowlton, N. *et al.* (2004) Predictions and tests of climate-based hypotheses of broad-scale variation in taxonomic richness. *Ecology Letters*, **7**, 1121–1134.

Damschen E.I., Haddad N.M., Orrock J.L., Tewksbury J.J., and Levey D.J. (2006) Corridors increase plant species richness at large scales. *Science*, **313**, 1284–1286.

Darwin, Charles (1839) *Journal and Remarks (The Voyage of the Beagle)*. Henry Colburn, London.

Davies K.F., Margules C.R., and Lawrence C.R. (2000) Which traits of species predict population declines in experimental forest fragments? *Ecology*, **81**, 1450–1461.

Davis B.N.K. (1975) The colonization of isolated patches of nettles (*Urtica dioica* L.) by insects. *Journal of Applied Ecology*, **12**, 1–14.

Deacon J. and Lancaster N. (1988) *Late Quaternary Environments of Southern Africa.* Clarendon Press, Oxford, U.K.

Delaney M.L. (1989) Extinctions and carbon cycling. *Nature*, **337**, 18–19.

D'Hondt S., Donaghay P., Zachos J.C., Luttenberg D., and Lindinger M. (1998) Organic carbon fluxes and ecological recovery from the Cretaceous–Tertiary mass extinction. *Science*, **282**, 276–279.

Diamond J.M. (1969) Avifaunal equilibria and species turnover rates on the Channel Islands of California. *Proceedings National Academy Sciences U.S.A.*, **64**, 57–67.

Diamond J.M. (1972) Biogeographic kinetics: Estimation of relaxation times for avifaunas of southwest Pacific islands. *Proceedings National Academy Sciences U.S.A.*, **69**, 3199–3203.

Diamond J.M. (1974) Colonization of exploded volcanic islands by birds: The Supertramp Strategy. *Science*, **17**, 803–806.

Dick C.W., Abdul-Salim K., and Bermingham E. (2003) Molecular systematic analysis reveals cryptic Tertiary diversification of a widespread tropical rain forest tree. *American Naturalist*, **162**, 691–703.

Doebeli M. and Dieckmann U. (1999) Evolutionary branching and sympatric speciation caused by different types of ecological interactions. *American Naturalist*, **156**, 77–101.

Donlan C.J. (2007) Restoring America's big, wild animals. *Scientific American*, June 2007.

Dyer L.A., Singer M.S., Lill J.T., Stireman J.O., Gentry G.L., Marquis R.J., Ricklefs R.E., Greeney H.F., Wagner D.L., Morais H.C. *et al.* (2007) Host specificity of Lepidoptera in tropical and temperate forests. *Nature*, **448**, 1–5.

Eldredge N. and Gould S.J. (1972) Punctuated equilibria: An alternative to phyletic gradualism. In T.J.M. Schopf (Ed.), *Models in Paleobiology*. Freeman Cooper, San Francisco, pp. 82–115.

Ehrlich P.R. and Murphy D.D. (2005) Conservation lessons from long-term studies of checkerspot butterflies. *Conservation Biology*, **1**, 122–131.

Erwin D.H. (1998) The end and the beginning: Recoveries from mass extinctions. *Trends in Ecology & Evolution*, **13**, 344–349.

Erwin T.L. (1982) Tropical forests: Their richness in Coleoptera and other arthropod species. *Coleopterists Bulletin*, **36**, 74–75.

FAO (2005) *Global Forest Resources Assessment*. Food and Agriculture Organization of the United Nations, Rome.

Fastovsky D.E. and Sheehan P.M. (2005) The extinction of the dinosaurs in North America. *GSA Today*, **15**, 4–10.

Feeley K.J., Wright J., Supardi M.N., Kassim A.H., and Davies S.J. (2007) Decelerating growth in tropical forest trees. *Ecology Letters*, **10**, 461–469.

Firestone R.B. *et al.* (21 co-authors) (2007) Evidence for an extraterrestrial impact 12,900 years ago that contributed to the megafaunal extinctions and the Younger Dryas cooling. *PNAS*, **104**, 16016–16021.

Fischer A.G. (1960) Latitudinal variations in organic diversity. *Evolution*, **14**, 64.

Fitzpatrick J.W, Lammertink M., Luneau Jr. M.D., Gallagher T.W., Harrison B.R., Sparling G.M., Rosenberg K.V., Rohrbaugh R.W., Swarthout E.C.H., Wrege P.H. *et al.* (2005) Ivory-billed woodpecker (*Campephilus principalis*) persists in continental North America. *Science*, **308**, 1460–1462.

France R. (1992) The North American latitudinal gradient in species richness and geographical range of freshwater crayfish and amphipods. *American Naturalist*, **139**, 342.

Freyer G. (1991) Comparative aspects of adaptive radiation and speciation in Lake Baikal and the great rift lakes of Africa. *Hydrobiologia*, **211**, 137–146.

Fuhrman J.A., Steele J.A., Hewson I., Schwalbach M.S., Brown M.V., Green J.L., and Brown J. (2008) A latitudinal diversity gradient in planktonic marine bacteria. *PNAS*, **105**, 7774–7778.

Fuller J.L. (1998) Ecological impact of the mid-Holocene hemlock decline in southern Ontario, Canada. *Ecology*, **79**, 2337–2351.

Galli A.E., Leck C.F., and Forman R.T.T. (1976) Avian distribution patterns in forest islands of different sizes in central New Jersey. *The Auk*, **93**, 356–364.

Galeotti S., Brinkhuis H., and Huber M. (2004) Records of post–Cretaceous-Tertiary boundary millennial-scale cooling from the western Tethys: A smoking gun for the impact-winter hypothesis? *Geology*, **32**, 529–532.

Gans M.W. and Dunbar J. (2005) Computational improvements reveal great bacterial diversity and high metal toxicity in soil. *Science*, **309**, 1387.

Gaucher E.A., Govindaran S., and Ganesh O.K. (2008) Palaeotemperature trend for Precambrian life inferred from resurrected proteins. *Nature*, **451**, 704–707.

Gentry A.H. (1988) Changes in plant community diversity and floristic composition on environmental and geographical gradients. *Annals of the Missouri Botanical Garden*, **75**, 1–34.

Gilbert F.S. (1980) The equilibrium theory of island biogeography: Fact or fiction? *Journal of Biogeography*, **7**, 209.

Gingerich P.D. (2006) Environment and evolution through the Paleocene–Eocene thermal maximum. *Trends in Ecology & Evolution*, **21**, 246–253

Gotelli N.J. and Colwell R.K. (2001) Quantifying biodiversity: Procedures and pitfalls in the measurement and comparison of species richness. *Ecology Letters*, **4**, 379–391.

Grassle J.F. (1989) Species diversity in deep sea communities. *Trends in Ecology and Evolution*, **4**, 12–15.

Grassle J.F. and Maciolek N.J. (1992) Deep-sea species richness: Regional and local diversity estimates from quantitative bottom samples. *American Naturalist*, **139**, 313.

Grice K., Changqun Cao, Love G.D., Böttcher M.E., Twitchett R.J., Grosjean E., Summons R.E., Turgeon S.C., Dunning, W., and Yugan Jin (2005) Photic zone euxinia during the Permian–Triassic superanoxic event. *Science*, **307**, 706–709.

Grime J.P. (2001) *Plant Strategies, Vegetation Processes, and Ecosystem Properties*. Wiley, New York, 417 pp.

Groombridge B. (1993) Global biodiversity. In B. Groombridge (Ed.), *Status of the Earth's Living Resources*. World Conservation Monitoring Center/Chapman & Hall, London, 605 pp.

Groombridge B. and Jenkins M. (2002) *World Atlas of Biodiversity: Earth's Living Resources in the 21st Century*. World Conservation Monitoring Center/United Nations Environment Program/University of California Press, 340 pp.

Grubb P.J. (1977) The maintenance of species richness in plant communities: The importance of the regeneration niche. *Biological Reviews*, **52**, 107–145.

Grubb P.J. (1987) Global trends in species-richness in terrestrial vegetation: A view from the Northern Hemisphere. In: J.H.R. Gee and P.S. Giller (Eds.), *Organization of Communities: Past and Present*. British Ecological Society/Blackwell Scientic, Boston, p. 576.

Guilday, J.E. (1984) Pleistocene extinction and environmental change. In R.S. Martin and R.G. Klein (Eds.), *Quaternary Extinctions: A Prehistoric Revolution*, pp. 250–258. University of Arizona Press, Tucson, AZ.

Guthrie, R.D. (1984) Mosaics, allelochemics, and nutrients: An ecological theory of late Pleistocene Megafaunal extinctions. In P.S. Martin and R.G. Klein (Eds.), *Quaternary Extinctions: A Prehistoric Revolution*, pp. 259–298. University of Arizona Press, Tucson, AZ.

Guthrie R.D. (1991) *Frozen Fauna of the Mammoth Steppe: The Story of Blue Babe*. University of Chicago Press, 323 pp.

Haffer J. (1969) Speciation in Amazonian forest birds. *Science*, **165**, 131–137.
Hammond P.M. (1990) Insect abundance and diversity in the Dumoga-Bone National Park, N. Sulawesi, with special reference to the beetle fauna of lowland rain forest in the Toraut region. In W.J. Knight and J.D. Holloway (Eds.), *Insects and the Rain Forests of South East Asia (Wallacea)*, pp. 197–254. Royal Entomological Society, London, U.K.
Hannah L. and Lovejoy T. (2005) Conservation, climate change and tropical forests. In M.B. Bush and J.R. Flenley (Eds.), *Tropical Rainforest Responses to Climate Change.* Springer/Praxis, Heidelberg, Germany/Chichester, U.K.
Hannah L., Betts R.A., and Shugart H.H. (2005) Modelling future effects of climate change on tropical forests. In M.B. Bush and J.R. Flenley (Eds.), *Tropical Rainforest Responses to Climate Change.* Springer/Praxis, Heidelberg, Germany/Chichester, U.K.
Hawksworth D.L. (1991) The fungal dimension to biodiversity. *Mycological Research*, **91**, 641–655.
Hecht A.D. and Agan B. (1972) Diversity and age relationships in Miocene and Recent bivalves. *Systematic Zoology*, **5**, 308–320.
Heliwell D.R. (1976) The effects of size and isolation on the conservation value of wooded sites in Britain. *Journal of Biogeography*, **3**, 407.
Hewitt G.M. (1993) Some genetic consequences of ice ages, and their role, in divergence and speciation. *Biological Journal of the Linnean Society*, **58**, 247–276.
Hill J.L. and Hill R.A. (2001) Why are tropical rain forests so species rich? Classifying, reviewing and evaluating theories. *Progress in Physical Geography*, **25**, 326–354.
Hodkinson D. and Casson D. (1991) A lesser predilection for bugs: Hemiptera (Insecta) diversity in tropical rain forests. *Biological Journal of the Linnean Society*, **43**, 101–109.
Hotinski R.M., Bice K.L., Kump L.R., Najjarand R.M., and Arthur M.A. (2001) Ocean stagnation and end-Permian anoxia. *Geology*, **29**, 7–10
Hubbell, S.P. (2001) *The Unified Neutral Theory of Biodiversity and Biogeography.* Princeton University Press, Princeton, NJ.
Hubbell S.P., Foster R.B., O'Brien S.T., Harms K.E., Condit B., Wechsler S.J., Wright S., and Loo de Lao (1999) Light-gap disturbances, recruitment limitation, and tree diversity in a Neotropical forest. *Science*, **283**, 554-557.
Huber M. (2008) A hotter greenhouse? *Science*, **321**, 353–354.
Huey R.B. and Ward P.D. (2005) Hypoxia, global warming, and terrestrial late Permian extinctions. *Science*, **308**, 398–401.
Hughes L., Cawsey E.M., and Westoby M. (1996) Climatic range sizes of *Eucalyptus* species in relation to future climate change. *Global Ecology and Biogeography Letters*, **5**, 23–29.
Huston M.A. (1980) Patterns of species diversity in an oldfield ecosystem. *Bulletin of the Ecological Society of America*, **61**, 110.
Huston M.A. (1993) *Biological Diversity: The Coexistence of Species on Changing Landscapes.* Cambridge University Press, Cambridge, U.K.
IPCC (2007) Australia and New Zealand. *IPCC Fourth Assessment Report*. Available at *http://www.ipcc-wg2.org/*
Jablonski D. (1993) The tropics as a source of evolutionary novelty through geological time. *Nature*, **364**, 142–144.
Jablonski D. (2002) Survival without recovery after mass extinctions. *PNAS*, **99**, 8139–8144.
Jackson J.B.C. and Johnson K.B. (2001) Measuring past biodiversity. *Science*, **28**(293), 2401–2404.
Jackson S.T. and Wang C. (1999) Late Quaternary extinction of a tree species in eastern North America. *PNAS*, **96** 13847–13852

James A.N. and Green M.J.B. (1999) *A Global Review of Protected Area Budgets and Staffing*. World Conservation Monitoring Center/World Conservation Press, Cambridge, U.K., 46 pp.

Janzen D.H. (1970) Herbivores and the number of tree species in tropical forests. *American Naturalist*, **104**, 501–528.

Jaramillo C., Rueda M.J., and Mora G. (2006) Cenozoic plant diversity in the neotropics. *Science*, **311**, 1893–1896.

John R., Dalling J.W., Harms K.E., Yavitt J.B., Stallard R.F., Mirabello M., Hubbell S.P., Valencia R., Navarrete H., Vallejo M., and Foster R.B. (2007) Soil nutrients influence spatial distributions of tropical tree species. *Proceedings National Academy Sciences U.S.A.*, **104**, 864–869.

Johnson C.N. and Wroe S. (2003) Causes of extinction of vertebrates during the Holocene of mainland Australia: Arrival of the dingo, or human impact? *The Holocene*, **13**, 941–948.

Kershaw S.P., van der Kaars S., and Flenley J.R. (2005) In M.B. Bush and J.R. Flenley (Eds.), *Tropical Rainforest Responses to Climate Change*. Springer/Praxis, Heidelberg, Germany/ Chichester, U.K.

King J.R. and Mewaldt L.R. (1987) The summer biology of an unstable insular population of white-crowned sparrows in Oregon. *Condor*, **89**, 549–565.

Kitagawa H. and van der Plicht J. (1998) Atmospheic radiocarbon calibration to 45,000 yr B.P.: Late glacial fluctuations and cosmogenic isotope production. *Science*, **279**, 1187–1190.

Kleidon A. and Mooney H.A. (2000) A global distribution of biodiversity inferred from climatic constraints: Results from a process-based modelling study. *Global Change Biology*, **6**, 507–523.

Kleidon A., Adams J.M., Breu B., and Pavlick R. (2009) Simulated geographic variations of plant species richness, evenness and abundance using climatic constraints on plant functional diversity. *Environmental Research Letters*, in press.

Klicka J. and Zink R.M. (1997) The importance of recent ice ages in speciation: A failed paradigm. *Science*, **277**, 1666—1669.

Koleff P. and K.J. Gaston (2001) Latitudinal gradients in diversity: Real patterns and random models. *Ecography*, **24**, 341–351.

Kortlandt A. (1984) Vegetation research and "bulldozer" herbivores in tropical Africa. In A.C. Chadwick and C.L. Sutton (Eds.), *Tropical Rain Forest*. Special Publications of the Leeds Philosophical and Literary Society.

Kreft H., Jetz W., Mutke J., Kier G., and Barthlott W. (2007) Global diversity of island floras from a macroecological perspective. *Ecology Letters*, **11**, 116–127.

Krings M., Stone A., Schmitz R.W., Krainitzki H., Stoneking M., and Pääbo S. (1997). Neanderthal DNA sequences and the origin of modern humans. *Cell*, **90**, 19–30

Kruess A. and Tscharntke T. (2000) Species richness and parasitism in a fragmented landscape: Experiments and field studies with insects on *Vicia sepium*. *Oceologia*, **122**, 129–137.

Kutschera U. and Niklas K.J. (2004) The modern theory of biological evolution: An expanded synthesis. *Naturwissenschaften*, **91**, 1432–1483.

Lack D. (1973) The numbers of species of hummingbirds in the West Indies. *Evolution*, **27**, 326–377.

Levy, S. (2006) Clashing with titans. *BioScience*, **56**, 292–298.

Liew P-M., Kuo C-M., Huang S.Y., and Tseng M.-H. (1998) Vegetation change and terrestrial carbon storage in eastern Asia during the Last Glacial Maximum as indicated by a new pollen record from central Taiwan. *Global and Planetary Change*, **16/17**, 85–94.

Little C.T.S. and Benton M.J. (1995) Early Jurassic mass extinction: A global long-term event. *Geology*, **23**, 495–498, June.

Looy C., Twitchett R.J., and Dilcher D.L. (1991) Life in the end-Permian dead zone. *PNAS*, **98**, 7879–7883.

Lubchenco J. (1978) Community development and persistence in a low rocky intertidal zone. *Ecological Monographs*, **59**, 67–94.

MacArthur R.H. (1972) *Geographical Ecology*, 269 pp. Harper & Row, New York.

MacArthur R.H and MacArthur M.W. (1961) On bird species diversity. *Ecology*, **42**, 595–599.

MacArthur, R.H. and Wilson, E.O. (1967) *The Theory of Island Biogeography*. Princeton University Press, Princeton, NJ.

MacGurran, A.E. (1988) *Ecological Diversity and Its Measurement*. Princeton University Press, Princeton, NJ.

Martens K. (1997) Speciation in ancient lakes. *Trends in Ecology & Evolution*, **12**, 177–182.

Martin P.S. (1989) Prehistoric overkill: A global model. In P.S. Martin and R.G. Klein (Eds.), *Quaternary Extinctions: A Prehistoric Revolution*. University of Arizona Press, Tucson, AZ, pp. 354–404.

Martin P.S. (2007) *Twilight of the Mammoth: Ice Age Extinction and the Rewilding of America*. University of California Press, Berkeley, CA.

Maslin M. and Burns S.J. (2000) Reconstruction of the Amazon Basin effective moisture availability over the past 14,000 years. *Science*, **290**, 2285–2287.

May R.M. (1990) How many species? *Philosophical Transactions of the Royal Society of London B*, **330**, 293–304.

May R. & J. Godfrey (1994) Biological diversity: Differences between land and sea. *Philosophical Transactions Royal Society*, **343**, 105–111.

McElwain J.M., Beerling D.J., and Woodward F.I. (1999) Fossil plants and global warming at the Triassic–Jurassic boundary. *Science*, **285**. 1386–1390.

Miles L., Grainger A., and Phillips O. (2004) The impact of global climate change on tropical forest diversity in Amazonia. *Global Ecology and Biogeography*, **13**, 553–565.

Miller G.H., Magee J.W., Johnson B.J., Fogel M.L., Spooner N.A., McCulloch M.T., and Ayliffe L.K. (1999) Pleistocene extinction of *Genyornis newtoni*: Human impact on Australian megafauna. *Science*, **283**, 205–208.

Mithren S. (1993) Simulating mammoth hunting and extinction: Implications for the Late Pleistocene of the central Russian plain. *Archeological Papers of the American Anthropological Association*, **4**, 163–178

Morecroft P.R., Hurtt G.C., and Pacala S.W. (2001) A method for scaling vegetation dynamics: The ecosystem demography model (ED). *Ecological Monographs*, **74**, 557–586.

Morley R.J. (2005) Cretaceous and Tertiary climate change and the past distribution of megathermal rainforests. In M.B. Bush and J.R. Flenley (Eds.), *Tropical Rainforest Responses to Climate Change*. Springer/Praxis, Heidelberg, Germany/Chichester, U.K.

Myers N. (1988) "Hot spots" in tropical forests. *The Environmentalist*, **8**, 1573–1580.

Myers N, Mittermeier C.G., DaFonseca G.A.B., and Kent J. (2000) Biodiversity hotspots for conservation priorities. *Nature*, **403**, 853–858.

Naniwadekar R. and Vasudevan K. (2007) Patterns in diversity of anurans along an elevational gradient in the Western Ghats, South India. *Journal of Biogeography*, **34**, 842–853.

Nelson B.W., Ferreira C.A.C., da Silva M.F., and Kawasaki M.L. (1990) Refugia, endemism centres and collecting density in Brazilian Amazonia. *Nature*, **345**, 714–716.

Newmark W.D. (1991) Tropical forest fragmentation and the local extinction of understory birds in the eastern Usambara Mountains, Tanzania. *Conservation Biology*, **5**, 67–78.

Newmark W.D. (1996) Insularization of Tanzanian parks. *Conservation Biology*, **10**, 1549–1556.

Novotny V., Basset Y., Miller S.E., Weiblen G.D., Bremer B., Cizek L., and Drozd, P. (2002) Low host specificity of herbivorous insects in a tropical forest. *Nature*, **416**, 841–844.

Novotny V., Drozd P., Miller S.E., Kulfan M., Janda M., Basset Y., and Weiblen G.D. (2006) Why are there so many species of herbivorous insects in tropical rainforests? *Science*, **313**, 1115–1118.

Odegard F. (2000) How many species of arthropods? Erwin's estimate revisited. *Biological Journal of the Linnean Society*, **71**, 583–597.

Orr J.C., Fabry V.J., Aumont O., Bopp L., Doney S.C., Feely R.A., Gnanadesikan A., Gruber N., Ishida A., Joos F. *et al.* (2005) Anthropogenic ocean acidification over the twenty-first century and its impact on calcifying organisms. *Nature*, **437**, 681–686.

Packer A. and Clay K. (2003) Soil pathogens and *Prunus serotina* seedling and sapling growth near conspecific trees. *Ecology*, **84**(1), 108–119.

Parker AG, Goudie AS, Anderson DE, Robinson MA, and Bonsall C. (2002) A review of the mid-Holocene elm decline in the British Isles. *Progress in Physical Geography*, **26**, 1–45.

Patrick, R. (1967) The effect of invasion rate, species pool, and size of area on the structure of the diatom community. *PNAS*, **58**, 1335–1342.

Patterson N., Richter N.J., Gnerre S., Lander E.S., and Reich D. (2006) Genetic evidence for complex speciation of humans and chimpanzees. *Nature*, **441**, 1103–1108.

Philips O.L., Lewis S.L. Baker T.R., and Malhi Y. (2007) The response of South American tropical forests to contemporary atmospheric change. In M.B. Bush and J.R. Flenley (Eds.), *Tropical Rainforest Responses to Climate Change*. Springer/Praxis, Heidelberg, Germany/Chichester, U.K.

Pianka E.R. (1988) The structure of lizard communities. *Annual Review of Ecology and Systematics*, **4**, 53–74

Poore G.C.B. and Wilson G.D.F. (1993) Marine species richness. *Nature*, **361**, 597.

Prance G.T. (1987) Biogeography and Quaternary history in tropical America. In T.C. Whitmore and G.T. Prance (Eds.), *Oxford Monographs on Biogeography No. 3*, pp. 46–65. Oxford University Press, New York.

Puerto A., Rico M., Matias M.D., and Garcia J.A. (1990) Variation in structure and diversity in Mediterranean grasslands related to trophic status. *Journal of Vegetation Science*, **1**, 445–452.

Pulliam H.R. (1988) Sources, sinks and population regulation. *American Naturalist*, **132**, 652–661.

Ralph C.J. (1985) Habitat association patterns of forest and steppe birds of northern Patagonia, Argentina. *Condor*, **87**, 471–483.

Rashit E. and Bazin M. (1987) Environmental fluctuations, productivity, and species diversity: An experimental study. *Microbial Ecology*, **14**, 1432.

Raup D.M. (1979) Size of the Permo-Triassic bottleneck and its evolutionary implications. *Science*, **206**, 217–218.

Raup D.M. and Sepkoski J.J. (1984) Periodicity of extinctions in the geologic past. *PNAS*, **8**, 801–805.

Reid, W.W. and Miller K.R. (1989) *Keeping Options Alive: The Scientific Basis for Conserving Biodiversity*. World Resources Institute, Washington, D.C.

Rettalack G.J. (2001) A 300-million-year record of atmospheric carbon dioxide from fossil plant cuticles. *Nature*, **17**, 287–290.

Rex M.A., Stuart C.T., Hessler R.R., Allen J.A., Sanders H.L., and Wilson G.D.F. (1993) Global-scale latitudinal patterns of species diversity in the deep-sea benthos. *Nature*, **365**, 636–639.

Richardson J.E., Pennington R.T., Pennington D., and Hollingsworth P. (2001) Rapid diversification of a species-rich genus of neotropical rain forest trees. *Science*, **293**, 2242–2245.

Ris Lambers H., Clark J.S., and Beckage B. (2002) Density-dependent mortality and the latitudinal gradient in species richness. *Nature*, **417**, 732–735.

Roberts D.L. and Solow A.R. (2003) Flightless birds: When did the dodo become extinct? *Nature*, **425**, 245.

Rodland D.L. and Bottjer D.J. (2001) Biotic recovery from the end-Permian mass extinction: Behavior of the inarticulate brachiopod *Lingula* as a disaster taxon. *Palios*, **16**, 95–101.

Rosenzweig M.L. (1971) Paradox of rnrichment: Destabilization of exploitation ecosystems in ecological time. *Science*, **171**, 385–387.

Rosenzweig M.L. (1995) *Species Diversity in Space and Time*. Cambridge University Press, Cambridge, U.K.

Royal Society (2005) *Ocean Acidification Due to Increasing Atmospheric Carbon Dioxide*, Policy Document 12/05, 60 pp. The Royal Society, London.

Sanderson M.J., Thorne J.L., Wikström N., and Bremer K. (2004) Molecular evidence on plant divergence times. *American Journal of Botany*, **91**, 1656–1665.

Sauer J.D. (1988) *Plant Migration: The Dynamics of Geographic Patterning in Seed Plant Species*. University of California Press, Los Angeles, p. 350.

Saxon E., Baker B, Hargrove W., Hoffman F., and Zganjar C. (2005) Mapping environments at risk under different global climate change scenarios. *Ecology Letters*, **8**, 53–60.

Scheffer M., van Geest G.J., Zimmer K., Butler M.G., Hanson M.A., Declerck S., De Meester L., Jeppesen E., and Sondergaard M. (2006) Small habitat size and isolation can promote species richness: Second-order effects on biodiversity in shallow lakes and ponds. *Oikos*, **112**, 227–231.

Schliewen T.K., Tautz D., and Pääbo S. (1994) Sympatric speciation suggested by monophyly of crater lake cichlids. *Nature*, **368**, 629–632.

Schweingruber F.H., Briffa K.R., and Jones P.D. (1993). A tree-ring densiometrric transect from Alaska to Labrador: Comparison of ring width and latewood density. *International Journal of Biometerology*, **37**, 151–169.

Sephton M.A., Looy C.V., Brinkhuis H., Wignall P.B., de Leeuw J.W., and Visscher H. (2005) Catastrophic soil erosion during the end-Permian biotic crisis. *Geology*, **12**, 941–944.

Sepkoski J.J., Bambach R.K., Raup D.M., and Valentine J.W. (1981) Phanerozoic marine diversity and the fossil record. *Nature*, **293**, 435–437.

Shackleton H.J. and Thomas E. (1996) The Paleocene–Eocene benthic foraminiferal extinction and stable isotope anomalies. *Geological Society London Special Publications*, **101**, 401–441, doi: 10.1144/GSL.SP.1996.101.01.20.

Shen Y., Zhang T., and Hoffman P.F. (2008) On the coevolution of Ediacaran oceans and animals. *PNAS*, **105**, 7376–7381.

Shipley B., Vile D., and Garnier E. (2006) From plant traits to plant communities: A statistical mechanistic approach to biodiversity. *Science*, **314**, 812–814.

Sigurdsson H., D'Hondt S., Arthur M.A., Bralower T.J., and Zachos J.C. (1991) Glass from the Cretaceous-Tertiary boundary in Haiti. *Nature*, **349**, 482–487.

Silman M.R. (2005) Plant species diversity in Amazonian forests. In M.B. Bush and J.R. Flenley (Eds.), *Tropical Rainforest Responses to Climate Change*. Springer/Praxis, Heidelberg, Germany/Chichester, U.K.

Simberloff D. and Abele L.G. (1982) Refuge design and island biogeographic theory: Effects of Fragmentation. *American Naturalist*, **120**, 41.

Simberloff D. and Wilson E.O. (1970) Experimental zoogeography of islands: A two-year record of colonization. *Ecology*, **51**, 934–937.

Smith V.H., Foster B.L., Grover J.P., Holt R.D., Leibold M.A., and deNoyelles F. (2005) Phytoplankton species richness scales consistently from laboratory microcosms to the world's oceans. *PNAS*, **102**, 4393–4396.

Soule P., Wilcox B.A., and Holtby C. (1979) Faunal collapse in East African game reserves revisited. *Biological Conservation*, **15**, 259–272.

Stanley S.M. (1986) Anatomy of a regional mass extinction: Plio-Pleistocene decimation of the western Atlantic bivalve fauna. *PALAIOS*, **1**, 17–36.

Steadman D.W. and Martin P.S. (2003) The late Quaternary extinction and future resurrection of birds on Pacific islands. *Earth-Science Reviews*, **61**, 133–147.

Steadman D.W., Pregill G.K., and Burley D.V. (2002) Rapid prehistoric extinction of iguanas and birds in Polynesia. *Proceedings National Academy Sciences U.S.A.*, **99**, 3673–3677.

Stebbins G.L. (1950) *Variation and Evolution in Plants*. Columbia University Press, New York.

Steele T.J., Adams J.M., and Slukin T. (1998) Modelling paleoindian dispersals. *World Archaeology*, **30**, 286–305.

Steuber T., Mitchell S.F., Buhl D., Gunter G., and Kasper H.U. (2002) Catastrophic extinction of Caribbean rudist bivalves at the Cretaceous–Tertiary boundary. *Geology*, **11**, 999–1002.

Stevens R.D. (2006) Historical processes enhance patterns of diversity along latitudinal gradients. *Proceedings of the Royal Society B*, **273**, 2283–2289.

Stewart J.R. and Lister A. (2001) Cryptic northern refugia and the origins of the modern biota. *Trends in Ecology & Evolution*, **16**, 608–613.

Stehli F.G., Douglas R.G., and Newell N.D. (1969) Generation and maintenance of gradients in taxonomic diversity. *Science*, **164**, 947–949.

Stringer C. and Gamble C. (1993) *In Search of the Neanderthals*. Thames & Hudson, New York, 247 pp.

Stuart A.J., Kosintsev P.A., Higham T.F.G., and Lister A.M. (2004a) Pleistocene to Holocene extinction dynamics in giant deer and woolly mammoth. *Nature*, **431**, 684–689.

Stuart S.N., Chanson J.S., Cox N.A., Young B.E., Rodrigues A.S.L, Fischman D.L., and Waller R.W. (2004b) Status and trends of amphibian declines and extinctions worldwide. *Science*, **306**, 1783–1786.

Svenning J-C. and Skov F. (2007) Could tree diversity pattern in Europe be generated by postglacial dispersal limitation? *Ecology Letters*, **10**, 453–460.

Tallis J.H. (1991) *Plant Community History: Long-term Changes in Plant Distribution and Diversity*. Springer-Verlag, New York, 398 pp.

Temple S.A. (1977) Plant–animal mutualism: Coevolution with dodo leads to near extinction of plant. *Science*, **197**(4306), 885–886.

Terborough J. (1973) On the notion of favorableness in plant ecology. *American Naturalist*, **107**, 481–501.

Theodorou G.E. (1988). Environmental factors affecting the evolution of islands endemics: The Tilos example for Greece. *Modern Geology*, **13**, 183–188.

Thomas E. and Shackleton N.J. (1996) The Paleocene–Eocene benthic foraminiferal extinction and stable isotope anomalies. *Geological Society, London, Special Publications; Global Perspective: Geochronology and the Oceanic Record*, **101**, 401–444.

Thuiller W., Lavorel S., Arajo M.B., Sykes M.T., and Prentice I.C. (2005) Climate change threats to plant diversity in Europe. *PNAS*, **102**, 8245–8250.

Tilman D. (1982) *Resource Competition and Community Structure*. Princeton University Press, Princeton, NJ.

Tilman D. (1994) Competition and biodiversity in spatially structured habitats. *Ecology*, **75**, 2–16.

Turner I.M., Tan Y.C., Ibrahim W.A.B., Chew P.T., and Corlett R.T. (2002) A study of plant species extinction in Singapore: Lessons for the conservation of tropical biodiversity. *Conservation Biology*, **8**, 705–712.

Turney C.S.M, Flannery T.F., Roberts R.G., Reid C., Fifield L.K., Higham T.F.G., Jacobs Z., Kemp N., Colhoun E.A., Kalin R.M. *et al.* (2008) Late-surviving megafauna in Tasmania, Australia, implicate human involvement in their extinction. *Proceedings National Academy Sciences U.S.A.*, **105**, 12150–12153.

Tzedakis P.C. (1993) Long-term tree populations in northwest Greece through multiple Quaternary climatic cycles. *Nature*, **364**, 437–444.

Vajda V., Raine J.I., and Hollis C.J. (2001) Indication of global deforestation at the Cretaceous–Tertiary by New Zealand fern spike. *Science*, **294**, 1700.

Vajda V. and McLoughlin S. (2004) Fungal proliferation at the Cretaceous–Tertiary boundary. *Science*, **303**, 1489.

Van Valen L. (1974) Molecular evolution as predicted by natural selection. *Journal of Molecular Evolution*, **3**, 1432–1441.

Verkeyen E., Salzburger W., Snoeks J., and Meyer A. (2004) Origin of the superflock of cichlid fishes from Lake Victoria, East Africa. *Science*, **300**, 325–329.

Via S. (2001) Sympatric speciation in animals: The ugly duckling grows up. *Trends in Ecology & Evolution*, **10**, 381–390.

Wallace A.R. (1869) *The Malay Archipelago*. Available at Googlebooks *http://books.google.com/books?id=geEBAAAAMAAJ&pg=PR11&dq=The+Malay+Archipelago*

Ward L.K. and Lakhani K.H. (1977) The conservation of juniper. *Journal of Applied Ecology*, **14**, 121–135.

Watkins Jr., J.E., Cardelús C., Moran R.C., and Colwell R.K. (2006) Diversity and distribution of ferns along an elevational gradient in Costa Rica. *American Journal of Botany*, **93**, 73–83.

Whittaker R.H. (1956) Vegetation of the Great Smoky Mountains. *Ecological Monographs*, **26**, 1–80.

Whittaker R.J. (2000) Scale, succession and complexity in island biogeography: Are we asking the right questions? *Global Ecology and Biogeography*, **9**, 75–85.

Wiggins D.A. (1999) The peninsula effect on species diversity: A reassessment of the avifauna of Baja California. *Ecography*, **22**, 542–547.

Wignal P.B. and Twitchett R.J. (1996) Oceanic anoxia and the end Permian mass extinction. *Science*, **272**, 1155–1158.

Wilf P, Johnson K.R., and Huber B.T. (2003) Correlated terrestrial and marine evidence for global climate changes before mass extinction at the Cretaceous–Paleogene boundary. *PNAS*, **100**, 599–604.

Wilf P., Labandeira C.C., Johnson K., and Ellis B. (2006) Decoupled plant and insect diversity after the end-Cretaceous extinction. *Science*, **313**, 1112–1115.

Willig M.R. and Lyons K.S. (1998) An analytical model of latitudinal gradients of species richness with an empirical test for marsupials and bats in the New World. *Oikos*, **81**, 93–98.

Woodward F.I. (1987) Stomatal numbers are sensitive to increases in CO_2 from pre-industrial levels. *Nature*, **327**, 617–618.

Wroe, S. and Field J. (2006) A review of the evidence for a human role in the extinction of Australian megafauna and an alternative interpretation. *Quaternary Science Reviews*, **25**, 2692–2703.

Wroe S., Field J., and Grayson D.K. (2006) Megafaunal extinction: Climate, humans and assumptions. *Trends in Ecology and Evolution*, **21**, 61–62.

Wroe S., Clausen P., McHenry C., Moreno K., and Cunningham E. (2007) Computer simulation of feeding behaviour in the thylacine and dingo: A novel test for convergence and niche overlap. *Proceedings of the Royal Society (London), Series B*, **274**, 2819–2828.

Zhang Y., Xu M., and Adams J.M. (in presss) Can Landsat imagery detect tree line dynamics? *International Journal of Remote Sensing*.

Index

Figure 1.15. Cloud layer at around 2,300 m in a tropical mountain area in northern Borneo. Areas regularly bathed by clouds are much damper and likely to support more species of ferns and amphibians (photo: Author).

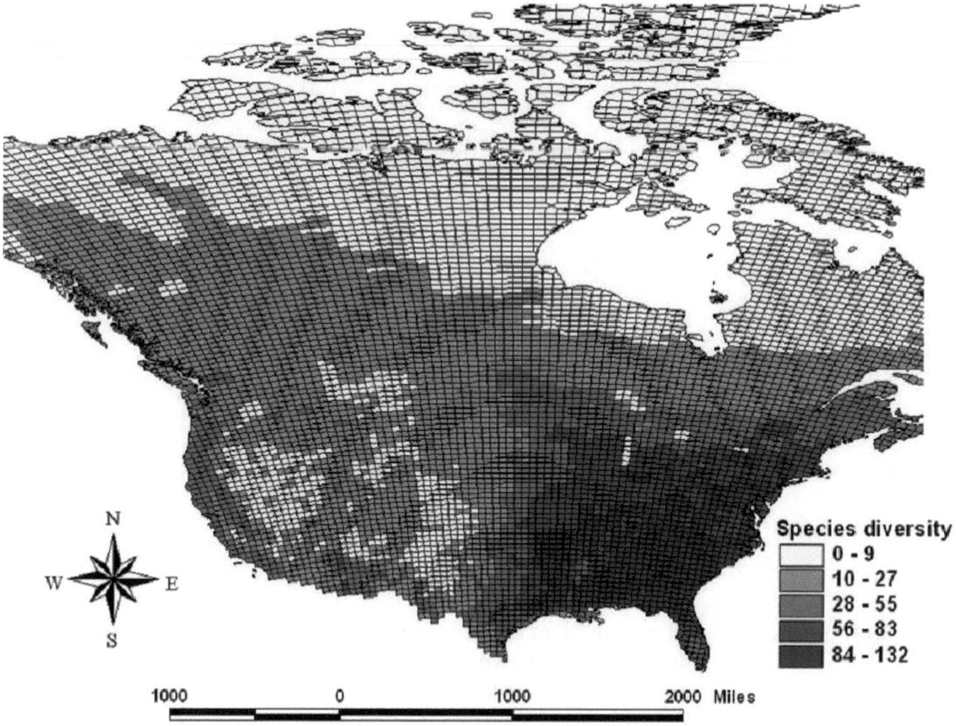

Figure 2.4. Tree species richness map of North America (Yangjian Zhang and J. Adams, unpublished).

Figure 2.7. The Florida plum yew, *Torreya taxifolia*, has a very restricted distribution in the southeastern U.S.A. This young specimen is from the Torreya State Park in Florida (photo: Author).

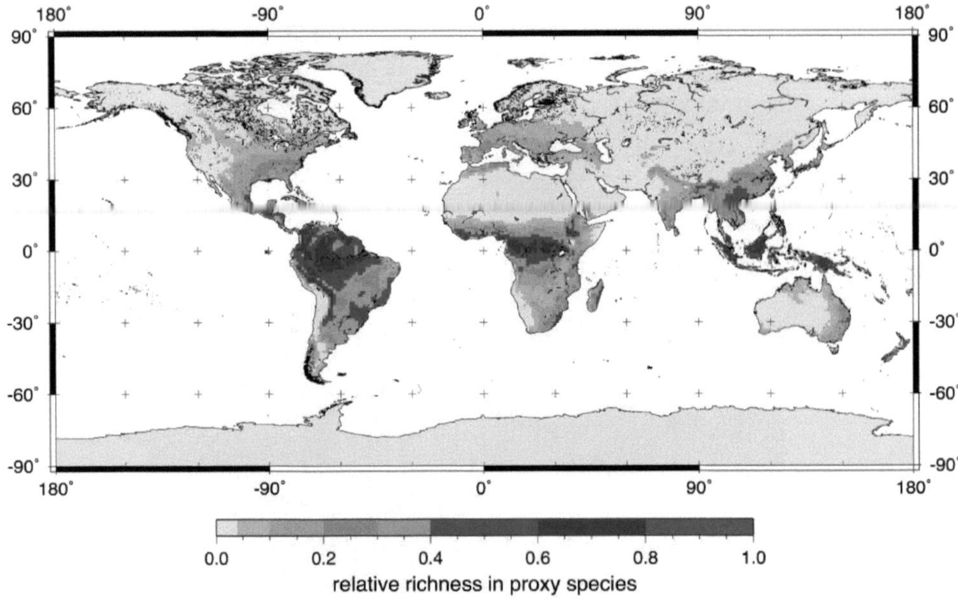

Figure 2.11. Predicted global map of plant species richness of the JeDi model. The patterns of species richness closely follow what is known for modern-day geographical patterns with latitude, and also manages to predict much of the east–west variation in species richness within regions (from Kleidon *et al.*, 2009).

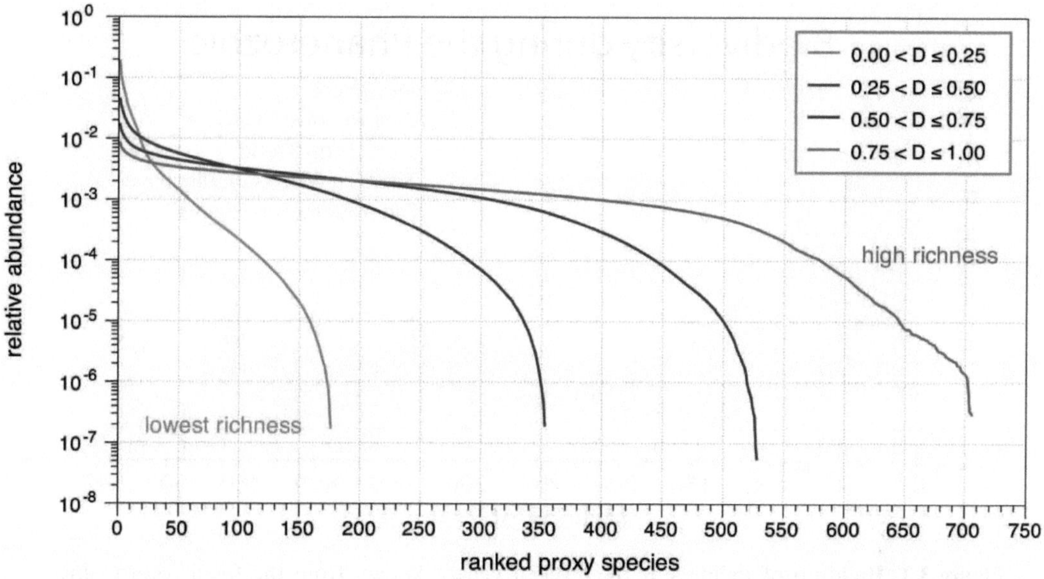

Figure 2.12. Relative abundance/evenness predicted in woody plant communities by the JeDi model. The least diverse communities in high latitudes are predicted to be dominated by a small number of species, with a small proportion of those being very abundant (green line). The most diverse low-latitude communities (red line) are predicted to have most of the species about equally abundant as one another (from Kleidon *et al.*, 2009).

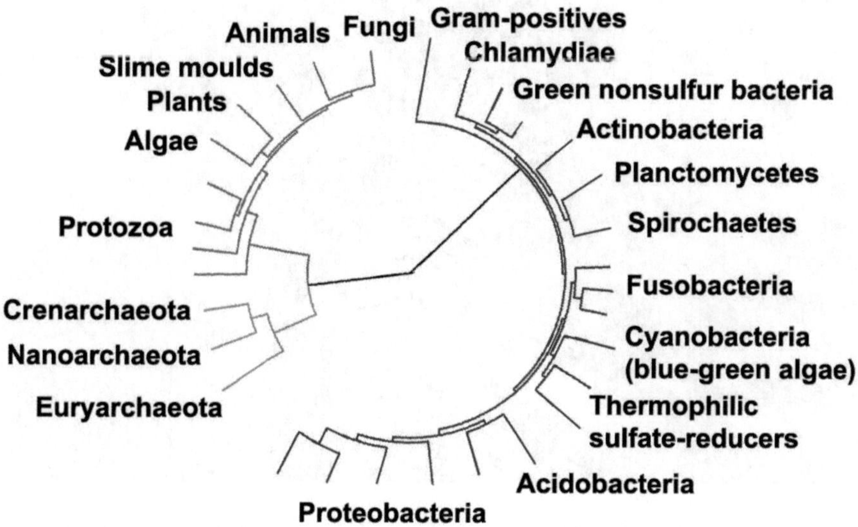

Figure 3.1. Family tree of all life on Earth, from comparing DNA sequences. Most of the broad- scale diversity of life is in the form of different groups of bacteria (blue) (source: Tim Vickers, Wikipedia Commons).

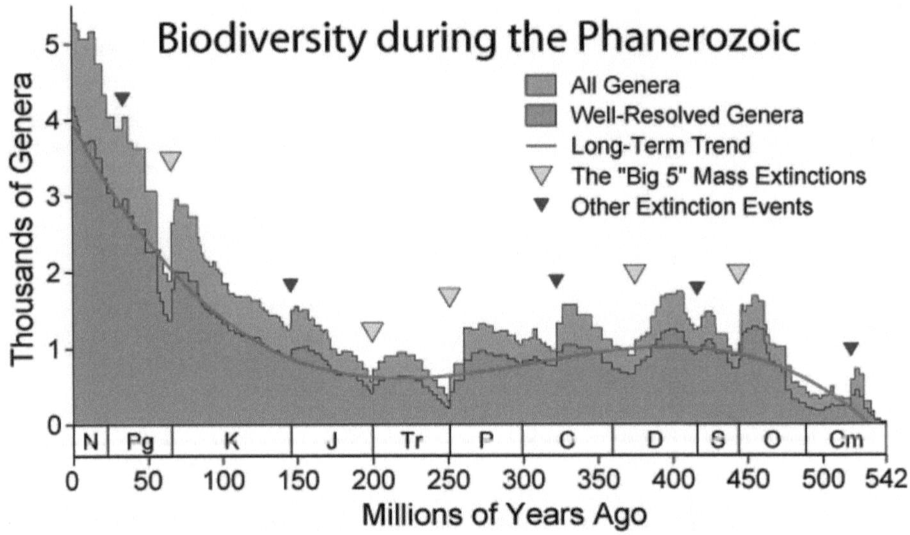

Figure 3.3. Build-up of richness at the level of genera known from the fossil record, since 540 million years ago (based on Alroy *et al.*, 2008, from Wikipedia Commons, diagram by DragonsFlight.

Figure 3.6. The tropical rainforest fern *Stenochlaena* blanketed North America for thousands of years after the end-Cretaceous mass extinction. In the present world, it only survives in Southeast Asia (photo: Author, northern Borneo).

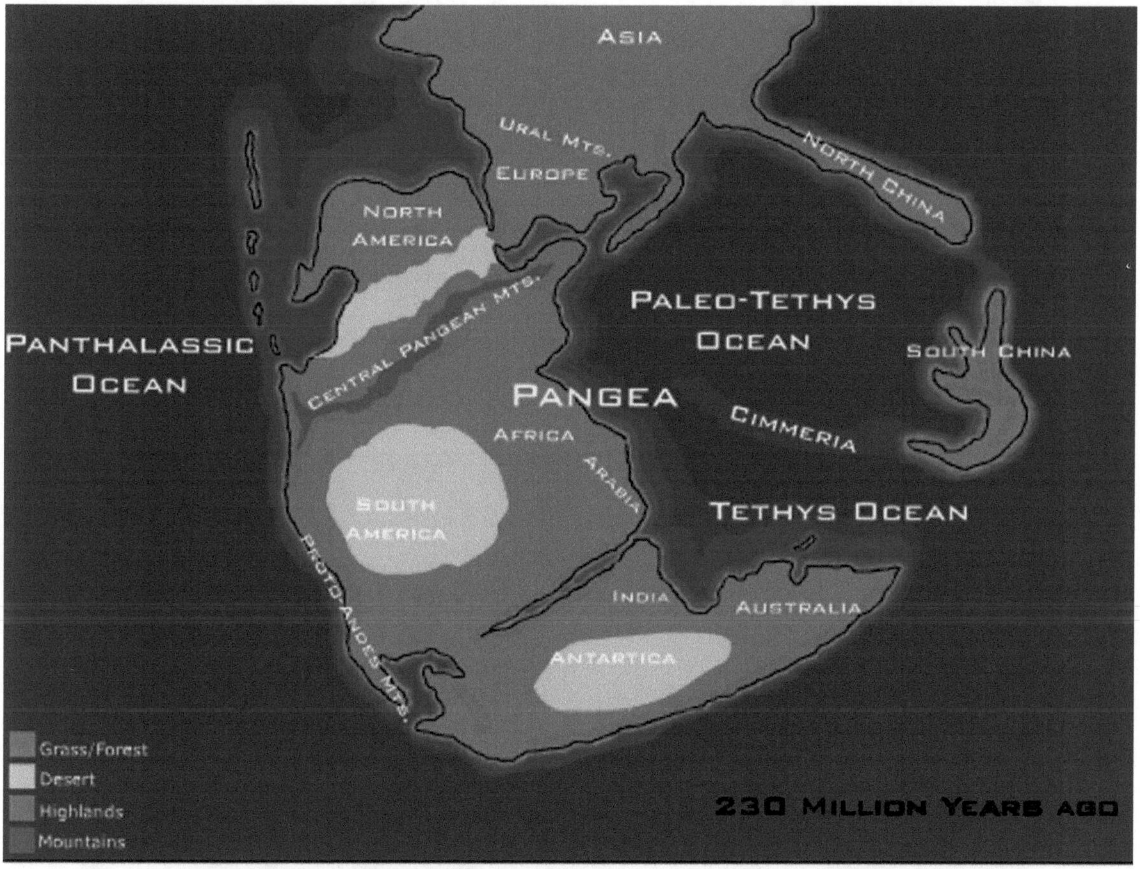

Figure 3.7. A map of the supercontinent of Pangea as it existed close to the time of the end-Permian extinction. Note how the lands seem to form a ring around the Paleo-Tethys Ocean: this might have been what set the scene for disaster, according to some hypotheses.

Figure 3.11. (Top) The maidenhair tree, *Ginkgo biloba*, was previously far more widespread in the northern temperate zone, and grows well when planted across its original range. (Bottom) The coast redwood *Sequoia* survives in a small area of northern California and adjacent Oregon in the U.S.A. Millions of years ago, it too was common and widespread around the northern hemisphere.

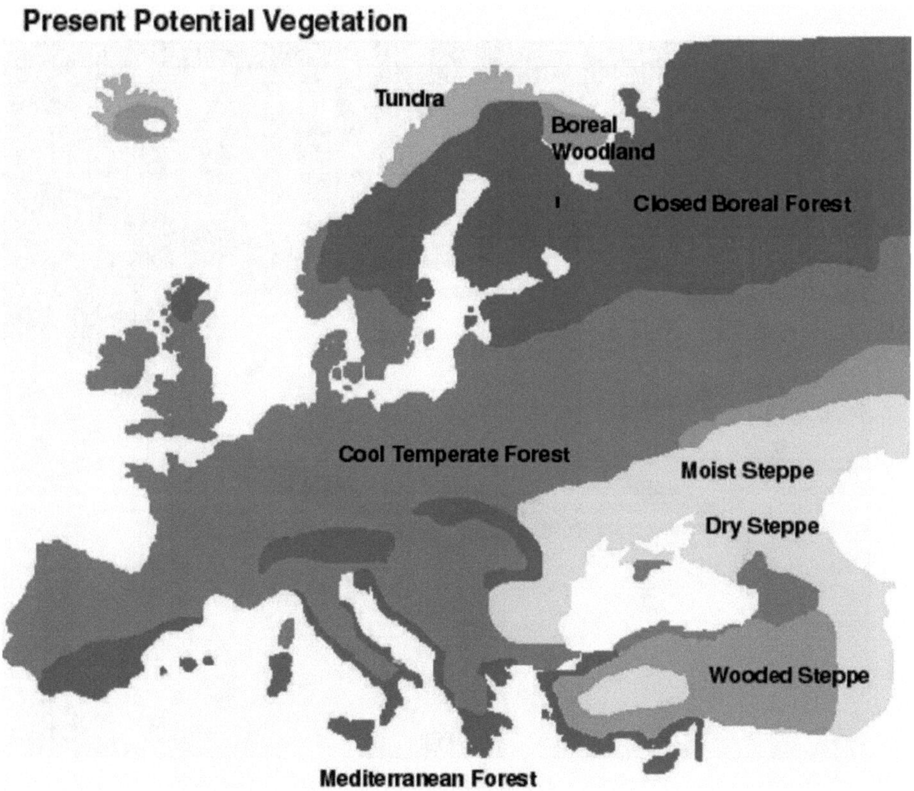

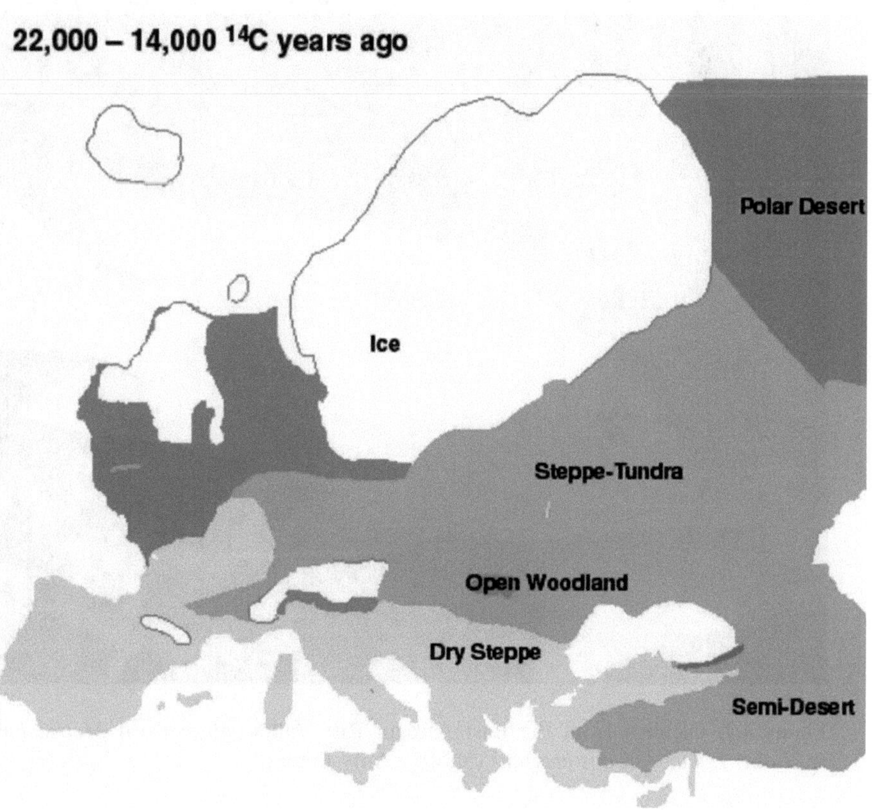

Figure 3.13. Reconstructed vegetation of Europe during (top) the present interglacial, before deforestation caused by farming, (bottom) during the last ice age. The ice vegetation was mostly devoid of forest cover due to a combination of cold and drought (source: Author).

Figure 4.2. Cichlids from the East African Rift Valley lakes: (top) *Metriaclima greshakei*, (bottom) *Labidochromis caerulus* (photos: Amy Lerch).

Figure 4.3. One of the several hundred species of heathers, *Erica*, endemic to the Cape (photo: Maille Neil).

Figure 4.4. A protea endemic to the Cape (photo: Maille Neil).

Figure 4.5. A western Australian mallee species of *Eucalyptus* (photo: Maille Neil).

Figure 4.6. Some examples of species from the daisy family, Compositae (source: Wikipedia Commons).

Figure 4.7. A member of the cabbage family Cruciferae: *Arabidopsis thaliana* (source: Wikipedia Commons).

Figure 4.8. Dipterocarp seeds, Malaysia (source: Author).

Figure 5.1. Cain kills his brother Abel, depicted in a 15th-century manuscript (by Butko, in Wikipedia Commons).

Figure 5.6. The Arctic rose, *Dryas octopetala*. A plant that thrives in the cold climates of tundra and high mountains, its pollen in lake beds in the European lowlands provides a sign of cooler conditions in the past (source: Wikipedia Commons).

Figure 5.7. The Carolina parakeet, extinct in 1918 (source: painting by Audubon).

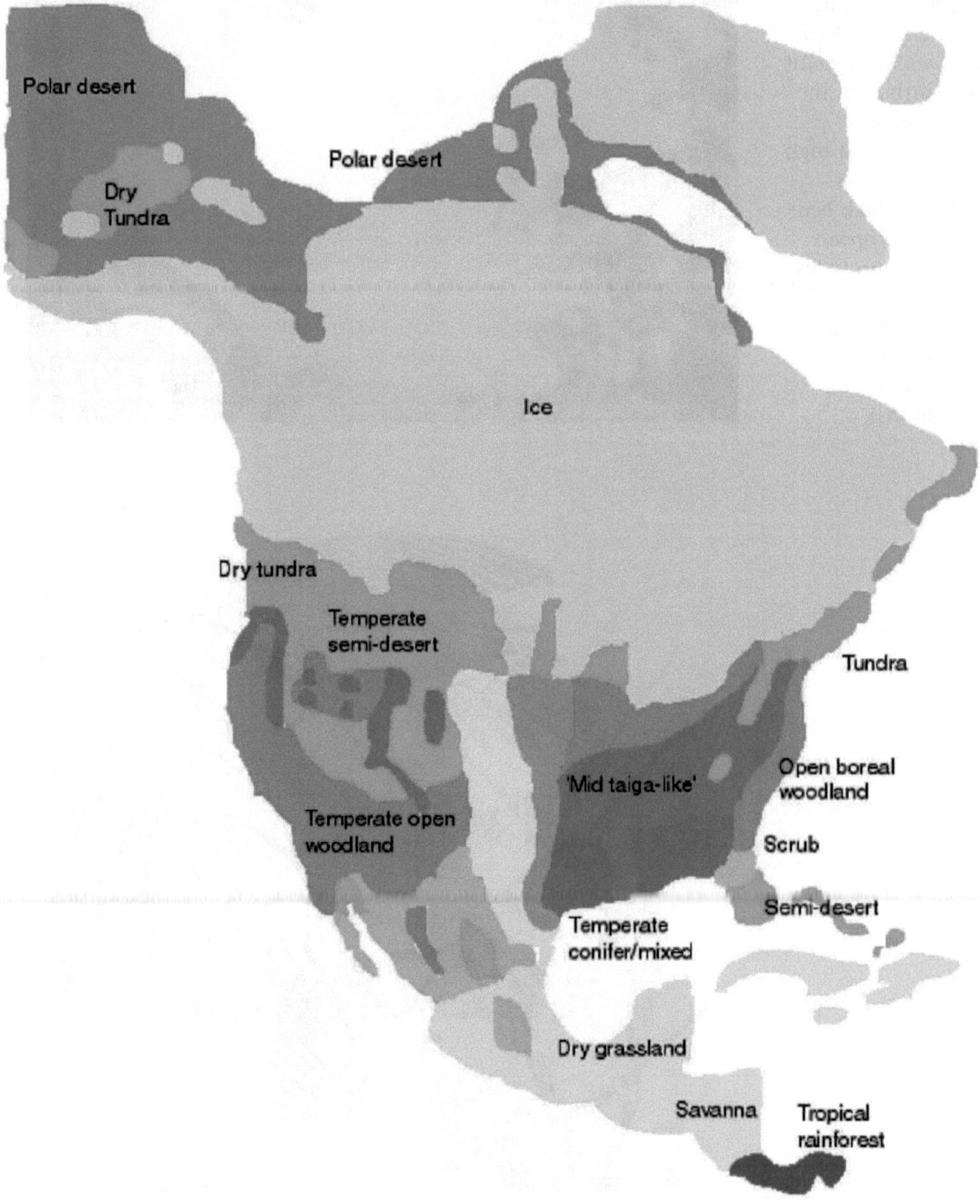
Polar desert
Polar desert
Dry
Tundra
Ice
Dry tundra
Temperate
semi-desert
Tundra
'Mid taiga-like'
Open boreal
woodland
Temperate open
woodland
Scrub
Semi-desert
Temperate
conifer/mixed
Dry grassland
Savanna
Tropical
rainforest

Figure 5.4. (Facing page) Last Glacial Maximum vegetation distribution, and (this page) the present potential vegetation without agriculture, for comparison (source: Adams and Faure, 1997).

Figure 6.3. The ivory-billed woodpecker. Extinct in the 1940s—or is it? (source: Audubon).

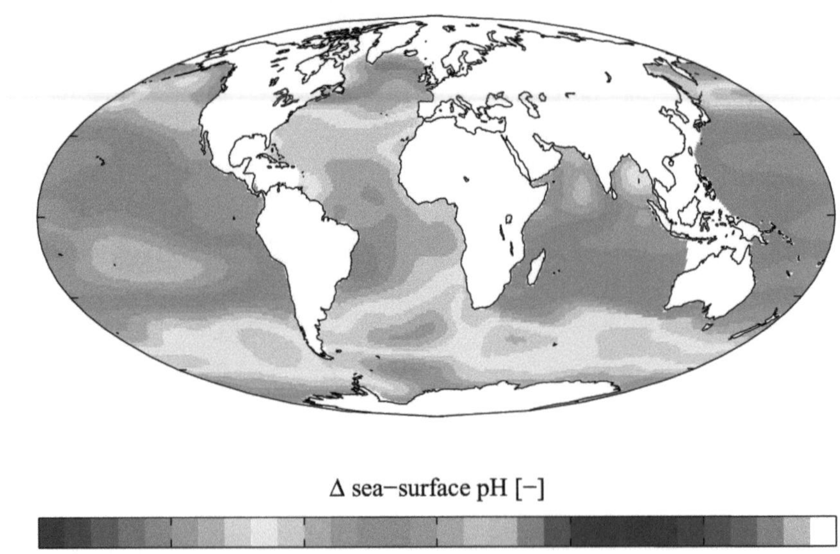

Figure 7.4. Estimated change in pH of ocean surface waters as a result of atmospheric CO_2 increase between the 1700s and the 1990s (source: "Plumbago", Wikipedia Commons).

MIX
Papier aus verantwortungsvollen Quellen
Paper from responsible sources
FSC® C105338

If you have any concerns about our products,
you can contact us on
ProductSafety@springernature.com

In case Publisher is established outside the EU,
the EU authorized representative is:
Springer Nature Customer Service Center GmbH
Europaplatz 3, 69115 Heidelberg, Germany

Printed by Libri Plureos GmbH
in Hamburg, Germany